From Autogiro to Gyroplane

FROM AUTOGIRO TO GYROPLANE

The Amazing Survival of an Aviation Technology

Bruce H. Charnov

Foreword by John Lienhard

Westport, Connecticut
London

Library of Congress Cataloging-in-Publication Data

Charnov, Bruce H.
    From autogiro to gyroplane : the amazing survival of an aviation technology /
    Bruce H. Charnov ; foreword by John Lienhard.
       p. cm.
    Includes bibliographical references and index
    ISBN 1-56720-503-8 (alk. paper)
    1. Autogiros. I. Title.
TL715.C43   2003
629.133'35—dc21   2002044972

British Library Cataloguing in Publication Data is available.

Library of Congress Catalog Card Number: 2002044972
ISBN: 1-56720-503-8

First published in 2003

Praeger Publishers, 88 Post Road West, Westport, CT 06881
An imprint of Greenwood Publishing Group, Inc.
www.praeger.com

Printed in the United States of America

The paper used in this book complies with the
Permanent Paper Standard issued by the National
Information Standards Organization (Z39.48-1984).

10 9 8 7 6 5 4 3 2 1

In Memory
Mel N. Morris Jones
1952–2002

Ken Brock
1932–2001

"There's nothing that's ever been made that can't be improved on."

"Don't tell me, show me."

"I've been in gyroplanes for quite a while. I don't know of anything else I'd rather be doing than what I'm doing today. I've had a lot of fun. It's taken Marie and me all over the world. We can't go to any spot in the world where we can't call up somebody we know."

Ken Brock, Bensen Days Seminar, Friday, April 16, 1993.

CONTENTS

FOREWORD

This child of the 1930s was hooked on airplanes. I built models, collected pictures, and created a vast mental inventory. Never in that time was I privileged to see an autogyro, but I knew they were out there, along with flying wings, Panama Clippers, and the *Hindenburg*—all the flying arcana that never quite made it over my rooftop in St. Paul, Minnesota.

There were two kinds of flying machines in those days: There were those that fate permitted me to see—DC-3s, Piper Cubs, blimps, and pontoon seaplanes. As World War II turned from a threat into reality, B-24s, P-40s, and all their kin joined that list. The other breed of airplane was that which I could only dream of seeing one day. By now, I have indeed seen many of those. The day this now-old man actually rode in a B-17 (such a reassuringly solid flying machine that was!) was the day childhood returned, not as it was, but as I had once wished it might be.

I knew then, and I still do, that I will go to my death without seeing a corporeal autogyro in flight. That mystic creature has long since become my personal *unicorn* among horses—a spectral creature destined to remain in an imagined memory.

Back in the late 1930s, one of the big tobacco companies, in its ongoing efforts to recruit addicts among children, began offering an airplane picture with each pack of cigarettes. It also provided albums to hold the pictures. I have three of those albums completely filled. My parents smoked the cigarettes for me, and both have long since died of heart and lung diseases. But, now and then, I pull the albums out to look at the airplane pictures. They are, I suppose, a costly part of my parental legacy.

Among such forgotten oddities as the flying automobile and the supposedly foolproof Stearman Hammond Safety Plane, is the Kellett Autogiro. It is described as "[a]n extremely useful ship for observation work because of its low minimum forward speed of 24 m.p.h."

That was 1940. Three years later, working as a bicycle delivery boy for a local drugstore, I was able to pay the full one-dollar cost of the *Aeronautics Aircraft Spotters Handbook.* If I'd felt there was a dimension of legend to the autogyro up to then, this compendium made it clear that it was real enough. The book displayed six very different autogyros along with two of the new helicopters.

Those helicopters would, of course, send the autogyro over into the land of unicorn-ness for a long time. One of them was the embryonic American Vought-Sikorsky VS300. The other was the German Focke-Wulf Fw-61. After the war, we would hear much about helicopters and almost nothing further about my unicorn autogyros.

Therefore I feel, as I read Bruce Charnov's extraordinary book, that I tread forbidden ground. It seems almost wrong to tear away the veil and reveal the flesh and blood, the wood and aluminum, that made the autogyro into a living beast of the forest.

Yet, as I read, Charnov systematically places the autogyro in direct contact with all I've known about early flight. I find Bert Hinkler test piloting one of my unicorns. He was the wild Australian pilot celebrated in a popular song as "Hustling Hinkler up in the Sky." Amelia Earhart turns out to have been deeply involved with popularizing the autogyro. Charnov shows us how the autogyro, once intended to bring flight into every backyard, actually played a real role in World War II. It represented a far greater presence in early-twentieth-century flight than we realize.

Charnov also reveals another idea that I was unprepared for. It is the ongoing reappearance of unicorns in the forest of new aircraft since World War II. The autogyro concept has been with us the whole while, despite its seeming displacement by the helicopter. In that, it reenacts an oft-repeated pattern in invention—particularly in invention as it relates to flight.

The pattern goes as follows: A technology arises and seems to offer a bright future. Then it is shelved because of some new gust of technological change. Finally, it reemerges, chrysalis-like, in an unanticipated reincarnation.

As an example, consider a technology that arose right alongside the new Cierva Autogiros. It was a means for achieving transatlantic service with the short-range passenger airplanes of the 1920s and early 1930s. The trick was to build floating airports, anchored like today's offshore oil rigs, so airplanes could hop their way across the ocean. Just as these were about to become a reality in the 1930s, they were preempted by a new generation of long-distance, ground-based passenger airplanes.

Now the Japanese have actually built a floating airport in the Bay of Tokyo—not to facilitate transoceanic travel, but to save precious land. They call it *Megafloat.* Thus a technology lingers as its purpose mutates.

The big transoceanic dirigible is part of a similar ongoing story. It seemingly died with the *Hindenburg,* but it is now being championed in a dazzling variety of new versions. None will replicate the great dirigibles, but all hope to transcend them. And, I believe, we will soon watch as a keeper emerges from that pack of contenders.

So it is with the autogyro. Charnov finishes with the rise of another flying machine, the *gyrocopter,* and its kindred forms of airplanes. A gyrocopter is an autogyro whose rotor blades can be powered like a helicopter's or can be permitted to freewheel like those of an autogyro while a regular propeller provides horizontal thrust. It is a composite between the older autogyro and the newer helicopter. As these gain in use and popularity, I may yet see my unicorn. I may yet even ride one as well.

Of course the gyrocopter is more nearly the spawn of a horse and a unicorn. But that's what all good technology is. Any machine that ultimately succeeds is an ongoing union of dream and reality. That's what I relearn from this scrupulous history of the corporeal autogyro—this reflection upon the creature of my nine-year-old imagination.

Any good engineer must keep the beings of his reveries alive, for the corporeal machine is merely the tangible shadow of that imagined reality. I read Charnov's history with wide-eyed fascination. But as I do, I protect my unicorn. I will not let this flood of reality blot out the autogyro of my child's imagination—still flitting there, amongst the treetops in my old backyard.

John H. Lienhard
Host and author of "The Engines of Our Ingenuity,"
broadcast nationally on public radio

PREFACE

In many ways this is the most personal preface I've written of the several books that have characterized my academic career, but it is only fitting, as this book has the most personal of origins. In November 1986 I awoke the morning of my fortieth birthday with the distinct feeling that somehow life was passing me by. Perhaps not an uncommon experience, my immediate reaction was to list those things I had always wanted to accomplish. Foremost were studying law and learning how to fly, and by the following August I was enrolled in Hofstra University's School of Law, a natural choice because I was already a tenured professor of management in its School of Business. For the record, the study of law was a magnificent undertaking, but the two years of subsequent legal practice while on leave from the university were of a different nature and did not result in new career directions, and I returned to the university in late 1992.

In 1999 I finally got around to flying and immediately was faced with the question as to *what* to fly. From my teen years I somehow dragged the long-forgotten memories of advertisements in *Popular Science* of something called a Gyrocopter and casually typed it into an Internet search engine. To my genuine surprise over 2,200 "hits" were referenced, and I started reading through them. It proved (and will continue to do so) an amazing experience. I quickly encountered references to the Spanish inventor Juan de la Cierva and descriptions of the first flight in 1923 outside Madrid, but also stunning photographs of the CarterCopter, the most technologically advanced autorotational aircraft flying and capable of doing what no other rotary-wing craft has ever done. Having approached this endeavor to determine what to fly, I found my focus shifting to the history of these amazing aircraft and soon realized that there was a mystery as to how Cierva's vision had survived to become the CarterCopter, and it is that mystery that became the genesis for this book.

Juan de la Cierva, son of a wealthy and politically prominent family, is considered the Father of Spanish Aviation as a result of having constructed the first local airplane, but his enduring genius came from his discovery (or rediscovery, according to some) of the aerial phenomenon of *autorotation*, namely that unpowered rotor blades will generate lift as long as air is flowing up through them. Cierva's inspiration was to see the aircraft wing differently than others—reasoning that aircraft *stall* arose when its speed fell below a critical point when the airflow over the wing could no longer generate sufficient lift, he advanced the then-radical idea that if the speed of the aircraft and the speed of the wing could be separated, lift could continue to be generated even at slower airspeeds—effectively creating an inherently safer means of air travel. His solution, developed in the 1919–23 time period, was to turn the wing into a four-blade, freely rotating rotor placed on *top* of the fuselage.

Cierva had relocated his aviation interests to England by 1925 and, calling his creation an Autogiro, trademarked the name. (Thus all Cierva and Cierva-licensed aircraft were/are correctly denoted as Autogiros. Rotary-wing craft that were either based on other technology or unlicensed copies of Cierva machines were called either *autogyros* or *gyroplanes*). I learned that this technology had been licensed by Harold Frederick Pitcairn and brought to America in 1928 and subsequently licensed by other American companies, including the Kellett Autogiro Company. But by end of World War II the Autogiro, displaced by the coming of the helicopter, had effectively disappeared. And so the question of how this technology survived came into sharp focus.

By this point in my inquiry I was beginning to encounter a fascinating cast of characters, notably a Russian immigrant named Igor Bensen, who had a fanatical zeal for his mid-1950s inventions, the names of which he also trademarked, the Gyro-Glider and Gyrocopter. During these early months of research I also found an enigmatic reference to a 1980s Afro-American autogyro designer, David Gittens, and a photograph of his Ikenga 530Z Autogyro being accepted into the Smithsonian National Air and Space Museum (NASM). I resolved to query the Smithsonian as to the uniqueness of this only Afro-American-designed aircraft in the national aviation collection, and this led directly to a most fortuitous encounter and the next step in this book's evolution.

I happened on a reference to the dedication in summer of 2000 of the oldest American-constructed Autogiro, the restored Pitcairn PCA-1B at the American Helicopter Museum & Education Center at the Brandywine Airport, West Chester, Pennsylvania (which also has the original Bill Parsons stretched Bensen two-place trainer, an aircraft as historic in its own way as the Pitcairn). Calling to get information, the docent who answered the phone took it upon himself to invite me to the dedication when he heard that I was a professor. At that ceremony I met Dr. Dominick Pisano and Russell Lee from the Smithsonian NASM and was introduced to Stephen "Steve" Pitcairn, Harold F. Pitcairn's son. Steve, unassuming and characteristically modest, asked me if I had read the book about his father and the history of the Pitcairn Autogiros. When I replied in the negative, he went into the museum shop and gifted me with a copy of Frank Kingston Smith's *Legacy of Wings: The Harold F. Pitcairn Story* (I would later learn that the Pitcairn family had funded the publication of this book). It was a generous and kind gift, and in many ways this book, in part, was inspired by Smith's work. For while it was immediately engaging and is required reading for anyone who wants to understand this facet of aviation history, it was evident that, although an informative volume handsomely done and with superb photographs, the view of the author was devotional and sometimes one-sided. It led me to seek other interpretations, which proved a decidedly difficult task. Peter Brooks's authoritative work *Cierva's Autogiros* is out of print but readily available from aviation used-book dealers, and although it was a scholarly undertaking, its coverage basically ended, except for some later comments on the post–World War II period, with the advent of the helicopter. Histories of the development of rotary-wing flight, basically chronicles of the evolution of the helicopter, consistently passed over the Autogiro and presented its history as a dead-end developmental path. No book covered the almost eight decades of Autogiro/Gyrocopter/gyroplane history nor was there any authoritative bibliography to fuel the scholarly inquiry necessary to answer the mystery of how the Autogiro technology had survived.

Through Steve Pitcairn I was introduced to one of his father's 1930s mechanics, George Townson (who argued with Amelia Earhart on April 8, 1931, when the famed American aviatrix set the first Autogiro altitude record of 18,415 feet in the Pitcairn PCA-2). George agreed to meet me at the Pitcairn hangar at the Robbinsville, New Jersey, airport to deliver a copy of his privately published book *Autogiro: The Story of the "Windmill*

Plane," a profusely illustrated, highly individual account of the American Autogiro, but the highlight of this first visit to the Pitcairn hangar was seeing the last flying 1930s Autogiro, the restored Pitcairn PCA-2 *Miss Champion.* I resolved to be there when she next flew, and that was to be in fall of 2000 at the American Helicopter Museum's Rotorfest 2000. And, watching this Autogiro fly by at twenty feet and twenty mph, I was fascinated and resolved to do the research, write the history, and produce the foundation bibliography for future readers/researchers. It was to prove an engaging task as I continually rediscovered an intriguing cast of genuine characters who had played unique roles in the development and preservation of Juan de la Cierva's vision.

In the course of my Internet search I had also encountered a new magazine, *FlyGyro!,* being published in England by Mel Morris Jones, and had immediately become a subscriber. Mel nurtured my interest and became a terrific and enthusiastic supporter of my research. Questions concerning Autogiro history were promptly answered even as future research directions were suggested. I sent him reports and photographs of the Rotorfest 2000 Fly-In and was thrilled when he put the photo of Steve Pitcairn flying *Miss Champion* on the cover. I asked him if he thought people might be interested in a reexamination of Amelia Earhart's involvement with the Autogiro on the seventieth anniversary of her world altitude record, and he quickly elicited my commitment to write the article, which was published in the February 2001 issue of *FlyGyro!* It was not only the first substantial result of my research to be printed, it proved to be exceptionally fortuitous and led to quite unexpected results that advanced the research in ways that could not have been envisioned.

In the article "Amelia Earhart and the PCA-2: A Re-evaluation of the First Woman Autogiro Pilot" I described how the famed aviator had been beaten cross-country in an Autogiro by the unknown John M. "Johnny" Miller from Poughkeepsie, New York. Shortly after the magazine appeared, I received a letter that began, "I'm the Johnny Miller you wrote about..." and continued for several single-spaced pages describing the transcontinental flight, including details that were available nowhere else. In the course of several subsequent conversations and letters Johnny provided me with a unique view of the early history of the Autogiro in America, including detailed descriptions of the transcontinental flight and the Eastern Airlines experimental Autogiro airmail route during the July 6, 1939, through July 5, 1940, period from the roof of the Philadelphia 30th Street Post Office and nearby Camden, New Jersey, airport. I finally met Johnny in the CarterCopter exhibit tent at the Experimental Aircraft Association (EAA) AirVenture Fly-In in July 2002 and consider the few hours we spent in conversation as one of the highlights of my Autogiro research; at ninety-seven he remains vital and vivid in memories and accounts of what happened—a national aviation treasure!

Through Russ Lee, curator of Rotary Aircraft at the Smithsonian, I had managed to locate David Gittens, and we exchanged many emails and much written material. And when I had finally completed the writing of David Gittens's gyroplane history (a version of which was subsequently adapted for publication by the PRA magazine *Rotorcraft*), he was kind enough to review what I had written and comment. I also count as a highlight of this research unexpectedly meeting David at the 2002 Popular Rotorcraft Association (PRA) Fly-In at Waxahatchie, Texas. As David resides in Sarasota, Florida, there was no logical reason to suppose that he would be in Texas, but there he was and we embraced as old friends—his Ikenga 530Z Gyroplane is a stunning adaptation of the earlier Cierva/Pitcairn "tractor" (engine in the front, like a conventional airplane) design. He gave me information on his particular venture and was genuinely enthusiastic and supportive as this book began to take shape. Written in nonsequential segments (as information became available) that were later to be placed in the proper position in the 1923–2003 history, the biographies of the living pioneers often were written (and rewritten) as the historical accounts of the early days were assembled, weighed, and analyzed and conclusions were drawn.

It has been a thrilling journey—as the eight decades of autorotational history assumed a form not previously available in a single work, relationships emerged to cast innovative, and sometimes startling, light on events and led to the new understandings and rediscovery of forgotten pioneers. A few examples illustrate each of these points: The conventional view was that Igor Bensen had, based on the earlier British Rotachute (a rotary kite) designed by Raoul Hafner for insertion of Allied agents into Europe during World War II, first designed the Gyro-Glider and later followed with the famed Gyrocopter, but this proved to be untrue. Although Bensen's accomplishments, discussed at length in the book, were many and varied, a forgotten inventor named Harris Woods independently came up with a design for a gyro-glider in 1945 based on an experience in the Kellett KD-1B Autogiro used in the earlier Eastern Airlines airmail route and perhaps also based on an artist's speculations accompanying the Ed Yulke article "Gyro Cars for Fun" in the November 1945 *Popular Mechanix*. Although Bensen's design was derivative from the Rotachute and the World War II German submarine Focke-Achgelis Fa-330 rotary kite, an obscure Californian named Arliss Riggs was building an independent series of Cierva/Pitcairn tractor autogyros in California in the 1950s and 1960s.

Understanding the derivative nature of Bensen's achievements helped explain why the Europeans, who had begun with his kits in the late 1950s, had so quickly modified Bensen designs and gone in different and sometimes radical directions. Mel Morris Jones had helped me contact the famed Finish designer/aeronautical engineer Jukka Tervamäki, who related to me details of his experience working in the Bensen Aircraft Company factory in 1958, and he too was kind enough to review what I had written and suggest how the text might be made better. And when I became aware that Jukka had, with mechanic Aulis Eerola, designed, constructed, and flown an autogyro (an ATE-3, OH-XYV) in the 1969 Finnish Spede Pasanen and Ere Kokkonen film *Leikkikalugangsteri* (Toy Gangster), I pestered him via email when I located a used copy in a video store in Helsinki until he was kind enough to get it and send it to me. By this time I had taken my first flight in an autogyro, a wonderful, state-of-the-art two-place Magni aircraft designed and sold as a kit through Greg Gremminger, American distributor for Vittorio Magni's Italian company (that same Magni with Greg is featured on the cover of the August 2001 *Popular Mechanics*), and I was hooked on flying, but the historical research continued at an increasing pace. Its own unique excitement was the encounter with historical figures, and none was more significant than the legendary Californian Ken Brock. It was to result, from a tragic turn of events, in the unexpected possession of and unanticipated obligations accompanying his legacy.

Concurrent with the work on this book, I had convinced Hofstra University senior administrators of the benefits of holding the first academic conference on the topic, "From Autogiro to Gyroplane: The Past, Present and Future of an Aviation Industry," and securing a 2003 date from the director of the Cultural Center, I embarked on securing the support and participation of those pioneers whose names were on my manuscript pages—one of whom was Ken Brock. He failed to respond to my mailings, so I brazenly walked up to his booth in an industry exhibits area in late July at the 2001 EAA AirVenture and said, "Ken—you don't answer my mail but I want you to come to my conference!" He was friendly and receptive and admitted that he had not looked at the mail, and almost immediately his wife and partner, Marie, briefly took me aside and told me to send mailings to *her* attention—she would see that Ken read them and they would both participate in the conference. I explained to Ken that I had written his history and would appreciate his comments. We agreed that I would send him my text, but he could not have guessed that behind my breezy demeanor I was quite nervous about this, for the interpretation I had reached was, I knew, quite at odds with the traditional view that Igor Bensen was the single most important moving force in the American gyrocopter movement. My research had strongly pointed to the interpretation that Brock had saved the PRA *from* the iconic cult of Igor Bensen, and Brock's long tenure (1972–88) as president of the organization had resulted in a creative burst of innovation. I was aware that no other writer had pictured him in this manner and was quite prepared to receive a denial or claim that I had unjustly slighted Bensen while aggrandizing Brock, but his reply was quietly eloquent and affirmative: "You've done a nice job ... ," he wrote on September 11, 2001.

I had additionally come to see the Brock KB-2 Gyrocopter as a truly historical aircraft—he and Igor Bensen had disagreed about that model in 1970, and Bensen, claiming that the KB-2 had been modified to such an extent that it could no longer be called a Gyrocopter, as that was a Bensen Aircraft–trademarked term. As a result, the KB-2 became the first modern *gyroplane*. I asked Ken if he would consider donating a KB-2 to the Smithsonian's National Air and Space Museum, and he replied that he would be humbled to have one of his models in the national aviation collection. In late August 2001 I was at the Smithsonian with Rod Anderson and Dr. Claudius Klimt of the CarterCopter organization and shared my research on Brock with members of the rotary aircraft group. It was well-received, and I was enthusiastically urged to pursue the donation of the KB-2 in 2004 after the NASM had completed and occupied its new exhibition space at Baltimore's Dulles Airport. I could almost imagine the dedication of the KB-2, with Ken and Marie looking on, but it was not to be. Ken Brock died in a crash of his private plane on October 19, 2001—the tail wheel collapsed upon landing and the plane skidded off the runway and flipped over, injuring Marie and fatally injuring Ken. So I found myself in possession of Ken's legacy and was immensely glad (and perhaps relieved) that he had approved it—his story is told here, but I was pleased to publish a modified tribute in the January 2002 issue of *Rotorcraft.* And it meant a great deal to me that Marie asked a friend to call me the day before Ken's funeral, at which hundreds of friends and admirers came to celebrate the life of this gyroplane pioneer, to tell me that the Smithsonian donation meant so much to Ken that she would put his KB-2 aside so that it could eventually be presented to the NASM. A few months later Mel Morris Jones passed away from a brain tumor—he had constantly encouraged this book and become a good friend. And it is with the express permission of the Brock and Jones families that I am honored to dedicate this book to these two gallant gentlemen who had become friends.

When I began this research I knew virtually nothing of the history of the Autogiro, but I knew, like many, only of the most famous modern auto-gyro, *Little Nellie,* in the 1967 James Bond film *You Only Live Twice.* That film's six and a half minutes of gyro flight and fight remain, based on forty-six hours of flying by Wing Commander Kenneth H. Wallis RAF (Ret), the most stunning international images of the autogyro, and I knew that it was a story essential in the modern history I was writing. Ken Wallis proved an engaging subject, and his history is told in detail, but there was to be much more. When Hofstra University had committed to the Auto-gyro/gyroplane conference, I had proposed that Ken Wallis receive an honorary doctorate in recognition of his life's achievements as an autogyro film star, designer, and pilot. That proposal had worked its way through the various committees and officials and was finally approved by the univer-sity's board of trustees after over a year's deliberations. I waited until the doctorate was approved to contact Ken—but he had heard of the confer-ence and let me know through a mutual friend that he intended to partici-pate, and he had reviewed what I had written about him. When I finally called his home, Reymerston Hall, late one evening and identified myself, Ken greeted me with, "Bruce, old man, how are you?" and we shared a long conversation about Autogiro/autogyro history.

It was apparent from my many contacts with living pioneers that there was a story to tell and those eager to tell it and to enthusiastically share in the recovered aviation history. And as the book assumed its first draft form, linkages became clear that found their way into the constant consul-tations that seemed to arrive daily via letters, email, fax, and the oc-casional telephone call. Roger Connor of the Smithsonian asked for information on the controversy concerning the first American Autogiro flight and the Henry Ford Museum–Greenfield Village called with a query as to where they might get a replacement tire for the Walter Scripts Pit-cairn PCA-2 Autogiro used by the *Detroit News* seventy years ago (I didn't know, but one call down to the Pitcairn Aircraft hangar in Robbinsville, New Jersey, and a short conversation with a member of Steve Pitcairn's restoration crew did the trick!).

This book also features a great number of acknowledgments—and the reader is invited to note those individuals cited. It is an accurate reflection of those who endorsed and participated in this project, enthusiastic in its vision of telling the eighty-year journey from Autogiro to Gyroplane. In significant measure those living individuals cited in the text were invited to comment on their stories, and in lesser measure, most did. This has resulted in a vibrant profusion of detail that hopefully tells a compelling story, while at the same time being backed up by extensive endnotes and documentation founded on an exhaustive research bibliography. I have endeavored to produce the volume that I had hoped to find when I got hooked by Autogiro history, and it is my hope that others reading this book will be intrigued and inspired to further the knowledge of this forgotten segment of aviation history. And if this book can serve as the foundation for such research or even just the popular retelling of the stories and personalities that have characterized Juan de la Cierva's vision, it will have succeeded in extending that history into a new century.

While preparing this preface, I read an article in the fall 2002 issue of *American Heritage of Invention & Technology* on vacuum tubes. Author Mark Wolverton engagingly described the history of the vacuum tube and the emerging technological innovations that may bring it back into a more general use in the twenty-first century. He concludes by asking, "Why abandon a technology just because it's old?...The stubborn longevity of [vacuum] tubes demonstrates that sometimes we may be a little too quick to discard the old in favor of the new. Once in a while even the obsolescent can have a few surprises and tricks in store." Such a question, asked about the Autogiro and Juan de la Cierva's autorotational technology, is an appropriate question for this book.

And with regard to the flying, I've never shaken the romance and sheer excitement of the original Pitcairn Autogiro. I recently took my twelve-year-old daughter, Jessica Lauren Charnov, down to visit Steve Pitcairn. I specifically asked if Jessica could see *Miss Champion,* as I wanted it to be part of her memories. Steve kindly asked if she would like to sit in the cockpit, which she happily did. And then I asked, "How about her father?" and Steve told me to go ahead. It proved an amazing experience—sitting there and taking in the controls, I thought of the 1932 flight in this very Autogiro by Lewis A. "Lew" Yancy to Havana, Cuba, and then on to the Yucatán, and the world's altitude record over Boston later that same year. In April I took Ken Wallis down to the Pitcairn Aircraft hangar, and although the weather did not allow flying, Steve introduced Ken Wallis to *Miss Champion*, an amazing coming together of two gyroplane pioneers. I've also contracted with Ron Herron of Little Wing Autogyros to "cut steel" and start welding a fuselage for me based on his tractor design. And

I have taken delivery of a fantastic seven-cylinder rotary engine being produced by Rotec in Australia that has a lineage stretching back to the original engines on the early Cierva and Pitcairn machines. The two together would result in a modern autogyro that strongly resembled the earlier Autogiros in appearance and performance. And what should its paint scheme be? Why, just like *Miss Champion,* of course!

ACKNOWLEDGMENTS

This book is the product of a great deal of help, encouragement, and shared enthusiasm. There are many who deserve thanks and acknowledgment and I can only hope that no one has been forgotten, for without their aid this project could not have been completed. I thank initially Steve Pitcairn, whose thoughtful, spontaneous gift of a copy of Frank Kingston Smith's *Legacy of Wings: The Harold F. Pitcairn Story* was where and when this book began, and who later so generously made available the Pitcairn Aircraft Association photo archives, and I also thank Sen. Juan de la Cierva, who kindly shared his memories of his uncle of the same name, inventor of the Autogiro in Spain in 1923. Historian/scholars Carl Gunther (Pitcairn Aviation), James G. Lear (American Helicopter Museum), Russell Lee, Roger Connor, Tom Deitz, and Dr. Dominick Pisano (Smithsonian Institution National Air and Space Museum), Dr. Bette Davidson Kalash (Jesse Davidson Aviation Archives), Glenn and Pam Bundy (Archimedes Rotary Aircraft Museum, Popular Rotorcraft Association), and pioneer editor/author/historian/pilot Paul Bergen Abbott were incredibly helpful, generous with their time, and genuinely tolerant of my many inquiries.

Rotary aircraft authors/journalists/publishers Mel N. Morris Jones *(Fly-Gyro!),* Katheryn Fields *(Rotorcraft),* Dan Leslie (TV journalist), Stephanie Gremminger (former editor, *Rotorcraft*), Ron Bartlett and Kay Verity *(Autogyro 1/4ly),* and the British Rotorcraft Association's T. P. "Ben" Mullett were amazingly helpful and shared enthusiasm for this project to present a comprehensive history of how the amazing autorotating vision of

Juan de la Cierva survived and, with photo/journalist Stu Fields, cannot be thanked enough for providing many of the stunning photographs that accompany the text. Legendary Autogiro/gyrocopter pioneers John "Johnny" M. Miller, Martin Hollmann, George Townson, Ken and Marie Brock, David Gittens, Italy's Vittorio Magni (who continues as a leading manufacturer), and Finland's Jukka Tervamäki were extraordinarily kind with their information (and photographs) and memories, and this book could not have been completed in the form that it assumed without their assistance. Nor could the story of the industry be told without the kind assistance of current industry leaders Ron Herron (Little Wing Autogyros), Greg and Steph Gremminger (Magni USA, LLC), Jim and Kelly Vanek (SportCopter), Donald and Linda (Haseloh) Lafleur (RAF), Robert E. Kopp (Gyro-Kopp-Ters), Don Bouchard, and many others who made time to talk to a stranger about their business. Special thanks are also due legendary pilot John T. Potter, who so generously shared his memories of Don Farrington. And Jay Carter Jr., Rod Anderson, Dr. Claudius Klimt, and Glenna Montgomery of CarterCopters, along with Jay and David Groen, Al Waddill, and James "Jim" P. Mayfield III of Groen Brothers Aviation willingly shared their vision of the future for this amazing aviation technology, and their enthusiasm about that future proved contagious.

I am also grateful for the support and encouragement of Tom Poberezny, president of the Experimental Aircraft Association, Ms. Ann Brown, executive director of the American Helicopter Museum, Hofstra University former and current presidents Dr. James M. Shuart and Stuart Rabinowitz, J.D., Provost and Executive Vice President Dr. Herman Berliner and Dean Ralph Polimeni, Ph.D., of the Frank G. Zarb School of Business, for their constant support. I also salute my wonderful administrative support team Geraldine "Geri" Woods and student research assistants Laura A. Barone and Audrey E. Damour, as well as administrators Anneliese Payne, Sherry M. Ross, Dr. David Klein, Barry Germond, and Dr. Lanny Udey of Hofstra University's Academic Computing services, and Brian Caligiure for their invaluable contributions in the area of photographic digitizing that made many of the book's illustrations possible.

I thank the editors at Greenwood/Praeger for their efforts, even as I fully accept responsibility for my errors.

As always, this is for my children Aharon Chayim Charnov, Miryam Esther Benovitz and her husband Don-E, and Jessica Lauren Charnov. And delightfully, for the next generation: Ariella Reva Benovitz, Yonaton Menachem Benovitz, and Mindel "Mindy" Leora Benovitz.

And for Ron Herron—now that the book is done, on to the Little Wing!

INTRODUCTION

A very interesting type of heavier-than-air craft is the Cierva autogyro. It is neither a helicopter nor an airplane. It consists of a body or fuselage very similar to that of an airplane. In its nose there is a motor and an ordinary type of propeller. Its small stubby wings and tail surfaces are somewhat similar to that of any airplane. The remarkable feature of this machine is the windmill, or rotary wing which is mounted over it. This windmill, or rotor as it is called, is not actuated directly by any power from the engine but is made to revolve by the airstream from the propellor striking against it. When an ordinary airplane takes off in flight, its propeller gives it speed through the air and makes air pass under the wing at such a rapid rate that the airplane rises. With the autogyro, instead of the wing being fixed in position, it rotates and goes through as many particles of air in a given time as the wing of an airplane does; but with the windmill, instead of a great deal of forward motion being required, it is accomplished in one place by rotation.... The machine can be made to hover over a certain place when climbing it upward. It can be brought straight down to the ground and landed on a place without any forward motion.... The autogyro has been very successful in its flights and promises a great deal for the future, where machines are required to land on the top of a building, a small field, in a forest or the top of a mountain.
—General William "Billy" Mitchell, *Skyways* (1930)

The PCA-2, flown by James G. "Jim" Ray, the first rotary-wing aircraft to land at the White House, just prior to the awarding of the Collier Trophy at the White House on April 22, 1931, to Harold F. Pitcairn and his associates by President Herbert Hoover. (Courtesy of Stephen Pitcairn, from the Pitcairn Archives.)

April 1931 saw what is undeniably the most dramatic moment in the history in America of the Autogiro, the 1923 invention of Spaniard Juan de la Cierva. On April 22, 1931, on the back lawn of the White House, President Herbert Hoover presented the prestigious Collier Trophy of 1930, for the greatest achievement in American aviation to Harold F. Pitcairn[1] and his associates Edwin Asplundh, Agnew E. Larsen, James G. "Jim" Ray, and Geoffrey S. Childs for "their development and application of the Autogiro and the demonstration of its possibilities with a view to its use for safe aerial transport."[2] Pitcairn thus joined the luminaries of aviation, many of whom were present for the ceremony, including Orville Wright and Senator Hiram Bingham, who had created the Air Commerce Act of 1926, the "cornerstone of modern aviation."[3]

Although Pitcairn sought to schedule the presentation so that Cierva could be present, the availability of President Hoover dictated the time and place of the ceremony. Hoover, previously secretary of commerce, whose jurisdiction had included the drafting of federal and state aeronautical safety regulations, had often consulted with Pitcairn, and he requested that the ceremony be held on the south lawn of the White House and that it commence with the landing of a Pitcairn Autogiro!

On April 21, pilot Jim Ray flew the Pitcairn PCA-2 from the factory in Willow Grove, Pennsylvania, to College Park, Maryland. The next morning, with newsreel cameras present, Ray landed the Autogiro on the south lawn of the White House, a landing strip of only two hundred feet, with the Pitcairn landing almost within its own length. It was the first rotary-wing aircraft to land at the White House,[4] and its takeoff was almost as dramatic—backed up close to the building, Ray took off in a crosswind and was climbing after a ground run of only forty-three feet. He then turned northeast ahead of an advancing squall that was already churning the Potomac River, a most effective demonstration of what *Fortune* magazine called "the only basic contribution to the art of flight since the Wright brothers rode a biplane into the air in 1903."[5]

A fixed-wing aircraft derives lift from its wings in accordance with Bernoulli's law, which states that because air has to flow faster over the curved top of a wing surface than under the flat bottom and that faster air exerts less pressure on the top of the wing, lift is produced.[6] The difficulty with a fixed-wing aircraft was that it could fall out of the sky when its speed was less than that necessary to keep sufficient air flowing over the wing—a condition known as a *stall.* Cierva's innovation was to separate the speed of the wing from the speed of the aircraft by letting the wing surfaces—here two to four rotor blades, each functioning as a separate wing—rotate freely above the aircraft. He named it an Autogiro, and it was the first aircraft that could not stall. The air flowing up through the rotor blades caused them to turn, a condition he called *autorotation,* and continually provided lift, even when the aircraft was descending.

The helicopter, which appeared after the Autogiro, flies with a powered rotor by "pushing" down. Should a helicopter lose power, its pilot immediately disconnects the rotor blades from its engine and allows free rotation, which lets the aircraft gently descend as it derives lift from autorotation; in such emergencies the helicopter becomes an Autogiro.

William E. Hunt later a pioneering rotary-flight engineer and an associate of Igor Sikorsky, described the dramatic impact of Ray's landing in the following manner:

In 1931, while I was attending a technical school in Washington, DC, taking a two-year course in aircraft design, aerodynamics and draughting, I happened to be passing the back of the White House when I became aware of an unusually loud aircraft engine noise above and behind me. Quickly

Amelia Earhart with the Beech-Nut Pitcairn PCA-2 Autogiro in 1931.
(Courtesy of Stephen Pitcairn, from the Pitcairn Archives.)

turning around, I saw it was an autogyro, and as it passed overhead it aimed
directly towards the Rose Garden, which it circled several times. Then I
noticed several people standing behind a table on the left side of the garden.

I ran across the street and stood peering over the iron fencing, hedge and
flower garden just as the autogyro made its final pass (years later, veteran
pilot Jim Ray told me he almost decided not to land) and turned directly
towards me, descending at a very steep angle. The aircraft made a quick
nose-up "flare-out," its whirling rotors tilting backwards to provide maximum lift, as Jim Ray with consummate skill made a perfect slow, soft landing on the lawn, the first and last time an autogyro would do so. He then
turned round and taxied back as far as he could, and turned round again, facing into the wind, keeping the rotor turning with the propeller wash, ready

Amelia Earhart with the Pitcairn factory PCA-2 Autogiro on April 8, 1931, at Pitcairn Field, Willow Grove, Pennsylvania, where she set a world altitude record of 18,415 feet. (Courtesy of Stephen Pitcairn, from the Pitcairn Archives.)

for take-off.... With wheel brakes hard-on, power was increased, causing the rotor blades to whirl faster, then the pilot simultaneously applied full power and released the wheel brakes into the wind. With another quick nose-up the 'gyro rose as if by magic at a steep angle to clear the tree-encircled lawn....Little did I realize at the time that this form of aircraft would lead me to the Sikorsky Division of the United Aircraft Corporation.[7]

This, then, was the best of times for Harold Frederick Pitcairn and the Pitcairn-Cierva Autogiro Company. Since acquiring the American manufacturing and licensing rights to Cierva's patents and inventions on February 14, 1929, for the significant sum of $300,000, Pitcairn had sold his transport airline and airmail service in June 1929 and embarked on a developmental program that resulted in the PCA-2, which had received its government certification on April 2, 1931.

Pitcairn needed to publicize his new model, and Amelia Earhart, ever mindful of current aviation developments, had been as eager to fly Pitcairn's new Autogiro as he apparently was for her to do so with the assured publicity. On April 8, 1931, she set an altitude record of 18,415 feet. On her first attempt, watched by over five hundred spectators, the news media, and cameras, she reached 18,500 feet but experienced fuel-line trouble. After a light lunch and a nap and remedial action by the Pitcairn mechanics, she made a second attempt, which took three hours, and returned at dusk with a National Aeronautic Association (NAA) sealed barograph attesting to the altitude record. The record would stand until broken by Captain Lewis L. A. Yancey, who reached 21,500 feet over Boston, Massachusetts, on September 25, 1932.

April 1931 was truly a month of triumphs for the Pitcairn-Cierva Autogiro Company as experienced by few aviation companies and catapulted the Autogiro into the American popular imagination. It was touted as the safest aircraft flying, increasing its appeal to a public recently battered by the death of noted coach Knute Rockne in the March 31, 1931, crash of a Fokker Tri-Motor (NC999E) at Bazaar, Kansas. But although these successes seemed to portend a bright future for the Autogiro, it was not to be, and Pitcairn would effectively be out of the business by 1941, the result of bad economic conditions, poor business decisions, and the advent of the helicopter.

Also, the fate of Autogiro development was indelibly impacted in England on the fog-shrouded morning of December 9, 1936. The KLM DC-2 (PH-AKL) flight, bound for Amsterdam from the airport at Croydon Aerodrome, London, was delayed beyond its 10:00 A.M. scheduled takeoff. Under the command of a Captain Hautmeyer, it finally made an instrument takeoff, made necessary by visibility that rarely exceeded 25 yards. The pilot, although experienced, inadvertently swung subtly off to the left from the white 2,100-foot takeoff guideline, which proved to be a fatal mistake to all but two of the sixteen people on board when the plane crashed—at the time, the worst air disaster in British history. Thus did Juan de la Cierva die—an ironic end to a man so passionately committed to developing a safe means of air travel. Stripped of his passion, the company he founded in England would quickly shift the focus of its efforts toward developing a helicopter. And while a few other Cierva licensees, notably the Kellett Brothers of Philadelphia, Pennsylvania,[8] would pursue Autogiro development, the technology would effectively disappear with the coming of the helicopter, which did the one thing that the Autogiro could not—hover over a place without vertical movement—an advantage that, particularly in the minds of senior military officers who procured the new craft, more than made up for increased cost and mechanical complexity.

Although it is tempting to assert that the Autogiro disappeared because of the shift of the world economies toward the waging of global war, this would be incorrect. For even though the United States minimally investigated application of the Kellett and Pitcairn Autogiros for specific combat missions, Japan, England, France, Sweden, Germany, and the Soviet Union regularly made use of Autogiro technology. But it was clearly the helicopter that had focused public and military attention on rotary-wing aircraft. The result was that all attempts to revive or even advance Autogiro technology, most notably by combining it with the helicopter in configurations that came to be known as a *compound helicopter* or *convertiplane,* came to naught.

Spectacular autorotational achievements such as Fairey Aviation's "Rotodyne" of England and its Russian counterpart, the Kamov Ka-22 Vintokrulya (Vintokryl) ("Screw Wing") were abandoned by the early 1960s, and it appeared that Cierva's technology would become so insignificant and irrelevant as to completely vanish from aviation sight. The impressive Umbaugh 18A (later, Air & Space 18A)[9] and McCulloch J-2 gyroplanes[10] of America, and the Canadian Avian 2/180[11] made little impact through the 1970s and soon faded from public view. Few recall that England's Wing Commander Ken Wallis RAF (Retired) flew his *Little Nellie* autogyro in the 1967 James Bond film *You Only Live Twice* (which was not a miniature helicopter as one character in the movie claims) or that Canadian Peter Rowland Payne designed the legendary Avian 2/180 gyroplane. If remembered at all, Payne is associated with his better-known inventions, notably the nonlethal rubber bullet (1967) and the crash-test dummy (anthropomorphic mannequin).

That this seeming inevitability of technological extinction did not occur was occasioned in part by the most unlikely combination: Igor Bensen, a Russian-immigrant engineer; a World War II German submarine rotary-wing kite invented by Professor Heinrich K. J. Focke; and the English 1942 Airborne Forces Experimental Establishment (AFEE) Rotachute designed by Austrian immigrant Raoul Hafner, Dr. J. A. J. Bennett, and O. L. L. Fitzwilliams. Although these obscure rotary-wing inventions had virtually no impact on the course of events for the combatants, Bensen's experiences with both would lead directly to the creation of the Gyrocopter, a term he would trademark just as Cierva had done with Autogiro thirty years earlier. The Gyrocopter, fondly known as the "flying lawnchair," was the unlikely vehicle that would enable autorotation technology to survive.

For a while in the middle of the twentieth century, it certainly appeared that Cierva's dream would survive *only* as an amateur project constructed in thousands of garages and backyards, but even that was not assured. For while Bensen preserved and nurtured it with missionary zeal, the Popular Rotorcraft Association, the movement he created, soon had to be rescued from his leadership lest it too fall victim to the very limitations embodied by that vision. And it was only during the subsequent leadership of Ken Brock, who never appears to have seen himself as the revolutionary that history reveals him to be, that the gyroplane movement has emerged as a true international endeavor.

Ultimately, the survival of this amazing aviation technology is to be found in its simplicity and safety—the very factors that characterized its early development. In many ways this has been a technology seeking a task and an effective form of business organization to achieve a successful commercial result, and that is only now happening. It is only in the eighth decade of the "windmill plane" that its industrial potential seems about to be realized, with the creation and marketing of the Groen Brothers Aviation (GBA) Hawk 4 series of gyroplanes to an international market. This success, which eluded so many previous entrepreneurs, is as much the product of the creation of an effective business organization and effective business decision making as it is the result of a superior gyroplane model.

This book presents the history and development of the autorotational aircraft. Whether called an Autogiro (trademarked by Cierva and referring to Cierva-licensed aircraft), Gyrocopter (the Bensen trademark until the company ceased operations), autogyro, or the general term *gyroplane,* there is a wonder as to the means and methods by which it survived bad business decision making, poor timing, the Depression, a world war, and bad luck. That the Autogiro survived is even more impressive given that for almost all of its first eight decades, it totally and completely failed to find a market! It was, in every real sense, a failure from the late 1930s onward. Its survival was not in any way assured, and in a dramatic turn of events, the most exciting developments are only now emerging at the start of the twenty-first century with the CarterCopter, which is poised to make aviation history in ways that Cierva, Pitcairn, and their contemporaries could not have anticipated but in a manner that they could immediately appreciate and applaud as the fruition of long-ago dreams of rotating wings. The story of the survival of this technology in the shadow of constant failure is a suspenseful drama with a fascinating cast of characters, but it is a worthy paradigm for technologies seemingly passed by. In this time ever-increasingly characterized as one of rapid change, there may be many such technologies that are worthy of current innovation. This book tells that story.

NOTES

1. The standard work on Harold F. Pitcairn is Frank Kingston Smith's *Legacy of Wings: The Harold F. Pitcairn Story* (New York: Jason Aronson, 1981). See also George Townson, *Autogiro: The Story of 'the Windmill Plane'* (Fallbrook, California: Aero Publishers; Trenton, New Jersey: Townson, 1985); Frank Anders, "The Forgotten Rotorcraft Pioneer: Harold F. Pitcairn," *Rotor & Wing International* 25, no. 5 (May 1990): 34–37; Frank Kingston Smith, "Mr. Pitcairn's Autogiros," *Airpower* 12, no. 2 (March 1983): 28–49; Carl R. Gunther, "Autogiro: Part 1 and 2," *Popular Rotorcraft Flying* 17, no. 5–6 (October–December 1979); George Townson, "History of the Autogiro," *American Helicopter Society Newsletter* 11, no. 4 (March 1961); George Townson, "General Information and History of the Autogiro," *American Helicopter Society Newsletter* 11, no. 4 (March 1961); George Townson and Howard Levy, "The History of the Autogiro: Part Two," *Air Classics Quarterly Review* 4, no. 3 (fall 1977): 4–19, 110–14.

2. Smith, *Legacy of Wings,* p. 184.

3. Smith, "Mr. Pitcairn's Autogiros," p. 36.

4. For photographs of the landing and takeoff at the White House, see Smith, p. 36.

5. "Autogiro in 1936," *Fortune* 13, no. 3 (March 1936): 88–93, 130–31, 134, 137; 88.

6. Francis Pope and Arthur S. Otis, *Aeronautics* (Yonkers-on-Hudson, New York: World Book Company, 1941) pp. 106–10.

7. See William E. Hunt, *'Heelicopter' Pioneering with Igor Sikorsky* (London: Airlife Publishing Ltd., 1998), pp. 35–36.

8. Philadelphian Roderick "Rodney" G. Kellett and his brother W. Wallace Kellett, along with C. Townsend Ludington and his brother Nicolas Townsend, had founded the Kellett Aircraft company in 1929 with a license from the Autogiro Company of America. The Kelletts had originally begun in the aviation business as dealers in 1923 operating out of the Pine Valley airport. Peter W. Brooks, *Cierva Autogiros: The Development of Rotary-Wing Flight* (Washington, D.C.: Smithsonian Institution Press, 1988) pp. 134–36. By 1936 Kellett Aircraft had invested $500,000 into the Autogiro business. See "Autogiro in 1936," p. 130 (photo caption).

9. Kas Thomas, "The Umbaugh Story: Rags to Riches (and Back?)," *Popular Rotorcraft Flying* 10, no. 2 (March–April 1972): 10, 23.

10. Bill Sanders, "The Rebirth of N4353G and N4364G," *Rotorcraft* 28, no. 4 (June–July 1990): 14–19; Robert Zimmerman, "A Reality at Last: The Family Autogyro," *Popular Mechanics* 131, no. 3 (March 1969): 112–13. For a critical appraisal of the J-2, see Peter Garrison, "Everybody Loves an Autogyro," *Flying* 88, no. 2 (February 1971): 66–68.

11. David Mondey (ed.), *The Complete Illustrated Encyclopedia of the World's Aircraft* (Secaucus, New Jersey: Chartwell Books, Inc., 1978), pp. 86–87.

Chapter 1

JUAN DE LA CIERVA

"It was still the only basic contribution to the art of flight since
the Wright brothers rode a biplane into the air in 1903."

"Cierva, the soft-spoken Spaniard with the bulbous intellectual forehead..."

"Autogiro in 1936," *Fortune*

Juan de la Cierva y Codorníu, "the greatest name in the history of rotary-
wing flight before Igor Sikorsky evolved the first widely used type of helicopter,"[1] has largely been forgotten.[2] Although his discovery (some would
say rediscovery)[3] of autorotational flight and development of the first functional rotary-wing aircraft—which he called the Autogiro and which
derived its lift from autorotation[4]—preceded Sikorsky by sixteen years, he
has largely been relegated to, at best, a curious footnote in aviation history,
and then only in reference to the development of the helicopter.

Autorotation is defined as "the process of producing lift with freely-
rotating aerofoils by means of the aerodynamic forces resulting from an
upward flow of air."[5] This means that as long as the aircraft is moving, lift
will be produced by the movement of air up through the rotating wing surfaces (called rotors). Thus the Autogiro's unpowered rotor lifts up, not
pushes down as does the helicopter's rotor. The Autogiro, the unique
spelling coined as a proprietary name by Cierva, was characterized in an
enduring descriptive manner as the "windmill plane"[6] and as a "'devil's
darning-needle,' a corkscrew plane, a dragonfly, a flapper flying machine...."

Spanish inventor Juan de la Cierva. (Courtesy of Stephen Pitcairn,
from the Pitcairn Archives.)

an intoxicated duck... [comparing] its method of making a turn in the air to
Charles Chaplin in his favorite fashion of turning a corner in a hurry."[7] Yet
these descriptions do not begin to capture the excitement that occasioned its
flight. Even today one marvels at the photographs and surviving films of the
Autogiros of the forgotten American pioneer Harold F. Pitcairn flying over
Manhattan, over the partially constructed George Washington Bridge, and
past the Statue of Liberty—and when his son Stephen "Steve" Pitcairn flies
one of the few surviving flying Autogiros,[8] *Miss Champion* (NC11609)[9] at
air shows, a hush falls over the crowd as aviation history flies by at eighteen
to twenty miles per hour, twenty feet off the ground! In what has been
termed the Golden Age of Aviation,[10] Cierva and Pitcairn's aircraft captured

the world's attention. And it all began with a young man, the son of an aristocratic family in Spain, who would first become, before inventing the Autogiro, the "Father of Spanish Aviation."

Juan de la Cierva was born in Murcia, Spain, on September 21, 1895, the first son of a privileged family. His father, Don Juan de la Cierva Peñafiel, made the family fortune from the practice of law and land ownership. After World War I the elder Cierva served in successive Conservative government administrations as a delegate to the national assembly, and as minister of education, the interior, war, the treasury, and development. He remained both active and identified with Conservative governments until he left public life with the formation of the Spanish Republic in 1931. He died in 1938, but unfortunately lived long enough to attend the funeral of his eldest son.

There was little in the education of the younger Cierva that would portend his interest in aviation. He attended local schools in Murcia and received private tutoring when the family moved to cosmopolitan Madrid in 1905. Although there is little mention of the impact of the Wright brothers first flight at Kill Devil Hill, North Carolina, in 1903 on the eight-year-old boy, while a student in Madrid Cierva begun studying the work of Samuel Pierpont Langley,[11] Sir Hiram Stevens Maxim,[12] Clément Ader,[13] Octave Chanute,[14] and other contemporary aeronautical writers.[15] He was particularly impressed with the scientific inquiries of Otto Lilienthal,[16] who had died in 1896, the year after Cierva's birth—and who had articulated principles by which manned flight could be made a reality. Cierva was also taken by the fact that Lilienthal had made more than two thousand gliding flights, a paradigm of practical and theoretical aviation that was to distinguish Cierva's own efforts.[17] Aviation achievements were regularly reported by the newspapers in Europe and the newly established aviation journals *Revista de locomocion aerea*[18] and *Aviacion,* and Cierva was certainly aware of the first officially observed European powered flight of Brazilian Alberto Santos-Dumont[19] in France on October 23, 1906, and of Wilbur Wright's demonstration flights in France in late 1908. Those flights led to the spread of "aviation fever."[20] He also could not help but be inspired by reports of the first crossing of the English Channel by Louis Blériot on July 25, 1909. By 1908–9, Cierva had decided to make aviation his career, later observing that he doubted that his parents looked favorably on his aeronautical endeavors. But he also noted that "boys find ways and means to satisfy their extraordinary impulses, and we spent much of our spare time and most of our spare cash in these primitive experiments in practical aeronautics."[21]

The first powered flights in Spain were made in Barcelona on February 11, 1910, followed by appearances in Madrid on March 23 by French pilot Julien Mamet in his Blériot. These later flights were witnessed by fourteen-year-old Cierva and his friends, who resolved to push ahead with their own design efforts. Beginning with models and larger kites, the aviation adventurers—Juan, his younger brother Ricardo, José "Pepe" Barcala Moreno, Pablo Diaz Fernández, Tomás de Martin-Barbadillo, Rafael Silvela Tordesillas, and Antonio Hernandez-Ros Murcia Codorníu (the mother's name following the father's, as was Spanish custom at the time)—advanced to gliders during the 1910–11 period, flying craft that were not very successful but gaining enough experience to convince Cierva to pursue aviation in his college education.

In 1911 Cierva enrolled at the Civil Engineering College of Madrid with his friend Pepe Barcala, and by 1912, along with Pablo Diaz, the young men were ready to build the first Spanish airplane. In 1912 the Spanish government purchased its aircraft from external sources, notably from France. A French pilot, Jean Mauvais, then living in Madrid, sold aviation supplies to the government and gave exhibition flights for the public in his Sommer biplane. His reputation was as a genial gentleman, and he had become friends with the young men, who spent spare time at the nearby Cuatro Vientos Airport, six and a half miles southwest of Madrid. The local fliers used to play a recurring practical joke on the French pilot by hiding his small Dion Bouton single-cylinder two-seat automobile at various places about the airfield—a ploy soon mastered by Mauvais as he regularly retrieved his car to drive back into Madrid at the end of the day. But on one occasion the locals were more inventive than usual, and it resulted in a lengthy search by the increasingly frustrated pilot. He was grateful for the help of Cierva and his friends, who eventually found the small automobile suspended by a rope from the ceiling inside the dirigible hanger! But it was Mauvais's misfortunate that would lead Cierva, Barcala, and Diaz to construct the first Spanish airplane.

Aviation was new and a relatively uncommon phenomenon—and although flying achievements and daring aerial feats were acclaimed, the public had not yet fully realized the dangers inherent in a speeding plane, with its various whirling blades and moving wings. Crowds regularly rushed onto runways to mob the successful aviator, as they would do to Charles Lindbergh when he landed at Paris on May 21, 1927. Famed woman early aviator Matilde Moisant's flying career would end on April 14, 1912 when she crashed after avoiding the spectators rushing on the runway at Wichita Falls, Texas. Pulled from the wreckage with her hair and clothes on fire, she was not seriously hurt, but her family's concern led her to end a promising aviation career at just twenty-six years old.[22] This rushing by the crowd happened to Mauvais at the conclusion of a demonstration flight at a Madrid race course. The spectators, having been amazed by the aerial spectacle, rushed the landing plane with tragic results: although the pilot tried to avoid the approaching crowd, several people were killed and the plane was wrecked. As a result, Mauvais, who had emerged almost unscathed, announced that he was quitting flying, and the wreckage of the Sommer was removed and unceremoniously heaped in back of the pilot's airport workshop.

The boys entered into an agreement with Mauvais to purchase the wreckage and, using what could be salvaged, began to construct a new aircraft. It was also agreed that Mauvais would act as the test pilot of the aircraft, as none of the boys could then fly. Almost twenty years later Cierva speculated that Mauvais "probably thought this an excellent joke; very likely he supposed the bargain was safeguarded by the likelihood that we would never complete our part of the contract."[23] The boys used their joint capital, approximately $60, and started construction in the workshop of Pablo Diaz's father at No. 10, Calle de Velázquez, Madrid. The boys, with their severely limited budget and fearing the disapproval of concerned parents, had to fabricate most of the parts that went into the rebuilding effort. A significant portion of their limited capital paid for the services of a carpenter at an hourly wage, but when it came time to replace the propeller that had been destroyed in the original crash, there was a genuine crisis. Cierva knew from his reading that propellers were crafted from seasoned wood, and he found it in the most unlikely of places for a young man from an aristocratic family—the barroom of a local inn![24] Cierva reasoned that the counter of the bar had been bathed in spilled drinks for years and that the constant exposure to alcohol would surely have seasoned the countertop; of perhaps equal importance, it was within the remaining funds, so they purchased it and from that countertop carved a propeller. Cierva had learned, by this point, enough of the requirements to design sufficient curve and balance that the propeller functioned, much to Mauvais's surprise!

It must be regarded as one of the ironies of Cierva's life and Spanish aviation that the propeller of the first aircraft constructed in Spain was crafted from a wine-soaked tabletop from a local bar. But although this failed to achieve mythic status, later Spanish lore would assert that Cierva's inspiration for the flexible rotor blade that made the Autogiro possible was occasioned during an operatic performance of *Don Quixote* when he saw the windmill onstage with its flexible blades![25] Although there is various evidence that Cierva was inspired at the opera, his intellect having been active in considering the matter of rotor design, actually there is agreement and a family memory that the opera was Verdi's *Aïda*.[26] So although invocation of a Spanish national literary hero is worthy of myth, propellers from local drinking establishments are apparently quite forgettable! But the aircraft constructed by the boys was not forgotten—using the first initials from each of their last names, the young builders dubbed their biplane the BCD-1, but it quickly became known as *El Cangrejo*—the *"Red Crab"*—because they had colored the wings and fuselage with aniline dye to a deep scarlet color.[27] Finally the day came when Mauvais took to the air, test-flying the plane as part of their deal—and it flew quite well! Although he had given up flying after the fatal accident that had rendered the original Sommer a pile of debris, he had helped the boys with advice and become interested in their plane's progress. Now he would perform a service that would echo in Cierva's life until his death in 1936—Mauvais often took Cierva flying.

Cierva flew in it many times, and Mauvais let the young man reach around from the rear passenger seat and hold the wheel. Cierva probably remembered those moments when he himself took flying lessons in 1927 so that he could fly his Autogiro! It was apparent that, despite some slight wobble and vibration, the problem of the propeller had successfully been solved. The wings were another matter. The boys had run short of money, and to cover the wing surfaces, they had purchased the cheapest canvas, which was stretched over the wooden framework of the wings and doped with a great deal of glue. Although the propeller lasted as long as the airplane flew, the wings became sticky when exposed to rain. Eventually the plane began to disintegrate. When the wings began to vibrate in flight, it was perceived as merely a minor annoyance. When parts of the aircraft began to fall off, however, it was an inglorious end to the *Red Crab,* Spain's first airplane.

The boys were emboldened by their success with the *Red Crab* and the subsequent benefit of the favorable impression that it had made upon their parents (including an increased allowance to spend on aviation projects). In 1913 they constructed a racing monoplane, its design probably inspired by the publicity that the French Nieuport was receiving in European air races. The new airplane, dubbed the BCD-2, was originally powered by a twenty-four horsepower Anzani engine, which was designed by Italian motorcycle builder Alessandro Anzani, then living in Courbevoie, France. But needing more power, the boys soon substituted a sixty horsepower Le

Rhône engine. First flown by pilot Julio Adaro Terradillos at Getafe airfield outside Madrid in December 1913, the aircraft proved unstable and soon crashed. Repaired with what the young men hoped would be sufficient corrections, the BCD-2 was transported to Cuatro Vientos outside Madrid where Mauvais agreed to be the new test pilot. Its development ended for good when it crashed again. But this setback did not dissuade the boys, and Cierva's next project was to prove far more ambitious although, unfortunately, no more successful. However, out of that experience would come the Autogiro, the most significant aviation innovation since the Wright brothers.

The young men then devoted their time to their studies—with Cierva concentrating on his classes at the Civil Engineering School in Madrid where he studied a course that resulted, in 1917, in graduation with the title of Ingeniero de Caminos, Canales y Puertos (Engineer of Roads, Canals and Ports). There were neither aeronautical schools nor courses of aeronautical study in Spain, and Cierva's academic achievement was then considered the highest engineering degree in Spain. His achievements, however, with the Autogiro were to be such that in 1930, when the first University for Aeronautics was established in Spain, he was awarded an honorary degree, granting him the additional academic title of Ingeniero de Construcciones Aeronáuticas.[28] By 1931 Cierva would become a permanent consulting member of the Junta Superior de Estudios y Pensiones para Extrajero, an institution for aeronautical research that awarded scholarships for study in foreign countries; a member of the Association of Spanish Civil Engineers and the Association of Aeronautical Engineers in Spain; an honorary member of the AIDA [a society for aviation engineering] in Italy; a member of the Société Française de Locomotion Aerienne in France; and a member of the British, German, French, Spanish, and Belgian Aero Clubs. Additionally, he would be made Chevalier of the Legion of Honor in France and holder of the Order of Leopold in Belgium, and awarded the Cross of Alfonso XII in Spain.

While a university student, Cierva followed the course of the war in Europe and read avidly of the rapid developments in air combat; spurred on by the needs of combat, the nature of the airplane was quickly changing as it developed into a lethal weapon. With such progress came scientific articles and treatises, which were devoured by young Cierva. Although he had ceased practical experimentation after the BCD-2, he immersed himself in the theoretical literature of wing curves and airfoil design, mathematically based subjects that were to furnish a theoretical foundation for the coming Autogiro development. However, one more practical excursion into airplane design was just around the corner in 1918.

Spain, which had remained neutral in the world war, now resolved to nurture local aviation talent to incorporate airpower into its armed forces, as the ongoing conflict had revealed such force to be a vital element of modern warfare. On September 5, 1918, the Spanish government announced an aviation design contest under the direction of Colonel Don Julio Rodríguez Mourelo, director of Spanish Air Services, for pursuit, reconnaissance, and bomber aircraft. Although the prize offered in each category was only 60,000 pesetas, equivalent then to about $9,600 ($60,000–$70,000 today), Cierva was undeterred. Even the possibility that there would be additional orders of the winning design was not a major incentive for the young graduate, as he felt it unlikely that the Spanish government would order many aircraft. Rather, he saw this as the beginning of his aviation career. Although he had involved himself in his family's business and political affairs after graduation in line with his father's expectations, his avid interest in aviation had continued and he could not let this opportunity pass. Perhaps because of ease of construction, the relatively smaller size, or the fact that pursuit and reconnaissance aircraft (out of which pursuit craft had evolved), there were several entries in those classes. But there was only one in the bomber category—Juan de la Cierva! His aircraft would have to be ready to fly in trials in March of the following year. The budget for this construction was set at 150,000 pesetas, or about $24,000[29] ($150,000–$175,000 today). This sizable amount was jointly financed by Cierva's father and a wealthy friend, Don Juan Vitórica Casuso, Conde de los Moriles.

The bomber, called the C-3, was a large aircraft, with an eighty-two foot, one-quarter inch wingspan, with a loaded weight of 11,000 pounds, and powered by three 225 horsepower Hispano Suiza 8Ba engines, produced in Barcelona by a nominally Spanish company (it was actually French). It was constructed with the help of his old friend Pablo Diaz in the Vitórica carriage workshops near Madrid, with Cierva noting, in what would become a lament of amateur builders everywhere, "we were obliged to tear down the walls of our workshop in order to get it out for its test flights."[30] After components had been assembled at Cuatro Vientos, it was an imposing aircraft indeed, a biplane designed to carry fourteen passengers or more than 2,000 pounds of bombs in addition to its two-person crew. Although the design followed generally accepted principles, it incorporated unique and innovative features, including three tractor (forward-facing) engines[31] and economical distribution of bracing struts between the wings which improved both speed and performance. Cierva also employed an original wing section, the design for which he had derived mathematically. These advanced design features were a result of Cierva's mathematical studies, an approach that would distinguish his efforts in all things aeronautical—first came the theory, then the application.[32] On February 13, 1930, in speaking to the Royal Aeronautical Society, a group whose opinions and approval he greatly valued and of which he was an associate fellow, Cierva summed up his method in this manner:

> My engineering theories, all based on energy equations since 1924 and very similar in general lines to that developed later by Mr. C.N.H. Locke, and published by the Air Ministry in the R. & M. 1127, in 1927 were not a useful guide to me until, in 1928, I succeeded in finding an analytic method of integrating the frictional losses of energy, when the aerofoil used in the Göttingen 429, which gives the average profile drag in any conditions and for any value of the parameters defining a rotor. The theory completed in this manner has allowed me to produce Autogiros with the correct proportions and I can safely say that the present results check with amazing accuracy the simple assumptions which form the basis of my theory.[33]

Even though the design was not destined for success in the bomber competition, several of Cierva's innovations were evident in the French Caudron C.25 transport plane exhibited in the 1919 Paris Salon de l'Aéronautique.

Cierva and his backers chose Captain Julio Ríos Argüeso, an experienced army pilot to fly the C-3. Captain Ríos was an experienced pursuit pilot who had been wounded by a sharpshooter while flying low against the Riffs in Morocco. He came highly recommended and was regarded as a thoroughly capable aviator. However, Ríos had never previously flown a large biplane, a lack of experience that proved his undoing on July 8, 1919, in its first flight at Cuatro Vientos. It was Cierva's conclusion, and that of the spectators to the first (and only!) brief flight of the C-3, that Ríos had initially been nervous and apprehensive but had soon gained an easy confidence in the manner in which it flew. The C-3 proved reasonably responsive to effective control—but then disaster! When Ríos, now apparently overconfident, acted as if he were flying a pursuit plane in a tight low turn, he caused the large trimotor bomber to lose its lift. All fixed-wing aircraft will lose essential lift if their speed falls beneath a predetermined level, and that loss of lift is called a *stall.* The result was a nonfatal crash that ended Cierva's hopes for the competition and, apparently, the chance for an aviation career.

Cierva, at his father's insistence and encouragement, again returned to family business affairs. Becoming manager of his father's agricultural and manufacturing interests, he also entered politics and became a member of the Spanish parliament in 1919, representing his native Murcia where the family still retained land. He was to remain a participating member of the Spanish government until 1923, and as we shall see, to play a significant if relatively unknown role in Spanish political affairs until his death in 1936. On December 10, 1919, Cierva married María Luisa Gómez-Acebo Varona. Their first child, also named Juan, was born on July 24, 1921. But of equal importance, he never stopped thinking about the crash of the C-3. Beginning late in 1919, just prior to his marriage, Cierva began a serious investigation into aviation theoretical literature as to how a stall-proof aircraft might be constructed. He had become convinced that his bomber design had been sound—it had flown successfully at its first trial—and that the pilot was not at fault. He stated that "[a] good airplane in the hands of a good pilot had no business to be turned so suddenly and conclusively into a useless tangle of fabric and machinery just because the pilot made a miscalculation of speed and distance."[34] Years later he stated that this thought was the "germ and genesis of the Autogiro."[35]

During his exhaustive search of the literature,[36] Cierva, now considered one of Spain's leading aeronautical engineers, became familiar with contemporary speculation about the helicopter. But he rejected this possibility as being too complicated and unlikely to lead to success,[37] but he became intrigued with the concept of a rotating wing as a means of dealing with airplane stall—the idea being that the rotating wings, or rotors, would always be in motion independent of the motion of the airplane itself. His C-3 had stalled, he reasoned, when the aircraft lost flying speed and fell below the limit at which sufficient lift was generated (the stall speed). He further reasoned that if the wing could be rotated and could generate lift, the speed of the aircraft would not be relevant. Cierva would call the phenomenon of a rotating wing generating lift *autorotation,*[38] and it led to the Autogiro. He had first demonstrated it with a helicopter toy launched from the balcony of his parent's Madrid home and watched it rotate as it descended. Each blade of the rotor, if canted at an appropriate angle, would function as a wing; in the case of autorotation, the air coming up through the rotor would cause it to turn and provide lift. All that was necessary was that the aircraft be powered to move forward so that the airflow up through the rotor could be maintained and lift provided by the unpowered, freely rotating blades. If the forward power failed, the rotor blades would still generate lift as air continued to flow up through the rotor even as the aircraft settled to the ground. And of greatest importance to Cierva, the autorotating aircraft could not stall.

Having arrived at the idea of autorotation, Cierva characteristically commenced in early 1920 an extensive scientific investigation of autorotation. Autorotation had been proposed in a paper in 1915 by Hodgson to the British Institute of Automobile Engineers and had even been described in a June, 1919, British patent application, No. 146,265, by the Argentinean of Italian descent, the Marquis Raul de Pateras Pescara, then living in France and active in helicopter development (although Pescara described autorotation, he never made use of it). Additionally, the French inventor Lucien Chauviére had claimed in 1917, based on the earlier writings of Russian D. P. Riabouchinskii, to have patented autorotation. There is no evidence that Cierva was aware of these individuals or their ideas, and he is honored for the discovery and application of autorotation. He himself did not cite any of the other inventors, but clearly understood and presented his discovery as something new. He wrote:

> My invention does not deny any of the advances of aeronautical theory, nor does it dispute the credit due to the achievements of the pass quarter century. In a thousand details it does not materially differ from the airplane. But it differs profoundly in a single important essential, which I have attempted to define as the basic principle of flight. It differs because it applies in a new way the idea of wings in motion as the essential of the flying machine—the law which permits man to fly in a heavy craft of wood, metal and fabric.[39]

He began with models and progressed to full-sized aircraft,[40] each of which was tested in the newly constructed wind tunnel at the Aeronautical Laboratory at Cuatro Vientos. Cierva, à la Thomas Edison before him, stated that "invention is more often in debt to persistence than to inspiration,"[41] and given that his first three Autogiro designs failed, his persistence stood him in good stead. Although little is known of the construction of his first model, dubbed the C.1, it is generally assumed that it was constructed at Getafe airfield near Madrid in the workshop of Amalio Diaz Fernandez, the older brother of Cierva's boyhood friend Pablo. Diaz was an experienced aircraft builder, having constructed a fighter for designer Julio Adara Tarradillos that had won a prize in the 1919 Spanish military competition. Cierva made use of old airframes and secondhand engines, primarily the fuselage, landing gear, and vertical tail surfaces of a 1911 French Deperdussin monoplane, powered by a sixty horsepower Le Rhône engine, readily available as salvage from World War I, as he did not care how the aircraft appeared—he was trying to prove his theories and show the world a new way of flying.

The 772-pound aircraft retained the wings and standard control surfaces of the monoplane, but mounted two counterrotating, rigidly braced rotors, one on top of the other, with a vertical control surface on the very top of the rotor pylon. It looked as if it would never fly, and looks were not deceiving—no matter how ingeniously its test pilot, Cierva's brother-in-law Captain Filipe Gómez-Acebo Torre, tried to coax it aloft, it refused to leave the ground. Analysis clearly demonstrated that the rotor blades interfered with each other as the lower rotor turned 50 rotations per minute (rpm), less than half of the top rotor's 110 rpm. The unceremonious result was an imbalance of lift that consistently rolled the aircraft on its side, with unpleasant but nonfatal results for the hapless but enthusiastic pilot. But Cierva was undeterred—the C.1 confirmed his theory of the autorotational abilities of an unpowered rotor.[42] Rejecting the suggestion that the two rotors be linked by gears to ensure identical rotation, Cierva did not hesitate to go back to the drawing board, dedicated to the creation of a safe aircraft that was less mechanically complex and therefore more reliable.

The C.2, based on models that had been tested, was constructed on an adapted Spanish Military Aircraft biplane fuselage in a carpentry shop established for that purpose and set up by Cierva, his brother Ricardo, and Pablo Diaz in Madrid. On November 18, 1920, the three builders agreed to establish a company for the manufacture of aircraft and general engineering products. The C.2 abandoned the two-rotor approach in favor of a large five-blade rotor, with blades rigidly braced by high-tensile steel wire. Power had been increased to 110 horsepower, metal fittings were subcontracted to the Industrial College, and duralumin spars for the rotor blades ordered from France. As the delivery time from France stretched into the spring of 1921, Cierva proceeded with the design of C.3, which was actually completed and tested before the C.2. This timing occasioned much subsequent confusion, as Cierva himself refers to the C.3 as his "second" and to the C.2 as his "third" Autogiro in his 1931 book.[43]

The C.3 was completed and ready for testing at Getafe airfield in June 1921. Based on a "Hanriot fuselage,"[44] it also employed a single rotor, but with only three large blades. The fabric-covered blades were very broad (had a wide chord) and could be controlled by a warping (twisting), which was achieved by the pilot, who could vary the incidence, or *tilt,* of each blade as it revolved by means of a coaxial shift that was attached to struts in each rotor blade. But these cantilever blades, as a result of shifts in the center of pressure as they revolved, could not achieve stability with the changing speed of the aircraft; C.3 never got more than a few inches off the ground before rolling on its side. Cierva was persistent—the machine, extensively tested by its new pilot, Lieutenant José Rodríguez Diaz de Lecea, was damaged and rebuilt several times, and was tried in nine different forms, to little success. Cierva had almost concluded that rigid blades would not work—in fact, the first rigid-rotor gyroplane would be fashioned by the forgotten American rotorcraft pioneer E. Burke Wilford ten years later, and effective control of rotor pitch would be proposed by Ralph H. Upson in 1931—but Cierva had one more aircraft to test.

The metal for the rotor blades finally arrived, and C.2 was tested in early 1922. Although it achieved several hops, it never got above six feet, and even though it seemed better balanced than C.1 or C.3, it clearly experienced a loss of control and a pronounced tendency to roll over. Rebuilt three times, it was finally abandoned in April of 1923. While the testing of C.2 proceeded in 1922, Cierva began designing C.4 with grant money from the Spanish government. Those funds allowed extensive wind tunnel testing of his models[45] at the Aeronautical Laboratory at Cuatro Vientos under the supervision of Lieutenant Colonel Emilio Herrera Linares.[46] Those rubber-band-powered models had their genesis in December 1920 and were based on the C.2 five-blade rotor configuration, but with one fortunate difference—the rotor blades had been constructed of thin, flexible rattan (palm wood) by Cierva, his brother Ricardo, and cousin Antonio Hernández Ros Murcia Codorníu—that model flew! And flew well! But the C.2 and C.3 did not, and this occasioned much thinking by Cierva as to why the small model worked but the larger ones did not. He eventually realized that the essential issue was *dissymmetry of lift*—that is, the unequal lift generated by the rotor blades as they move around the rotor disk (the circular route of the blades). The rotor blades that are *advancing* (moving in the same direction as the aircraft) move faster than those blades that are *retreating* (moving in an opposite direction from that of the aircraft). The problem, then, was that the blades generated more lift while advancing than while retreating, and this dissymmetry of lift caused each of Cierva's first three models to turn over as the aircraft twisted and turned in a gyroscopic effect to balance the lift.

The moment of genuine revelation when Cierva realized the answer, as discussed earlier, came at the Theater Royal in Madrid during an opera performance of *Aïda*[47] in January 1922. Cierva suddenly realized, that the model flew precisely because the blades were made of a flexible material that allowed lift to be balanced, effectively dealing with dissymmetry. It was a remarkable rediscovery of the concept of the flexible, freely flapping blade, an idea that had been first mentioned by Charles Renard in 1904 and later patented[48] on October 29, 1908, by the French aviation pioneer Louis Bréguet and in 1913 in Germany by Hungarians Max Bartha and Josef Madzsar (patent number 249702).[49] There is little doubt that Cierva independently arrived at autorotation with flexible blades, the result of his deep and insightful thought, engineering inspiration, and thorough grounding in the mathematics of aeronautical science. For that he was and continues to be justly honored, and he was granted Spanish patent No. 81,406 on November 15, 1922.

Cierva, seeking to protect his discovery of the flapping rotor blade, secured several foreign patents, including the following:

France	No. 562,756, granted September 14, 1923
United Kingdom	No. 196,594, granted June 30, 1924
Germany	No. 426,727, granted July 27, 1925
United States	No. 1,590,497, granted June 29, 1926

As important as these patents were, Cierva limited them to the Autogiro and lost all protection they might have afforded in England for subsequent rotary-aircraft development. As observed by Cierva associate Dr. J.A.J. Bennett:[50]

The Autogiro principle, which was concerned basically with blade autorotation at a positive pitch angle, was covered by Cierva's first application for a patent but, with little knowledge of patent law at the time, he allowed this early parent application to lapse in favour of the other relating to a single rotor with articulated blades. Convinced that no Autogiro could be successful without this particular feature, he abandoned the possibility of securing a very broad patent on the basic principle of the Autogiro. In trying to reclaim later what he had lost, most of his subsequent British patents were restricted in their application to rotorcraft with autorotative blades, Cierva having established a clear distinction between the Autogiro and helicopter. *In the United States, however, no such distinction was made and most of Cierva's American patents were considered to be applicable to helicopters as well as Autogiros.* (emphasis added)

But Cierva had not waited for the patent—construction of the C.4 began in March 1922 even as C.2 continued to roll over. The C.4 had a four-blade rotor (Cierva had been experimenting with three- through five-blade rotors but may have selected four blades due to the greater ease of balancing lift, as each blade had one opposite in the rotor disk), with the individual blades attached at the root (base) with hinges that allowed them flap (i.e., to move up and down as they rotated). As Cierva anticipated, the blades assumed different angles to the horizontal rotor disk as they rotated, in effect seeking a position that equalized lift and effectively equalized the dissymmetry previously observed with blades rigidly fixed to the hub. They were braced from the top of the hub with steel cables and upward from the bottom with rubber shock absorbers, restraints that allowed flapping within a safe range. It was not a handsome machine, with the fuselage and motor probably coming from a scrap Sommer monoplane[51]—there is even some speculation that C.4 was cobbled together from the remains of C.3. Testing began upon its completion in April or May of 1922, and for a while it looked no different than its predecessors. During the testing period, which extended from June through January of 1923, pilots José María Espinosa Arias (July–August 1922) and Lieutenant Alejandro Gómez Spencer (June 1922 and September 1922–January 1923), C.4 crashed several times, as fifteen different configurations were constructed and tried with constant lesser modifications. Although the rotors worked and did not force the aircraft to turn over, control now became an issue. The solution to the difficulties was to prevent pilot manipulation of the rotor and rely on the standard fixed-wing controls on the wings and tail rudder. Cierva was optimistic that he had solved all the problems, but C.4 protested one last time, when it rolled over on January 10, 1923—but at least it rolled in the opposite direction than all other similar occasions!

NOTES

1. Peter W. Brooks, *Cierva Autogiros: The Development of Rotary-Wing Flight* (Washington, D.C.: Smithsonian Institution Press, 1988), p. 14.

2. "A half-forgotten phase in the history of aviation is the story of the unique, even mysterious, Autogiro." Charles Gablehouse, *Helicopters and Autogiros: A History of Rotating-Wing and V/STOL Aviation,* rev. ed. (Philadelphia: Lippincott, 1969), p. 35.

3. See, for example, John Fay, *The Helicopter,* 4th ed. (New York: Hippocrene Books, 1987), pp. 126–27; Charles Gablehouse, *Helicopters and Autogiros* (1969), pp. 1–7. Gablehouse further observes:

Oddly, the principle of autorotation seems to have been known long before the first Autogiro took to the air. It is thought that, as early as the Middle Ages, the masters of windmills understood they could get the wind wheels to turn into the airflow, rather than with it, by setting the sails at a very flat angle to the wind. Another example, further removed from the dream of rotating wings, is the ability to tack a sailing ship well up into the eye of the wind and still have the ship move forward; the wind can actually be striking the sail from an angle to the front, and yet the ship will be moving at an angle more or less *against* the force of the wind. And still another model is to be found in nature itself, in the whirling flight of the maple-leaf (or sycamore) seedlet, which has the form of a beautifully shaped single-blade rotor. (p. 36)

4. For an extensive theoretical discussion and mathematical modeling of autorotation, see Wayne Johnson, *Helicopter Theory* (Mineola, New York: Dover Publications, 1980), pp. 10, 101, 105–14, 132–33, 282, 286, 295–96, 325–30; Gablehouse, *Helicopters and Autogiros* (1969), pp. 36–38.

5. Fay, p. 80.

6. See, for example, George Townson, *Autogiro: The Story of "the Windmill Plane"* (Fallbrook, California: Aero Publishers; Trenton, New Jersey: Townson, 1985); Cierva himself accepted the popular *windmill* designation. Juan de la Cierva and Don Rose, *Wings of Tomorrow: The Story of the Autogiro* (New York: Brewer, Warren & Putnam, 1923), pp. 112–14; see also Gablehouse, "A Spanish Windmill," in *Helicopters and Autogiros* (1969), pp. 34–67.

7. Cierva and Rose, pp. 112–13.

8. For a photograph of a recently restored Kellett K-2 Autogiro, see Michael O'Leary, "It's a Kellett!" *Air Classics* 38, no. 6 (June 2002), pp. 68–72.

9. Federal registration designation, in most cases N-numbers, for each Autogiro (the Cierva-licensed models), Gyrocopter (as proprietary to the Bensen models), autogyro (non-Cierva-licensed aircraft), or gyroplane (dating from Ken Brock's break with Igor Bensen in the early 1970s) will be given wherever possible. Note that while the later "experimental," amateur-built category requires such registration, "ultralight" aircraft, weighing 254 pounds or less, do not.

10. See Frank Kingston Smith, *Legacy of Wings: The Story of Harold F. Pitcairn* (New York: Jason Aronson, 1981); Walter J. Boyne and Donald S. Lopez, *Vertical Flight* (Washington, D.C.: Smithsonian Institution Press, 1984); R. A. C. Brie, *The Autogiro and How to Fly It* (London: Sir Isaac Pitman & Sons, 1933); Derek N. James, *Westland Aircraft Since 1915* (London: Putnam Aeronautical Books, 1991: Annapolis, Maryland: Naval Institute Press); Brooks; Cierva and Rose; John W. R. Taylor and H. F. King, *Milestones of the Air: JANE'S 100 Significant Aircraft* (New York: McGraw-Hill Book Company, 1969); Michael J. H. Taylor and John W. R. Taylor, *Encyclopedia of Aircraft* (New York: G. P. Putnam's Sons, 1978); Townson.

11. As a young man Samuel Pierpont Langley, an American, had been a railway surveyor and civil engineer. He would later achieve fame as an astronomer, but in the 1880s he experimented with model wings while at the Smithsonian Institution.

12. Sir Hiram Stevens Maxim was a "colorful American who, while resident in England, developed the renowned machine-gun bearing his name. In England, too, he built an amazing airplane.... the man must be admired not only for his determination but for the brilliant engineering achievement represented by the

very light steam engine he developed for his aeroplane." John W. R. Taylor and Kenneth Munson, *History of Aviation* (New York): Crown Publishers, Inc. 1972), p. 41. See also Edward Jablonski, *Man with Wings* (Garden City, New York: Doubleday & Company, 1980); Philip Jarrett, *Ultimate Aircraft* (London and New York: Dorling Kindersley 2000), p. 157.

13. Adler was an electrical engineer and inventor, and his name endures in history not only for his pioneering flights, or hop, which he claimed to have made in secret on 9 October 1890, but for his work in developing the telephone. Taylor and Munson, p. 43.

14. Cierva's reference to and reliance on Chanute is not surprising. Octave Chanute, an American railway engineer, published a series of 1894 articles that were reprinted in a book entitled *Progress in Flying Machines*. It is generally acknowledged that it was this book that stimulated the Wright brothers! Although he built successful gliders, Chanute's greatest contribution to aviation was as a collector and disseminator of information.

15. See Cierva and Rose, p. 24.

16. The glider movement founded by Lilienthal swept Germany and would, after the defeat of Germany in World War I, become the foundation for pilot training of the Luftwaffe prior to the German elections of 1933 that thrust Adolf Hitler into power. C. R. Roseberry, *The Challenging Skies: The Colorful Story of Aviation's Most Exciting Years 1919–39* (Garden City, New York: Doubleday & Company, 1966). p. 213; see also Taylor and Munson, p. 43.

17. "The name of Lilienthal stands as the greatest in the history of practical flight before the Wright brothers, to whose achievements, as already noted, his work led directly." Taylor and Munson, p. 43.

18. P. T. Capon, "Cierva's First Autogiros: Part 1," *Aeroplane Monthly* 7, no. 4 (April 1979): 200–205, 201.

19. "Santos-Dumont caught the imagination of the air-minded world when, in his 110-ft (33.5 m) airship No. 6 [balloon], he flew from St. Cloud to Paris, round the Eiffel Tower and back in 29 and 1/2 minutes on 19 October 1901. Blériot later wrote to Santos-Dumont, "For us aviators your name is a banner. You are our Pathfinder." Taylor and Munson, p. 53 (photo caption).

20. Taylor and Munson, p. 238.

21. Cierva and Rose, p. 26.

22. See Jablonski, p. 102.

23. Cierva and Rose, p. 34.

24. For a discussion of the originals of the first propeller, see Leo J. Kohn, "Mr. Cierva and His Autogiros," *Air Classics* 15, no. 6 (June 1979): 87–93, 87.

25. Brooks claims, with reference to the January 1922 opera, that "[a] windmill on the stage had hinged blades" (p. 40); even though Cierva relates the same opera incident, he does not cite the stage windmill but merely states that

I was attending the opera in Madrid when I suddenly realized why the model would fly so well and the big machines badly or not at all. My wife assures me that my excitement at the moment was a little out of place; I doubt that I heard the end of the performance. For it dawned upon me that the construction of the model's rotor blades held the secret of success for the Autogiro. They were not rigid, but built of such flexible material that they would bend easily in flight. In no other respect was there any important or essential difference between the model and the full-sized machines. (Cierva and Rose, p. 97)

26. This is also documented without further references in Warren R. Young, *The Helicopters* (Alexandria, Virginia: Time-Life Books, 1982), p. 57, and Richard Aellen, "The Autogiro and Its Legacy," *Air & Space Smithsonian* 4, no. 5 (December 1989/January 1990): 52–59, 54.

27. For a rare photograph of the BCD-1, see Capon, "Part 1," p. 201.

28. Cierva and Rose, p. 39; Cierva, Juan de la, "The Autogiro," *The Journal of the Royal Aeronautical Society* (November 1930): 902–21.

29. In his 1931 book with Don Rose, Cierva himself estimated total cost at $32,000 (p. 45).

30. Cierva and Rose, p. 46; for photographs of Cierva's bomber, see Capon, "Part 1," p. 202.

31. Cierva himself claimed that the C-3 was the second trimotor plane to be built in the world, the first being an Italian Caproni, which flew earlier but was powered with two tractor motors, with the third motor being a "pusher" type. Cierva and Rose, p. 46; Brooks, p. 16.

32. In the current world of the gyroplane, this engineering approach is represented by the computer modeling methods of American Martin Hollmann, Finland's Jukka Tervamäki, France's Jean Fourcade, Scotland's Dr. Stewart Houston, and to a more limited extent in the 1950s, that of Igor Bensen. In the Gyrocopter movement created by Bensen, there was and continues to be a great deal of "fly by the seat of your pants" experimentation with little formal engineering analysis. Chuck Beaty, participant in the American gyroplane movement for over three decades and writer of gyroplane engineering and technical analysis, has observed that "[t]he main problem is that so few qualified engineers have taken an interest in gyroplanes." Chuck Beaty, "Gyro Stability: Understanding PIO, Buntover, and How Gyroplane Rotors Work," *Rotorcraft* 33, no. 5 (August 1995): 18–23, 23.

33. Cierva, "Autogiro, " p. 964.

34. Cierva and Rose, p. 60.

35. Ibid., p. 61.

36. Devon Francis, *The Story of the Helicopter* (New York: Coward-McCann, Inc., 1946), pp. 46–47.

37. Cierva stated: "From my study of aerodynamics I knew so far no practical method had (or has) been found to drive these horizontal windmills by an engine. Mechanical difficulties have been insuperable, and even when they are solved, very much greater power is needed to drag a weight vertically as compared to pulling it horizontally while its wings ride on a cushion of air." Ibid., p. 46.

38. For a graphic illustration of the autorotational forces that lift the Autogiro, see Gablehouse, *Helicopters and Autogiros* (1969), pp. 36–38.

39. Cierva and Rose, p. 74.

40. Hollmann states that Cierva "built test facilities with wind tunnels. Hundreds of tests were performed, but most were disappointing." Cierva cited such tests from mid-1921 on, although his first Spanish patent application was applied for on July 1, 1920, and granted on August 27, 1920. Although construction of the closed-circuit wind tunnel at the Aeronautical Laboratory at Cuatro Vientos was begun in 1919, it was not commissioned until 1921, well after the granting of Cierva's first patent. See Martin Hollmann, *Flying the Gyroplane* (Monterey, California: Aircraft Designs, Inc., 1986), p. 7; Cierva and Rose; Brooks, p. 357 n. 5. From Francis, p. 48, quoting Cierva:

> We built small models mounted on ball bearings and placed them in a wind tunnel, in winds up to forty-five miles an hour. We found that the rotating windmill offered definite resistance to forward motion. Then we inclined the windmill slightly, so that the plane of rotation was more in the direction of the wind. We found a definite and measurable tendency for the model to rise. This was highly important, for it proved I was on the right track. The lifting force of the wind, I now knew, acted on the rotating windmill much as sit does on the slightly inclined fixed wings of an ordinary airplane.

41. Cierva and Rose, p. 96.

42. For drawings of Cierva's first four models, C.1–C.4, see Fay, figs. 139–42; Brooks, pp. 35–45 (photos on 37–41, 43–44, and 46 and accompanying text); Capon, "Part 1," pp. 200–205.

43. Cierva and Rose, pp. 92–94.

44. Although Brooks reports this to be "possibly the modified fuselage of an old 1911–12 French Type E Sommer monoplane, probably originally imported into Spain by Jean Mauvais in 1913" (p. 38), Cierva reports this as a "Henriot fuselage" (Cierva and Rose, p. 92).

45. For a photograph of Cierva holding a model of the C.3 used in wind tunnel testing, see Capon, "Part 1," p. 205.

46. P. T. Capon, "Cierva's First Autogiros: Part 2," *Aeroplane Monthly* 7, no. 4 (May 1979): 234–240, 236.

47. Young, p. 57.

48. French patent No. 395,576.

49. German patent No. 249702. E. K. Liberatore, *Helicopters before Helicopters* (Malabar, Florida: Krieger Publishing Company, 1998), p. 95.

50. J.A.J. Bennett, "The Era of the Autogiro (First Cierva Memorial Lecture)," *Journal of the Royal Aeronautical Society* 65, no. 610 (October 1961).

51. Cierva himself states that the C.4 was based on a "Henriot fuselage" (Cierva and Rose, p. 92), an obviously incorrect reference, and is corrected by others to *Hanriot*, which clearly refers to the French firm of Aeroplanes Hanriot et Cie. See David Mondey (ed.), *The Complete Illustrated Encyclopedia of the*

World's Aircraft (1978; updated by Michael Taylor, *The New Illustrated Encyclopedia of Aircraft,* Secaucus, New Jersey: Chartwell Books, Inc., 2000), p. 283, where it would seem, based on the description of the firm's World War I activities, that Cierva in Spain would hardly make use of a plane that had been rejected by the French but was primarily used by Italian and Belgian services. The Sommer story is lent credence in that the BCD-1, Cierva's first plane, was clearly based on a reassembled Sommer flown by Jean Mauvais. Gablehouse, *Helicopters and Autogiros* (1969). The previous edition (*Helicopters and Autogiros* [1967], p. 38) repeats the "Hanriot scout biplane used by the Allies in World War I" version, but it is obvious from simultaneous references to the Cierva and Rose book that Gablehouse is merely repeating what Cierva has said. However, Brooks in his book and Capon ("Part 1" and "Part 2") state that the C.3 and the C.4 utilized a Sommer monoplane fuselage. Gablehouse (p. 41) states of the Cierva C.1–C.4 series, "A number of test beds were built, using parts of World War I airplanes such as the Hanriot scout, in order to try out various rotors, blade settings, and configurations," thus lending credence that the C.4 was a composite of several surplus planes and casting doubt on Cierva's dubious assertion of the "Henriot" origins of the C.4. In a slightly later passage in his coauthored 1931 book Cierva states ambiguously, "This fourth Autogiro was a very simple machine. Its fuselage and motor were an old airplane assembly." Cierva and Rose, p. 102.

Chapter 2

CIERVA'S AUTOGIRO

To land in a vertical descent! Think of it! And not roll a foot forward...

Captain Frank M. Hawks, "a colorful pilot in the leather breeches and silk scarf tradition," after flying an Autogiro

Cierva states[1] that the first flight of C.4 was on January 9, 1923, at Getafe airfield, when (cavalry) Lieutenant Alejandro Gómez Spencer, "a Spanish gentleman whose surname and appearance both indicate an English ancestry...one of the best known Spanish fliers," guided the craft in taxi tests during which the craft became airborne. But some historians[2] maintain that the first observed (and filmed) flight of C.4 took place on January 17, 1923, when Gómez Spencer flew six hundred feet at a steady height of thirteen feet across the field. Additional flights took place on January 20 and 22 before assembled dignitaries, including official observers from the Spanish Royal Aero Club and high-ranking military officers, such as General Francisco Echague Santoyo, then Spanish Director of Air Services and Don Ricardo Ruiz Ferry, president of the Spanish Royal Aero Club Commission. The January 20 flight was to prove of particular significance—the C.4's engine failed at twenty-five to thirty feet while rising and with its nose slanted upward, a dire circumstance in any conventional aircraft. But the Autogiro merely descended vertically, a slow settling to the ground, and was undamaged due to its autorotating blades. On January 31, Gómez Spencer flew the C.4 at the Cuatro Vientos military airfield over a circular course of two and a half miles before an even larger delegation of military officers. It was an unqualified success, and the Autogiro was on its way. That way would not be free from misstep, however, and both commercial and military success would eventually elude its hopeful advocates.

Cierva funded the construction of his next aircraft, the C.5, in the workshops of the Industrial College, which had done some of his previous subcontracting in early 1923, soon after the C.4 had established autorotational flight. He was anxious to explore different configurations, and this model was larger and employed a three-blade rotor. Completed in April of that same year, it was flown successfully by Gómez Spencer, but it proved overly sensitive in its control system and was destroyed while on the ground. The rotor blade, a different section (shape) than used previously, suffered from metal fatigue and failed. Cierva would not regularly employ a three-blade rotor again until 1931; although he would experiment with both two and three blades in 1927, he adopted four blades on all subsequent designs, perhaps because this allowed for a balanced allocation of stress.

By May of 1923, Cierva's wife Mária Luisa had borne the second of his eventual seven children—and although his C.4 Autogiro was receiving much notice and growing public acclaim, he was still being funded by his wealthy father and living off a family allowance. It was evident that he would continue in his aviation endeavor, but it was also readily apparent that it would have to evolve into a business. He could not continue to spend family resources in development of the Autogiro; another source of funding would have to be found. Cierva wrote the Westland Aviation Works at Westland Farm, Yeovil, Somerset, in the fall of 1923, but although the company would build Cierva Autogiros under license in the 1930s, it expressed no interest in the letter from an unknown Spanish inventor. The Spanish government, however, had previously provided funding for the testing of one-tenth-scale models at the Aeronautical Laboratory wind tunnel, and now General Francisco Echague, who had been an impressed observer of the C.4 flights on January 22, 1923, offered additional government funding for construction of the C.6. It was later suggested that the general had been misled by the wind tunnel testing of the model, which inaccurately suggested that the new Autogiro could theoretically have a much greater speed than a comparable fixed-wing airplane. But if the general made such a mistake, it surely was an inadvertent result of his misinterpretation and not an intended misrepresentation by Cierva. Even so, the decision was fortunate, for the C.6 was to deliver the best performance to-date.[3]

Even though 1923 was exceptionally exciting, busy, and productive for Cierva, he still found time to answer a handwritten letter from an American third-year high school student. John M. "Johnny" Miller, reading the newspaper accounts of the new Autogiro, wrote the inventor with little expectation of a reply, but much to his amazement, a reply did indeed come. Miller, writing almost seventy years later, observed that the letter was in perfect English, leading him to erroneously assert that Cierva had "been educated in Oxford."[4]

Although Miller is certainly wrong about Cierva's attending Oxford and it is not possible to ascertain who actually translated the letter he received, the significance of an English reply should not be missed. Cierva's ability to speak English would not be sufficient for a public address on the principles of the Autogiro until 1930, but he either had the ability to write perfect English or was willing to have his letter translated. In either event, Miller received two letters from the Spanish inventor "explaining his autogyro[5] in detail, including its aerodynamics and its possible development into a future helicopter."[6] Those letters would make an indelible impression on the young Miller, leading him to an engineering education and a flying career that began in 1924 and extended into the twenty-first century. He would develop a lifelong fascination with the Autogiro, would become the first private individual to place an order when they became commercially available, and would be famous for his Autogiro flying exhibitions in the 1930s.[7]

As the C.6 was a military project, it was constructed in the Spanish Military Aircraft Works located at Cuatro Vientos under the supervision of Captain Luis Sousa Peco. Utilizing a surplus fuselage, tail, engine, and modified landing gear of a British A. V. Roe & Co. Ltd. ("Avro") 504K World War I trainer, this was Cierva's first experience with the English company that would eventually manufacture his Autogiros. The first flights were made in February 1924 by Captain José Luis Ureta Zabala of the Aeronautical Laboratory's Experimental Squadron. The military evaluation was quite positive, as the C.6 showed great improvement over its predecessor, and the British finally began to take notice. In April 1924 Captain Oliver Vickers, representing British aircraft manufacturers Vickers Ltd., visited Spain and watched a C.6 demonstration. His report was favorable, and Vickers wrote to Cierva in May offering to do wind tunnel testing of a model C.6. Cierva provided specifications, but the English testing at Weybridge was to yield less favorable results than the previous Spanish research. Testing of the C.6 slowed in late 1924, when Cierva traveled to England on family business and pilot Ureta was sent to a new assignment, but Captain Joaquín Loriga Taboada arrived in August and became enthusiastic about the new aircraft. Following Cierva's return to

Spain, new testing successfully commenced on December 9, after a short pilot briefing (there was no one to show him how to fly the C.6). An additional flight was made on the 11th, but the third sortie, on the 12th, was to prove a milestone, when Loriga flew from Cuatro Vientos to Getafe airfield. The seven-and-a-half-mile trip, completed in eight minutes twelve seconds (average speed forty-eight miles per hour), was the first cross-country Autogiro flight. Cierva, who would not gain his pilot's license until 1927, followed as a passenger in Fokker C.IV.

Word of the Spanish inventor's extraordinary aeronautical innovation was spreading, and it burst upon the international aviation community at the Ninth Paris Salon Aéronautique[8] in December. Loriga presented a film of the C.6 to a meeting of the Société Française de la Navigation Aérienne (SFNA), to considerable interest. Captain George Lepère, who had already achieved a reputation as an aeronautical engineer, was particularly enthusiastic. He would play a significant role in the French development of the Autogiro with others at Lioré et Olivier.[9] The English returned for another observation on January 16, 1925—a decidedly serious effort led by Vickers Aviation Department chief designer Rex Pierson. During that flight Loriga experienced engine failure but was able to turn back to the airport and safely land. Rather than being dismayed by the unexpected landing that damaged a rotor blade, the observers were decidedly impressed by the safety inherent in the free-spinning rotor, as such a maneuver would have been fatal in a fixed-wing aircraft. Even though Vickers was not to consider acquiring an Autogiro license for another ten years,[10] news of the Autogiro had reached England from Paris and from the local Spanish media, notice that would bring Cierva to England by the end of the year and shift the focus of Autogiro development permanently out of Spain.

Just prior to the arrival of the second Vickers delegation, the British Commercial Attaché in Madrid had written to Don Juan de la Cierva, inquiring about his son's invention and requesting performance data of the latest flights. The elder Cierva, being an experienced senior government official, immediately recognized this for what it was—the first official British interest in the Autogiro. The British government was the first to express such an interest, but the rest of the aviation world was also expressing interest. The C.6 had been damaged in March of 1925, and while it was being repaired, Cierva went to France to speak to the SFNA and to show the C.6 films as Loriga had previously done. Returning to Spain via England, he again met with Vickers but made no further inroads. Even though his contacts and discussions with the British civil aviation executives continued to bring disappointment, the relationship with the official government aviation authorities was assuming a critical mass. Cierva met with H.E. Wimperis, director of research at the British Air Ministry, who expressed serious interest and solid encouragement. Of greater importance, The Royal Aeronautical Society (TRAS) requested that he lecture on the Autogiro. The aristocratic Cierva knew well the aviation acceptance and status inherent in TRAS; he eventually would give several lectures and be named an associate fellow of the society. France was not long in expressing its interest: before reaching Spain, Cierva again returned to Paris to meet with Victor Laurent-Eynac, French undersecretary of state for aeronautics, the most significant government contact to date, and with General Fortant, director of the Service Technique de l'Aéronauique, which produced a request for an official demonstration at Villacoublay.

That the military authorities were interested in the Autogiro is not surprising. The great innovation in aviation in the previous decade had clearly been occasioned by the world war. Civil aviation was still in its infancy—routes and mechanisms for profit had yet to be established in America and most pilots earned their livelihood by barnstorming, putting on exhibitions, taking people for rides, and an occasional advertising assignment. The Air Commerce Act of 1926 that would bring order to the route structure in America was still over a year away, and the plane that was to revolutionize passenger service, the Ford 4-AT Tri-motor,[11] would first fly on June 11, 1926. Given that, Cierva's interest in the military is also not surprising, and he immediately sought permission from the Spanish military authorities to use the C.6 for demonstration flights in England and France in May and June—but this was not to be. C.6 had been rebuilt after its March accident, was committed to exhibit at the Fourth Automobile and Aero Show at Montjuich (near Barcelona) the last ten days of May, and is never known to have flown again.

An improved model, the C.6A, flew for the first time in early June (the records indicate two dates for the first flight, June 6 and 8). The technology was improving, as the slightly larger rotor ran smoother, an arrangement of wooden pegs on the bottom of the blades allowed for a starter rope to be used to *spin up,* or prerotate, the blades to assist in the takeoff,[12] and the aircraft incorporated wider landing gear and a higher rotor pylon. Cierva had begun to confront the issue early on of how to get the rotor blades spinning to the necessary rpm to allow takeoff. While it was true that air flowing up through the rotor blades, given the cant (tilt) of the individual blades, would cause them to autorotate, this was accomplished

Harold F. Pitcairn and Juan de la Cierva, 1929. (Courtesy of Stephen Pitcairn, from the Pitcairn Archives.)

by running the aircraft up and down the runway. Cierva would try a manual rope spin-up and also prerotation by use of a *box tail* that directed the flow of air from the propeller upward to the rotor blades to cause them to spin. Eventually, Harold F. Pitcairn would channel power from the engine to the rotor by means of a clutched gearbox on the PCA-2—but that solution was still over five years away.

The C.6A was fine-tuned and demonstrated before King Alfonso XIII, government officials, and foreign attachés at Cuatro Vientos on June 24, 1925. Even though the king was suitably impressed with pilot Loriga's flight, subsequently appointing Cierva "Caballero of the Civil Order of Alfonso XII" for his aviation achievements, and even though the Spanish government subsequently allocated an additional $34,000 toward Autogiro development, events were even then occurring that would shortly take such activities out of Spain.

Harold Pitcairn initially believed that Cierva was in England, perhaps based on the comments of British aeronautical engineer W[ynn] Laurence

LePage, who had apparently spoken to Pitcairn about the wind tunnel testing of Cierva models performed by Vickers at Weybridge in January and February of 1925. LePage, then associated with England's National Physical Laboratory, was on loan to the Massachusetts Institute of Technology (MIT) to assist them in developing similar testing programs, and Pitcairn had contracted with MIT for testing of his fixed-wing aircraft designs. But the British officials were mystified as to why the Americans had come to Weybridge seeking Cierva, who was not nor had ever been resident in England. They promptly informed the Americans of Cierva's address in Spain, and Pitcairn promptly wrote Cierva on April 25, 1925, requesting an appointment.[13] This was not a casual contact—Pitcairn was in England with his friend and chief engineer Agnew Larsen, *and* he had a letter of introduction from Heraclio Alfaro, a "distinguished Spanish engineer who had learned to fly in France in 1911" who was well-known to Cierva.[14] Alfaro, destined to play a minor role in development of the Autogiro in America, had designed a fighter entered in the 1919 Military Aircraft Competition that had seen the crash of Cierva's C-3 bomber.

The meeting between Pitcairn, Larsen, and Cierva took place in May 1925 in Madrid. An interpreter was present, as Cierva did not yet speak English with fluency. Captain Frank T. Courtney, who was to have an exceptionally long and distinguished flying career and who would become Cierva's English military test pilot, described the Spaniard's ability with the English language at that time thusly:

> In London a few weeks later I had a phone call that Señor de la Cierva would like to have a talk with me. At his hotel the next day, after the usual preliminary chit-chat, we settled down to a discussion. Cierva's English and my Spanish didn't meet at any useful point, but we got along fine in French. (He later learned to speak excellent English.)[15]

The discussion during the brief meeting between Pitcairn, Larsen, and Cierva was of a general nature as it passed through a translator. Also, the American visitors were not able to see the Autogiro in flight other than the C.6 film, as C.6 was then under repair and C.6A was still being constructed. Pitcairn and Larsen were very impressed with the films of level, controlled flight—but, as suggested by Pitcairn's biographer, they may have felt that Cierva was deliberately being evasive. Later, when they had grown close and Cierva conversed in excellent English with only the slightest hint of an accent, Pitcairn apparently wrongly concluded that the use of an interpreter in May of 1925 had been a close-to-the-vest ploy, failing to recognize that

Cierva C.6A demonstrated at Farnborough in England by pilot
Captain Frank T. Courtney on October 30, 1925. (Courtesy of Dr.
Bette Davidson Kalash, from the Jesse Davidson Aviation
Archives.)

Cierva had managed to improve his English-language skills. The Americans
returned home to the business of developing Pitcairn Aviation's aircraft.
They were convinced that Cierva had an innovative aviation achievement
but were unsure of what direction it would take. They resolved to watch for
new developments coming from the Spanish inventor.

The French government had postponed the projected demonstration of
the C.6A, preferring to wait for a more powerful model powered by the 180
horsepower engine manufactured by the Soc. Hispano-Suiza at Levallois-
Perret, Paris, in its large factory Bois-Colombes, an idea that was perhaps

suggested by Cierva himself. So Cierva turned his attention to the English demonstration, which had been scheduled for October. The C.6A, having been completed and shipped from Bilbao to London, was ready, but the need arose to replace Captain Loriga, who had contracted pleurisy.

The C.6A has been reassembled at the Royal Aircraft Establishment at Farnborough, but there was no one to fly it. Cierva spoke with H. E. Wimperis, who suggested Captain Frank T. Courtney. Courtney, then a test pilot for the British de Havilland Aircraft Company, was highly respected for his flying skill and was known as "the Man with the Magic Hands."[16]

Courtney had already seen the C.6 while delivering a fighter plane to Cuatro Vientos earlier in the summer of 1925. Viewing the C.6 in a hanger, he was highly dubious of its flying ability. Assured that it had previously made several flights, he could not imagine, nor could any aviator present, how such a feat was possible. The officers related to him that the C.6 was the invention of "an eccentric character named Juan de la Cierva."[17] Courtney later wrote of his first impression of the Autogiro: "Well, anyway, lots of funny and useless things had been coaxed into the air at one time and another—this was probably just another one!"[18] But after meeting with Cierva and finding him a "cultured, charming, and serious gentleman of about my own age," he listened for almost three hours while Cierva explained the aerodynamic principles of the Autogiro. Courtney then concluded that Cierva was "no crazy inventor, but an imaginative engineer who had braved repeated disappointment and employed persistent ingenuity in making the most important breakthrough in aerial locomotion since the days of the Wright brothers."[19] He agreed to be the test pilot and familiarized himself with the aircraft on October 10, during which he spent about thirty minutes practicing taxi runs and short hops. The first demonstration flights took place on October 14, 1925, before Sir Philip Sassoon, undersecretary for air. It was successful, and a second demonstration was conducted on October 19 before Secretary of State for Air Sir Samuel Hoare, Chief of the Air Staff General Sir Hugh Trenchard, Director of Civil Aviation Sir Sefton Brancker, and other interested dignitaries. Brancker would have his first flight in an Autogiro, a C.19 Mk.II (G-AALA) piloted by Arthur Rawson, at Heston on January 7, 1930. He reportedly was enthusiastic, but it would not benefit the cause of the Autogiro, as he died in the crash of the British dirigible R 101 on a trip to India.[20] And the queen of Spain came to see what had become Cierva's triumph on October 28.

Cierva also gave the lecture he had previously been invited to give before a large audience at The Royal Aeronautical Society on October 22,

1925. As he felt that his English was not adequate, his paper was read in translation by TRAS Chairman Sir Sefton Brancker.[21] It was well-received and evoked great interest, particularly since the demonstration flights had taken place on October 14 and 19. Test-flights continued, usually with Courtney as pilot,[22] but on at least one occasion with RAF Squadron Leader Rollo A. de Haga Haigh, evidencing an increasing interest by the military. The military testing increasingly stressed the aircraft with steeper angles of descent until it was damaged in a rough landing at the end of October.[23] Courtney later recounted that the hard landing, the product of a near-vertical descent, had collapsed the landing gear and left the Autogiro "right side up however, and the rotor continued to revolve tranquilly overhead. Through some stroke of luck, the Autogiro had 'landed' behind a shallow hill, and only the rotor remained visible from the official observation location."[24] Courtney had actually been thrown from the cockpit and found himself sitting unceremoniously but safely on the ground. He was able to stand and regain his composure so that by the time the official observers appeared, he was able to "charge the whole thing off to landing gear failure!" Repaired and modified with stronger oleo landing gear adapted from an Avro 504N, the Autogiro was soon ready for the tests before representatives of the British Patent Office.

As the Autogiro was an entirely new flying craft, the British Patent Office was initially skeptical and requested that tests take place before its official observers.[25] Cierva was keen to gain patent protection—he would even trademark the term *Autogiro*—so he readily acquiesced to the need for a demonstration on November 11, 1925. It was a success, and he was granted a patent. The impression on the military observers was favorable, with senior officials calling in late November for the English government sponsorship of further development of three new types: (1) a more powerful version of the C.6A, utilizing a locally produced Armstrong Siddeley Lynx engine and a mechanical device to spin up the rotor blades in place of Cierva's rope system; (2) a small, agile, single-seat model; and (3) a research vehicle, the design of which was to be developed by the Aeronautical Research Committee. This represented a major endorsement of the Autogiro, and the British government guided Cierva into a licensing agreement with the significant English aviation company, A. V. Roe & Co. Ltd. In January 1926, both Cierva and the Air Ministry ordered a C.6A from A. V. Roe & Co. Ltd., the start of that company's involvement with the Autogiro.

Cierva also sought out the French government regarding the postponed demonstration. Courtney and two French pilots, Ingenieur Cousin and

Adjudant Moutonnier, flew the C.6A at the Villacoublay airfield near Paris, to mixed response in late January and early February in 1926. On January 27 Courtney had the misfortune of being blown over by a forty-two mph wind as he landed in a muddy field. Later, while watching the newsreel films of the accident, he stated that it looked like "a giant golfer gone mad, the whirling blades were shown smashing into the ground, distributing huge divots of wet earth and mud all over our VIPs."[26] The impressions of the crash were indelible and, even though the flight demonstrations were judged as good or better than in England, the French government decided against any further involvement. Even though Cierva was awarded the Grand Prix Scientifique de l'Air for 1925 by the Société Française de la Navigation Aérienne, it would take three years before the French would again become involved with the Autogiro. The C.6A returned to England, where it was subsequently demonstrated before Americans Harry Guggenheim and Rear Admiral Hutchinson Cove during March and April.

One of those watching the October 1925 demonstration flights at Farnborough was the holder of Aviator's Certificate No. 24, who had soloed in a Blériot monoplane on November 8, 1910, and who had later been secretary of state for air—James G. Weir.[27] He had been favorably impressed by the C.6A demonstrations and brought it to the attention of his brother, Viscount William Weir of Eastwood. They consulted with Frank Courtney and with a banker who was to become instrumental to bringing Cierva to England—Sir Robert M. Kindersley[28] of the London commercial banking firm Lazard Brothers & Co. Ltd. Cierva's father had previously dealt with the English banking firm in Madrid and undoubtedly knew Sir Robert's son Hugh K. M. Kindersley, who worked in the Madrid office and was fluent in Spanish. Don Juan de la Cierva now took it upon himself to write personally to Sir Robert, suggesting that the banking firm could profit from consideration and backing of the Autogiro's commercial potential. The English group, negotiating with Cierva's father through the younger Kindersley, soon reached acceptable terms and established the Cierva Autogiro Company Ltd. on March 24, 1926. The goal of the company was not manufacturing but development and exploitation of Autogiro developments through licensing, selling patents, and royalties. James Weir was named chairman, Cierva became technical director, and the Spanish-speaking Hugh Kindersley served as a director. Although Lord Weir refrained from financial participation, fearing that such activity would compromise his governmental influence, thirty thousand pounds was put up by James Weir, Hugh Kindersley, John Jacob Astor, and several participating financial institutions. For his patents Cierva was allotted twenty thousand preferred shares of stock in the new company. The future development of the Autogiro was now firmly in England, and it would remain the locus of autorotation development until the establishment in 1929 of the Autogiro Company of America by Harold F. Pitcairn.

But first, the Air Ministry took delivery of the improved Autogiro built by Avro. As the C.6C designation[29] suggests, this model was based on the C.6A but in a single-seat configuration and with a more powerful engine. Although senior officials had recommended an English engine, a locally manufactured French Clerget rotary was employed. Cierva also received an Avro aircraft, the C.6D. It was a two-seater more closely resembling the C.6A. The military was enthusiastic about testing their new Autogiro and arranged for the newly established Cierva Autogiro Company to begin flying the C.6C in June 1926 at Hamble. The pilot was, of course, Frank Courtney, newly appointed technical manager. The flight-test went well, with performance that improved on the earlier Spanish model—so well, in fact, that the Royal Air Force decided to advance their cause with a demonstration before King George V and Queen Mary, the king and queen of Spain, and assembled dignitaries at the Seventh RAF Display at Hendon on July 3, 1926.

This was *the* RAF showcase, and its senior officers surely counted on a dramatic impression being made. It almost didn't happen. Three days before the display, one of the cables that held the rotors up when at rest on ground broke while in flight.[30] Upon landing after a test-flight, Courtney was almost decapitated when the rotating blades sank lower and lower until they hit the rudder in the aircraft's tail and the prop in the front, finally slamming against the side of the plane. The quick-thinking pilot had saved himself by leaping from the Autogiro as the whirling blades came to rest. The Autogiro was repaired in time for its royal performance and fully justified the expectations of the RAF—it was publicly acclaimed, with commentators even projecting commercial applications that were as yet unproven. All-in-all, it was an auspicious beginning to the Autogiro's English career, but the C.6C would not have such an end—it crashed again in September 1926 and then went through a refit, with the long booms, which extended out from the fuselage ending with the aileron control surfaces, being replaced with short "stub" wings.

While this change may seem insignificant, it was the result of Cierva's ongoing research that was now both theoretical and practical. He had taken his first Autogiro ride in the C.6D with Courtney on July 30, 1926, fittingly enough, the first passenger in rotary aircraft in history. The second passenger was H.J.L. (Bert) Hinkler, who would become Cierva's test pilot after Courtney's departure. The flying experience gave Cierva a foundation of experience that would both contribute to his theoretical considerations and lead to his becoming a pilot. He gained the Royal Aero Club Aviator's Certificate (No. 8077) on January 20, 1927, and his "A" license (No. 1035) on February 29, 1927, flying the Autogiro a few months later.

Cierva, ever seeking to improve flight efficiency, had calculated that an Autogiro with fixed wings generating 30 percent of the aircraft's lift would have a constant rotor speed through much of the aircraft's speed range and be more efficient. Although this innovation was apparently successful and influenced future designs, the C.6C crashed for the final time on February 7, 1927, when a rotor blade fell off in flight and a second became detached just before the aircraft hit the ground.[31] Although Courtney was fortunate (again) to suffer only slight injury when the Autogiro was destroyed, it was the last straw—the two men had come to a parting of the ways. Courtney later claimed his difficulties with Cierva had begun the previous September 5, during a demonstration of the C.6D before German aviation officials and dignitaries at Tempelhof, Berlin. During that demonstration a German journalist named Kleffel was given a ride in the Autogiro, no doubt an attempt to generate favorable publicity. It was a public-relations technique later adopted by Harold F. Pitcairn in America, who arranged for a ride for the first aviation editor of the *Washington Daily News,* Ernie Pyle, at the Washington National Airport. Pyle's testimonials would appear in Pitcairn advertisements. Pyle would later become one of the most famous war correspondents during World War II. He was genuinely mourned by all of America when he was killed on Ie Shima Island just off Okinawa in April 1945.

Courtney, examining the aircraft that had been reassembled after being transported to Berlin, observed a distortion in the metal at the base of the metal core of the rotor blade (the "root" of the "spar"). He claimed that Cierva "declared flatly that this could not possibly have been caused in flight by aerodynamic loads; it must have been done when the blades were crated in England."[32] But the pilot ventured the opinion that the mechanical design was lacking in that although the current flexible horizontal arrangement allowed for the rotor blades to "flap" up and down, producing equalized lift as the blades rotated, there was no provision for the horizontal stress on the blade roots as they moved around the rotor disk. Courtney felt that this stress was causing the observed distortions in the blade root, a metal fatigue that could potentially result in the blade breaking off. The obvious solution was vertical drag hinging, which would allow the blades to move slightly back and forth in a lead-lag motion as they rotated, effectively equalizing the fore-and-aft forces under drag loads as the blades went from forward to rearward motion around the rotor hub.

Cierva, however, would hear nothing of lead-lag drag hinges and apparently did not take kindly to this suggestion by Courtney—leading the pilot to conclude either that since the idea did not originate with the proud inventor or that he had never considered lead-lag drag blade hinging, it was unacceptable. And indeed this may have been the case, as Cierva did not consider modification of his rotorhead necessary, and he would continue to design Autogiros without lead-lag drag hinges until 1928.[33] He could see the pilot's agitation and offered him the opportunity to abandon the German demonstration, but Courtney, installing a spare set of blades that he figured would last for the Templehof flights, triumphed before a crowd estimated in excess of two hundred thousand people, resulting in enormous public enthusiasm.

Courtney remembered what he had observed, however, and on his own began to discuss with trained engineers the necessity and feasibility of lead-lag drag hinges. Based on his discussions he attempted to introduce Cierva to the engineers, but the inventor would have nothing of it, apparently concluding that such an arrangement would introduce additional mechanical complexity into the Autogiro. So the stage was set for the final crash of the C.6C in early February of 1927 when a blade did indeed fall off during the aircraft's landing. Courtney estimated that had a second blade come off ten feet higher, he would have died instead of finding himself in a hospital with only shock, concussion and some broken ribs. He left the company and shortly thereafter moved to America. He would next encounter Cierva nine years later when, while on a visit to London, he ran into him in the Hungaria restaurant on the evening of December 8, 1936. The former aviation colleagues shared some drinks and conversation and, by Courtney's account, parted as friends. The next morning, shortly after 10:00 A.M., Cierva would die in the crash of a Dutch airliner. As Courtney ironically observed thirty-six years later, Cierva "had devoted his life to the creation of an aircraft that could not stall, and he lost it in an airplane that stalled on takeoff."[34]

Bert Hinkler, Avro's chief test pilot,[35] had assumed official responsibility for testing Cierva's Autogiros after Courtney's departure, but by the latter half of 1927, there it no doubt that much of the flying was actually being done by Cierva himself even though he was probably not accepted by the Air Ministry as an approved test pilot until, at the earliest, mid-1928. The Air Ministry, in response to Courtney's final crash of the C.6C, grounded all Autogiros and asked Cierva to improve the rotor by incorporating vertical lead-lag drag hinges. They were installed in the Avro C.6D model, thereafter known as the C.8R, which was the first Autogiro to receive a civilian registration—G-EBTW.[36] On August 2, 1927, Cierva made his first flight in the C.8R as pilot, and he was to fly a total of thirty-five hours of Autogiro test-flights that year.

England was not, however, the only location that saw autorotational development. Under a Cierva license, the Spanish government had in 1925 awarded a sizable grant of two hundred thousand pesatas to Dr. Jorge Loring Martinez, a noted builder of aircraft, to construct a larger, heavier, and more powerful Autogiro. That aircraft, dubbed the C.7, was completed in October of 1926, in time to be exhibited at the First Madrid Aero Show. Between October 27 and November 7, thousands of spectators viewed the C.7 and expressed much admiration. It was an impressive sight, promising even more outstanding rotary-wing achievements. Unfortunately, when flown for the first time eight days after the close of the Aero Show, the C.7 proved disappointing. With all its impressive changes, this two-seat tandem open-cockpit aircraft was found to be tail-heavy, necessitating readjusting the weight distribution. Additionally, although the initial construction lacked lead-lag drag hinges, Cierva subsequently modified the C.7's rotor after Courtney's accident to incorporate such hinges and installed two-wheel landing gear. It first flew in modified form on May 19, 1927, piloted by Reginald Truelove. He was an English pilot who had moved to Spain and become Loring's test pilot. Of greatest importance, Cierva flew as a passenger on several occasions, making firsthand observations as to rotor vibration, stability, and flight performance. Although the Spanish government, disappointed in the C.7's performance, apparently canceled an order for a second aircraft, its significance should not go unappreciated. This aircraft, far from England and the attention of the Air Ministry, was the test platform for the development of lead-lag drag hinges—a vital component on all future rotary aircraft, including the helicopter.

The government's disappointment signaled, however, almost the end of Autogiro development in Spain, as Cierva shifted his efforts to England. H.M. Yeatman had designed a special test platform at Hamble where rotors could be mounted on top of a tower designed to facilitate rotation in the wind for testing, effectively anticipating the testing rigs later to characterize helicopter development. The Spanish government would make a final development effort in 1929, when Loring constructed the C.12 under a Cierva license and under contract from the Aeronáutica Militar. That model used an American Wright R-760 Whirlwind J-5 engine, first placed in an Autogiro in late 1928 at the request of Harold Pitcairn. Cierva used the C.12 in Spain to develop the deflector tail for slip-stream rotor starting, a method by which the prop flow was directed upward to spin the rotor in preparation for takeoff. But the pattern of using Spanish models for research and development would come to an end with the C.12. England would be the center of Cierva's efforts.

In England Cierva continued to improve the Autogiro, now incorporating lead-lag drag hinges, first suggested by Frank Courtney in 1925, into the rotor system of a subsequent series of models and creating more effective systems for bracing the rotors and achieving flight control. Becoming more skilled as a pilot, he actively participated with Hinkler in flight-testing, and on September 30, 1927, he made the first cross-country flight in an Autogiro in England, flying the forty-four miles from Hamble to Farnborough via Worthy Down. Notably, Cierva had an accident in a model C.8V in February of 1928 while acting as test pilot, as Hinkler[37] had resigned the previous month as Avro chief test pilot. Although attributed at the time to insufficient rpm of the rotors prior to takeoff, it was probably due to a gust of wind catching the rotor blades and tipping the aircraft. But it did serve to highlight ongoing stability problems and the issue of spinning the rotor blades up to takeoff rpm—a problem that would be solved by Harold Pitcairn. He would be the first American to fly an Autogiro, initially in England, and then in Pennsylvania. But the first person to fly this aircraft *in* America would be the pilot who replaced Hinkler, RAF Flight Lieutenant H.C.A. "Dizzy" Rawson, who flew the C.8R from Hamble to the Royal Aircraft Establishment on June 2, 1928. Arthur Rawson would remain with the Cierva company as chief test pilot until 1932, during which time he would become the first person to accumulate one thousand Autogiro flying hours.

Between Cierva and Rawson, 1928 was a banner year for Autogiro development, testing, and notable achievements. Cierva, ever mindful of publicity, had previously made sure that he was photographed showing the aircraft to one of Britain's famous female aviators, Lady Mary Heath,[38] and offering rides to journalists such as well-known aviation writer Major C.C. Turner. But Cierva's greatest triumph occurred when, on September 18, 1928, he flew from London's Croydon airfield to Le Bourget in Paris with Henri Bouché, editor of *l'Aéronautique,* as passenger. With stops at Saint-Inglevert and Abbeville, it was the first international flight by an Autogiro and secured for Cierva the 1928 Grand Prix de l'Académie des Sports and the 1928 Lahm Prize presented by the Aéro Club de France. These two prestigious aviation awards totaled forty-five thousand francs, and the Union pour la Sécurité en Aéroplane awarded an additional twenty thousand francs for the feat, but the international public notice was worth even more. The most significant event in the development of the Autogiro, however, was undoubtedly the return of Harold F. Pitcairn in July of that year.

NOTES

1. Juan de la Cierva and Don Rose, *Wings of Tomorrow: The Story of the Autogiro* (New York: Brewer, Warren & Putnam, 1931) p. 104; see also P. T. Capon, "Cierva's First Autogiros: Part 1," *Aeroplane Monthly* 7, no. 4 (April 1979): 200–205; Ibid., "Cierva's First Autogiros: Part 2," *Aeroplane Monthly* 7, no. 5 (May 1979): 234–40.

2. See, for example, Peter W. Brooks, *Cierva Autogiros: The Development of Rotary-Wing Flight* (Washington, D.C.: Smithsonian Institution Press, 1988), p. 43; Charles Gablehouse, *Helicopters and Autogiros: A History of Rotating-Wing and V/STOL Aviation,* rev. ed. (Philadelphia: Lippincott, 1969), p. 38.

3. For a description of the C.6, see "Cierva C.6 Autogiro," *Rotorcraft* 28, no. 4 (June–July 1990): 53; Brooks, pp. 46–48; Capon, "Part 2," p. 238.

4. John M. Miller, "The First Transcontinental Flights with a Rotary-Wing Aircraft 1931," *Popular Rotorcraft Flying* 30, no. 5 (August 1992): 11–19, 11.

5. Miller uses the term *autogyro* incorrectly. Although Cierva originated the term *Autogiro,* Miller incorrectly asserted in a letter to the author of February 28, 2001, that "[w]hen I use the term autogiro I refer to the Pitcairn products. That name is a copyrighted name by Pitcairn. The generic name is autogyro."

6. Miller, "First Transcontinental Flights," p. 11.

7. See Brooks, p. 130.

8. For a description of the Paris Aeronautical Salon, see "Salon Stars: A Selection of Types at the November 1936 Paris Salon Aéronautique," *Air Enthusiast,* no. 91 (February–January 2001): 2–6 (note the Lioré et Olivier C.34 Autogiro in the background of the photo on the lower part of p. 5).

9. For a description of his career Brooks, pp. 48, 89–91, 129, 164, 206.

10. This may have been due to the differences in the results between the Spanish and English wind tunnel tests of Autogiro models. The Spanish results of the tests at the wind tunnel at Cuatro Vientos were suspected of overstating the potential of the C.4 and were more optimistic than the same tests performed by Vickers at Weybridge in January and February of 1925 under the direction of Captain P. D. Acland, managing director of the Vickers aviation department. Brooks, p. 46 (Brooks attributes the misstated Spanish results to "scale effects" which led to exaggerated projections of Autogiro speed).

11. Fredric Winkowski and Frank D. Sullivan, *100 Planes, 100 Years: The First Century of Aviation* (New York: Smithmark Publishers, 1998), pp. 48–49. See also *Aircraft of the National Air and Space Museum,* 4th ed. (Washington, D.C., and London): Smithsonian Institution Press, 1991), under "Ford 5-AT Tri-Motor."

12. For a picture of the C.6A (also known as the C.6bis) showing the men who would run with the rope to prerotate the rotor blades at its English demonstration at Farnborough, see Peter Almond, *Aviation: The Early Years (The Hulton Getty Picture Collection)* (Köln, Germany: Könemann Verlagssgesellschaft mbH, 1997), pp. 342–43; see also Frank T. Courtney, *The Eighth Sea* (New York: Doubleday & Co., 1972; also published as *Flight Path,* London: William Kimber), pp. 159–60.

13. Frank Kingston Smith, *Legacy of Wings: The Story of Harold F. Pitcairn* (New York: Jason Aronson, 1981), p. 67.

14. Brooks, p. 139.

15. Courtney, p. 162.

16. Warren R. Young, *The Helicopters* (Alexandria, Virginia: Time-Life Books, 1982), p. 58.

17. Brooks states that Courtney was told that the C.6 had been "built by a nut called Cierva," but this is clearly an unjustified exaggeration; nor does Courtney refer to the C.6 as "crazy looking." Compare Brooks, p. 53, with Courtney, p. 162. It is difficult to understand the source of Brooks's material, yet it is an understandable exaggeration—Courtney was skeptical and would never establish other than a minimally cordial working relationship with Cierva that would not last beyond February 1927. And Courtney would continue to bemoan the fact that Cierva only mentions him once in his 1931 book (see Cierva and Rose).

18. Courtney, p. 162.

19. Ibid., p. 163.

20. C.R. Roseberry, *The Challenging Skies: The Colorful Story of Aviation's Most Exciting Years 1919–39* (Garden City, New York: Doubleday & Company, 1966), pp. 2–4, 354.

21. Brooks, p. 55; Cierva would read his own paper in English on February 13, 1930, before TRAS.

22. For a photograph of Juan de la Cierva, Frank Courtney, and British Air Minister Sir Samuel Hoare at Farnborough in 1925, see Courtney, p. 160.

23. After the Autogiro had experienced its rough landing, and that incident was obscured in the successful completion of the tests, Cierva presented Courtney with an expensive gold cigarette case inscribed with "... recuerdo de unos experimentos que nos han hecho amigos para siempre." Courtney, p. 166.

24. Bill Hannon, "Those Infuriating 'Palm Trees,'" *Popular Rotorcraft Flying* 7, no. 6 (November–December 1969): 30.

25. Although Cierva is silent about this in his 1931 book with Rose, Courtney lectured on it almost forty years later, claiming that the patent office had initially refused Cierva's application, with the observation that the Autogiro "couldn't possibly work." Hannon, p. 30.

26. Courtney, p. 169. The assembled dignitaries included Pierre Etienne Flandin, the Spanish ambassador, and the Belgian and Italian military attachés.

27. He was a prominent Scottish industrialist who had risen to the rank of brigadier general at the end of World War I.

28. Sir Robert later became Lord Kindersley and served as a director of the Bank of England.

29. The C.6C was also given a non-Cierva manufacturer's designation of Avro Type 574 and was given the military serial J8068.

30. Frank Courtney, writing in his autobiography almost forty-six years later, claims that this accident happened the *day before* the royal demonstration, but given the damage, it is unlikely that the necessary repairs could have been made by the next morning. Brooks's account of a three-day interval between accident and display is far more credible. Compare Courtney, p. 170, with Brooks, pp. 62–63.

31. For a photograph of the destroyed C.6C, see Courtney, p. 161.

32. Compare Courtney, pp. 170–71, with Brooks, pp. 171–72.

33. The C.10 and C.11 designs of 1927 and 1928 did not have any vertical hinges.

34. Compare Courtney, p. 170, with Brooks, p. 175.

35. For a description of Hinkler's test-flying, see A. J. Jackson, *AVRO Aircraft Since 1908* (London, England: Putnam & Company, 1962) under each Avro Cierva Autogiro model.

36. For a photo of G-EBTW in flight, see Gablehouse, *Helicopters and Autogiros* (1969), p. 46 (Gablehouse identifies G-EBTW as a C.8, and this is certainly correct, as the C.6D had by then been fitted with lead-lag drag hinges and had been redesignated C.8R). See also Brooks, p. 69.

37. For a rare picture of Hinkler flying with a passenger in the C.8V, see Juan de la Cierva, "The Autogiro: Its Future as a Service Aeroplane," *United States Naval Institute Proceedings* 54, no. 8 (August 1928): 696–701, 697.

38. That photo is reproduced in Almond, p. 343. Lady Heath is unidentified, but see Roseberry, p. 427, where Lady Heath is identified and shown wearing the same hat and with Cierva. She completed her record flight from Cape Town to London in 1928 and thoroughly charmed the public by refusing to wear masculine attire like Amelia Earhart did, but flying in an afternoon gown and high heels! The most famous English aviatrix, she was the wife of Lord James Heath and held the first transport license ever granted to a woman in England. Her husband's status and position had undoubtedly helped change the law specifically for her. See Susan Butler, *East to the Dawn: The Life of Amelia Earhart* (New York: Da Capo Press, Inc., 1999), p. 212.

Chapter 3

CIERVA AND HAROLD F. PITCAIRN

The Navy is very interested in the possibilities of the autogiro. We have ordered one so that experiments may be carried out toward determining its adaptability to naval needs. The ability of the autogiro to land within a limited space its ability to hover over one point should make it extremely useful for reconnaissance work over bad country where adequate landing fields do not exist. There can be no doubt but what the development of the autogiro is the outstanding achievement in aviation during the past year.

> Rear Admiral W. A. Moffett, Chief of the Bureau of Naval Aeronautics, speaking at the White House at the ceremony awarding the Collier Trophy on April 22, 1931

Harold F. Pitcairn returned to England in July 1928 with the aim of more closely examining the Autogiro, which had, according to its press notices, advanced considerably since his last encounter with Cierva in 1925. Both the daily newspaper and, perhaps of greater importance, the American military journals began to take notice of Cierva's aerial achievements from the first demonstrations in England.[1] Although Pitcairn's aviation company was successful, with an airmail route spanning the Atlantic coast from the northeast to Florida and westward to Atlanta, Georgia, he was aware that there was an ongoing consolidation that would leave Pitcairn Aviation vulnerable to one of the nascent industry giants: the Aviation

Company of America, North American Aviation, General Motors, and the Ford Motor Company. The larger companies were buying up smaller air-mail carriers and preparing to make the massive investment necessary to initiate passenger service. Because he had always been interested in rotary-wing flight and had in fact received the first of his patents for rotary-wing aircraft in 1925 and because he had been experimenting with rotors powered by compressed air, Pitcairn decided to come to Europe to see the new Autogiros for himself. He was then able to fly the C.8L-II after only a short briefing by Cierva's chief test pilot, Arthur Rawson. Pitcairn made several landings, including steep descents and short takeoff rolls; he was impressed and was also amazed at how slowly the craft could fly under complete control, speeds at which a fixed-wing airplane would stall and fall from the sky. He immediately offered to purchase a C.8, providing that it could be *Americanized* with an American-built Wright 220 horse-power J-5 engine that would turn the propeller in a clockwise direction, perhaps influenced by Charles Lindbergh's use of the Wright engine in the *Spirit of St. Louis*. Assured by the engineers that the rotor would function regardless of propeller spin, the order was made and he committed to sending a Wright engine for installation. Pitcairn, the businessman and aviation strategist, proceeded to engage Cierva and his associates, including James Weir, in discussion as to what form an American license might take. Even though his brief C.8 flights had gone extraordinarily well and he was genuinely enthusiastic, his aviation and business experience suggested a cautious approach. He negotiated an option for the American licensing and manufacturing rights to Cierva's inventions and patents and proposed a reciprocal licensing of all future Pitcairn patents on rotary-wing inventions to the English company. Further, Pitcairn also proposed that his American company would now be named the Pitcairn-Cierva Autogiro Company of America.

Cierva agreed to the option, and Pitcairn returned to America to arrange for the shipment of the Wright engine. Of greater importance, he engaged in extensive consultation with his attorneys, the Philadelphia patent-law firm Synnestvedt & Lechner, who were tasked with an in-depth investigation of the current status of rotary-flight patents in America. Reassessing the state of his business affairs and evaluating the productive capacity of his factory, Pitcairn hired Edwin Asplundh, who had gained a reputation as a production genius, in preparation for expansion of the facilities in Bryn Athyn, Pennsylvania. Pitcairn, accompanied by patent lawyer Ed Davis, returned to England to finalize the licensing and partnership arrangements. They returned on December 11, 1928, on the SS *Aquitania* with signed

Harold F. Pitcairn flies the Cierva C.8W over Pitcairn Field. The Bryn Athyn Cathedral is in the upper right. The C.8W, now in the collection of the National Air and Space Museum of the Smithsonian Institution, was the first rotary aircraft in America. (Courtesy of Stephen Pitcairn, from the Pitcairn Archives.)

agreements calling for a payment of $300,000 and, of equal importance, the crated C.8W (the *W* referring to Wright engine) and Cierva pilot Arthur Rawson. Pitcairn had flown the C.8W in England before it was disassembled and crated, and he was anxious to fly it in America. The Autogiro was even then gaining additional notoriety in England, when Cierva, who had gained his pilot's license in 1927, flew to Paris on September 18, 1928, with passenger Henri Bouché, editor of *L'Aéronautique*—the first rotary-wing crossing of the English Channel. The Autogiro would soon gain similar notice in America.

Reassembled at Bryn Athyn and given the registration number NC418, the C.8W was first test-flown by Rawson on December 18, 1928, and was then flown by Pitcairn, although there is some confusion as to when the Pitcairn flight actually occurred. Pitcairn company historian Carl Gunther claims[2] that Pitcairn flew immediately after Rawson on the December 18, but the Autogiro Company of America's 1932 publication maintained that the first flight occurred[3] on *both* December 18 *and* 19! Assuming that the 18th is correct, it was twenty-five years and a day since the Wright brothers first flew—the age of rotary-wing flight had come to America. And America noticed—the flights from the Pitcairn airfield in Bryn Athyn gained increased public notice of the strange aircraft, and the army and navy, well-familiar with the dispatches describing Cierva's English achievements published in the United States Naval Institute *Proceedings*,[4] requested demonstration flights. Additionally, interest was expressed by the Post Office Department, the Department of Commerce, and the National Advisory Committee for Aeronautics (NACA).[5] But Harold Pitcairn resisted all entreaties—he wanted to gain more experience and make a decision whether to exercise his option. He was also painfully aware that should anything happen to the single flying American Autogiro, it would certainly retard, if not actually doom future development.

Flying the Autogiro,[6] Pitcairn made comparative flight-tests[7] against his successful Pitcairn PA-5 Mailwing—flown by Pitcairn Aircraft chief pilot James G. "Jim" Ray—in takeoff, climb, speed, and control aspects. Although the C.8W had generally good handling characteristics, Pitcairn and his associates immediately noted that at high diving speeds the rotor slowed and developed excessive and rough blade flapping. The tests were analyzed by Pitcairn, Ray, and associates Agnew Larsen and Paul Stanley, and they began suggesting ways to alleviate the problem (which was solved by the end of 1929 by changing the wing configuration) and improving the Cierva model. Pitcairn was in a precarious situation—the clock was running on the Cierva option, the definitive legal opinion from Synnestvedt & Lechner was not yet ready, and he was being pressed by Clement Melville Keys to sell the Pitcairn Aviation interests. But it all came together—the opinion letter from the attorneys, carefully worded and even more carefully researched, stated in unambiguous terms that the Cierva patents were favorable for going ahead with the business deal. It was what Harold Pitcairn had been hoping to hear—by transatlantic calls and telegrams, negotiation began between the Cierva and Pitcairn groups. It was a heated exchange, with significant differences of opinion regarding royalty payments for use of patents and corporate governance issues, but in the end it was obvious that both parties wanted it to work. Cierva sent his managing director, Colonel John Josselyn, to negotiate the final terms; Josselyn arrived at Bryn Athyn with an unrestricted power of attorney, and it was obvious to all that he would not leave without concluding the deal. The final agreement was signed on February 14, 1929, by which Pitcairn acquired the American rights to all Cierva's patents and inventions for $300,000. Pitcairn became a director of the English company, and Cierva became a director of the Pitcairn-Cierva Autogiro Company of America. It was intended that the Pitcairn-Cierva Autogiro Company, like the Cierva Autogiro Company Ltd., would be a research and development company that would hold and license the use of Pitcairn-Cierva Autogiro patents for manufacturing by others.

Harold Pitcairn undoubtedly took a deep breath. Along with his close associates, he had estimated the first year's budget for the new company—it was a staggering *one million dollars,* and that did not even include the payment to Cierva. It was obvious that money would have to be raised to nurture this new enterprise. Pitcairn's answer was to enter into serious negotiations with C.M. Keys to sell Pitcairn Aviation's eighteen hundred miles of federally subsidized mail routes. Those routes were continuing to grow and Pitcairn had just successfully exhibited his new PA-Super Mailwing airplane at the National Air Show in Detroit, and rumors were already circulating of a Keys-Pitcairn merger. Throughout these sensitive negotiations Pitcairn successfully concealed his deal with Cierva and interest in the Autogiro, but official pressure proved too much and he felt he could no longer ignore the requests for public demonstration. On May 13, 1929, Pitcairn flew the C.8W from Pitcairn Field at Bryn Athyn to Langley Field, Virginia, via Washington, D.C. He demonstrated the Autogiro before the annual NACA conference and then went on to the United States Naval Base at Norfolk, Virginia, where he showed the aircraft's unique flying characteristics before an enthusiastic audience of senior military officers. But of particular note was the reaction of those who observed this unique machine flying over Philadelphia, Wilmington, Baltimore, Washington, and Richmond—everywhere Pitcairn flew people stopped what they were doing, came outside, and gazed up at what local papers termed the "wonder-plane." It was the first cross-country flight in America and the longest Autogiro flight to date. To top it off, Orville Wright inspected the Autogiro and expressed his admiration. It was, by all accounts, a triumph, and it was soon repeated. After returning to Bryn Athyn to let the engineers inspect the aircraft, Pitcairn was off again to Washington, D.C., to demonstrate its flying abilities before the Spanish ambassador and dignitaries, government officials, and members of Congress. His demonstrations were met with public acclaim and served to reinforce his decision to go ahead with Autogiro development, but the problem of financing such an undertaking still remained.

Harold Pitcairn, ever the sharp businessman, decided to move the negotiations with Keys along by nurturing the rumors then rife that he was considering a merger with one of Keys's competitors. Merger-mania was then sweeping American aviation, and Pan America, Texas Air Transport, St. Tammany, and Gulf Coast, Colonial were all rumored to be considering joining with Pitcairn's Atlantic coast routes, and he did nothing to deny the circulating stories. He deliberately let slip in conversation with Keys that his new aircraft design, the PA-6 Super Mailwing, was selling well and that even more powerful planes were on the drawing board and that he intended to exhibit the PA-7 Super Mailwing at the 1929 National Air Show in Detroit. The seemingly casual remarks were intended to make Pitcairn Aviation even more desirable. But Keys and his associates were delaying, forcing Pitcairn to be even more creative in his attempts to move the negotiations to a successful conclusion. He asked his close associate Geoffrey M. Childs and his wife to take a European vacation, and he made sure that pictures of the smiling, confident executive were published in the business press with captions proclaiming that the Pitcairn Aviation official was going to Europe to ascertain new developments in passenger service that could be adapted to America. The final straw for Keys was the Post Office Department report for May, which indicated that Pitcairn's routes had made a $34,000 profit. After consulting with his associates, Keys requested a meeting with Pitcairn.

Keys took no chances—he invited Pitcairn to come to his New York office and there, surrounded by the evidence of his success and corporate power, told his guest that the board had authorized an offer of $2,500,000. Pitcairn put on his best poker face. It was a spectacular sum—not more than the airline routes were worth for future expansion, but certainly more than he expected from a sharp businessman such as Keys. Not wanting to appear overly eager, Pitcairn departed for the return trip to Philadelphia. Although he stated that he needed to consult with his board, he had already decided to accept. The deal was finalized less than a week later on June 12, 1929, with the signing of documents and a certified $500,000 check. Final payment was made in mid-September—seven weeks before the October 29 stock market crash on "Black Friday." He had the funding for the Pitcairn-Cierva Autogiro Company.

That Harold Pitcairn would become involved in aviation was not obvious. His father, John Pitcairn, was born in Johnstone, Scotland, on January 10, 1841. His family immigrated to America five years later. Growing up in the industrial town of Pittsburgh, John left school by fourteen and went to work at the Pennsylvania Railroad with his older brother Robert. He rose quickly and was soon in charge of the Philadelphia branch, a position in which he became friends with such business leaders as John D. Rockefeller, Richard Mellon, Henry Clay Frick, and Andrew Carnegie. In midlife he became a devoted follower of the Swedish revelator, Emanuel Swedenborg and the New Church, which embodied Swedenborg's teachings of meditation on the word of God, responsibility to family, hard work, and society. Above all, the New Church emphasized the proper use of one's God-given gifts. John Pitcairn founded the Pittsburgh Plate Glass Company with Captain John B. Ford on the nearby Allegheny River and served as its president from 1896 until 1906. Under his leadership the company experienced great growth, and by 1900 it produced 65 percent of American plate glass. At his death in 1916 he left an estate with an estimated value of between $60 and $270 million dollars[8] for his widow and three surviving sons. Two sons had died in infancy, and his only daughter, Vera, had died six years earlier at the age of twenty-three. Also at the time of his death, the New Church Cathedral, which he had endowed and for which he had donated land in Bryn Athyn, was in the early stages of construction and would be completed by his oldest son.

His oldest son, Raymond (1885–1966), was a successful lawyer, businessman, musician, civic leader, and noted Abraham Lincoln scholar. In 1914 he set aside his legal career to complete the New Church Cathedral. It was an original, innovative, beautiful, and inspirational building, so much so that Raymond was elected to the American Institute of Architects on its merits. The middle brother, Theodore (1893–1973), inspired to live a life of religious pursuit, theological studies, missionary work, and the ministry, became a priest in the New Church. Theodore's wife, Maryke Urban, whom he had met while on an overseas mission, aided his ministry as he preached the Swedenborg faith in Basutoland, South Africa. He returned to Bryn Athyn to the cathedral but eventually fell out with its leaders and doctrines. Breaking away, he took his followers with him and founded a new church on his own property, funding it with proceeds from the sale of his art collection. His Monet, which he had purchased for $11,000, sold for $1.4 million, establishing the record for such art at the time.

The youngest of John Pitcairn's children was Harold Frederick Pitcairn, born in 1897. Harold took an early, and not entirely approved of, interest in aviation. Inspired by the first flight of the Wright brothers in 1903, he began flight training as an air cadet in the last days of World War I, and he would eventually earn a pilot's license signed by Orville Wright. During his flight training he had met a talented engineer, Agnew Larsen. They had combined talents to produce "a series of clean, efficient aircraft that were to be of great significance to air transportation."[9] This was the Mailwing series that Pitcairn would tout to Keys and for which he is justly famous,

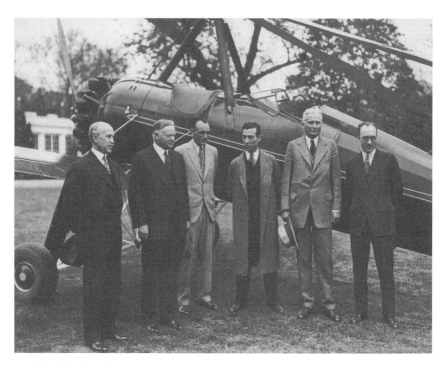

Awarding of the Collier Trophy at the White House on April 22, 1931, to Harold F. Pitcairn and his associates by President Herbert Hoover for the most "significant contribution to aviation" in 1930. The PCA-2, flown by James G. "Jim" Ray, was the first rotary-wing aircraft to land at the White House. Left to right: Orville Wright; President Herbert Hoover; James G. "Jim" Ray; Colonel Clarence A. Young, Assistant Secretary of Commerce for Aviation; Senator Hiram Bingham; president of the National Aeronautical Association, Harold F. Pitcairn. (Courtesy of Stephen Pitcairn, from the Pitcairn Archives.)

as it was ideal for the emerging government-subsidized mail route system, with a baggage compartment economically sized for the small mail loads that initially characterized such service. The construction was also innovative, with the unique use of stronger square tubing for the fuselage and quick-change engine mounts. Pitcairn Aviation's mail route system was efficiently planned and well-staffed, and it provided quality flight training. Even if Pitcairn had not gone on to the Autogiro, the Mailwing and the airmail route system he created would have earned him inclusion in the Aviation Hall of Fame. But after selling Pitcairn Aviation's airmail routes, plans, and airports to Keys, he threw himself into the development and promotion of the Autogiro. The results of the next sixteen months would earn him and his associates the Collier Trophy,[10] the most prestigious award for the greatest aviation achievement for the year.

Dr. Charles Abbot, secretary of the Smithsonian Institution, accepting the donation of the Cierva C.8W, the first rotary aircraft in America, from Harold Pitcairn and shaking hands with James G. "Jim" Ray, Pitcairn pilot on July 22, 1931. Left to right: Dr. Charles Abbot, Harold F. Pitcairn, Jim Ray, Geoffrey Childs, Colonel Clarence Young, William P. McCracken, and George Lewis. (Courtesy of Stephen Pitcairn, from the Pitcairn Archives.)

The Pitcairn-Cierva Autogiro Company's first licensee was Pitcairn Aviation Inc., which was to develop and manufacture Autogiro models based on both Cierva and Pitcairn patents. The latter company, owned by Pitcairn, immediately began an evaluation of the C.8W with the aim of developing an American model, as the Guggenheim Safe Airplane competition had been announced for the fall of 1929 and Pitcairn thought that an American Autogiro had a good chance of winning. Paul E. Garber, curator of aviation artifacts for the Smithsonian and after whom its complex at Silver Hill, Maryland, would eventually be named, wrote Pitcairn. Citing the importance of the first Autogiro in America, Garber requested that it be donated to the national aviation collection. Pitcairn was flattered and immediately accepted, recognizing that such an accolade acknowledged the important nature of his aviation endeavor and placed the C.8W with Lindbergh's *Spirit of St. Louis* and the other historic American aircraft.

When the evaluation of the C.8W was complete, Pitcairn presented it to the National Air and Space Museum of the Smithsonian Institution on July 17, 1931. It was accepted by Dr. Charles Greely Abbot, secretary of the Smithsonian Institution, when Jim Ray landed on the National Mall on July 22, 1931.[11]

The initial intention of the Pitcairn design team was much like that for the earlier Cierva designs, to modify an existing fixed-wing configuration. That aircraft, the PCA-1, employed the Mailwing welded chrome-molybdenum, steel square-tube fuselage that had proven itself on the demanding night airmail flights with a widened landing gear to increase stability. It also had plywood-covered strut-braced thirty-three-foot wings with upturned tips, and a boxlike tail derived from the Cierva designs (known as a "Scorpion tail"), a tiltable horizontal system constructed to channel the prop wash up into the rotor blades to aid in gaining the rpm necessary for takeoff. It was, as were the mail planes, an open-cockpit, two-seat, three-passenger design with a rotor pylon above the forward seat consisting of four tubular struts extending almost to the engine mount and anchoring just in back of the forward cockpit. It was larger than any Autogiro built, and the passengers rode in the forward cockpit while the pilot flew from the rear. The RB-55 rotor blades were supplied by the Cierva Autogiro Company and had been manufactured by Avro. Although Pitcairn remained optimistic that the PCA-1 would be completed in time to enter the Guggenheim Safe Aircraft competition on October 10, it proved more difficult than anticipated, and the aircraft was not ready. It was an ambitious attempt to create a large aircraft and was considerably larger than even the C.19 Mk.II (G-AAKY) that Cierva himself brought to America in August for the Cleveland Air Races of 1929. The new American PCA-1 had been covered with doped linen and then hand-rubbed to a glossy black finish. It was an impressive, innovative undertaking—one that unfortunately did not fly nearly as well as it looked.

Early in the PCA-1's ground testing, it was evident that the larger scale of this first American prototype produced serious vibration problems because of an unbalanced rotor configuration. But the very complexity of the rotor hub and suspension systems of the individual blades made diagnosis of the problem difficult. As Pitcairn and his development team solved one vibration problem, another would emerge. Still it looked as if the aircraft might be ready for the Guggenheim competition, until a new difficulty was recognized—the flight requirements for the competition had been written for an eight-to-one angle of glide (i.e., go forward eight feet for each foot of descent). Such was the characteristic of a fixed-wing aircraft, conditions totally inapplicable to the Autogiro, but the competition committee steadfastly refused to alter the requirement. Had the prototype been ready to fly, Pitcairn intended to enter it anyway, hopeful that its unique rotary-wing performance would impress the judges, but that never happened. The vibration problems with the rotor blades had clearly not been solved as the date of the competition approached, forcing Pitcairn to withdraw his entry, but he continued experimentation on the prototype. New rotor designs were tried, weights were adjusted to fine-tune the center of gravity of the fuselage, and various blade lengths were tried. It was an intensive effort that relied in part on Cierva's mathematical models and theoretical analyses, but the pragmatic experimentation was leading Pitcairn and his team to formulate new and innovative theories. There was a growing confidence that the American group was gaining a serious, in-depth understanding of this new aircraft as each problem was solved, but the feelings were clearly mixed when Cierva demonstrated the C.19 Mk.II in Cleveland before wildly enthusiastic crowds.[12] Even as he cheered the Spanish inventor, Pitcairn quietly vowed that next year, it would be an American-created and registered Autogiro that would thrill the spectators—but it would not be the PCA-1. In early October, with a more conventional tail structure and other improvements, the prototype was crash-landed by Cierva a few days after its first flight. It was moved to the old Pitcairn factory in Bryn Athyn for reconstruction, but it was consumed in a fire on November 18, 1929, which gutted the factory just three weeks after the stock market crash. It was an inglorious end to the first American Autogiro design, but its legacy and lessons had already been incorporated into the next Pitcairn model, the PCA-1A.

The PCA-1A incorporated a lighter duralumin tube construction with a stronger, braced wing structure. The landing gear was redesigned with low-pressure Akron tires. That this model was ready for flight-testing in late October was a clear indication that the Pitcairn team was successfully integrating the ongoing experience, the result being increasingly confident construction. Cierva, still in America, flew the PCA-1A in late October. The PCA-1A's end was far better than its predecessor's, as Harold Pitcairn presented it to the Franklin Institute in Philadelphia, and it hung in exhibition for twenty years in the institute's museum. Finally removed from exhibition in 1954 and crated, it was presented to the Smithsonian Institution and placed in the Paul E. Garber Preservation, Restoration and Storage Facility at Silver Hill. It was loaned to the American Helicopter Museum and Education Center in West Chester, Pennsylvania, with the understanding that it would first be restored by Harold's son Steve Pitcairn.[13] That restoration effort, led by Mike Posey, who was assisted by Steve Pitcairn, James Cole, and Joseph Rommel, was completed in March 2000, and the PCA-1A was finally placed on exhibition on October 4, 2000. At the dedication, Dr. Dominick Pisano, chairman of the National Air and Space Museum, stated that it was the finest restoration of an Autogiro he had ever seen. Viewed by thousands of visitors, it occupies an honored position as the oldest surviving American Autogiro.

The PCA-1 was flown for several months as a test platform. Pitcairn never applied for an Approved Type Certificate (ATC), as it was never intended for production. The three Pitcairn Autogiro pilots, Jim Ray, Jim Faulkner, and Harold Pitcairn, analyzed the flight experience and suggested improvements and modifications. The results of those lessons were incorporated into the ongoing development of a commercial Autogiro, but first a third experimental aircraft was constructed, the PCA-1B. It was similar to the earlier prototype aircraft, incorporating a light but strong duralumin tube fuselage and cable-braced fabric wings with the now-familiar upturned tips for increased stability. The box-deflector tail of the earlier models was replaced by a more traditional vertical tail unit. Pitcairn characterized the deflector box tail as "un-American," which would lead to the distinctly American development in 1930–31 of a mechanical system clutched to the engine for prerotation. Called the *Black Ship,* the PCA-1B was flown by Cierva before he returned to Europe in the first part of November 1929, and it was later demonstrated at the 1930 National Air Races at the Curtiss-Reynolds Airport in Chicago. Although the British Cierva C.19 Mk.II was also flown, Harold Pitcairn could not have but felt pride and a sense of accomplishment when the PCA-1B and prototype PCA-2 thrilled the crowd—it had been almost a year since he had vowed to have an American Autogiro flying. There was a certain irony in that the Pitcairn aircraft had flown to Chicago from Willow Grove, 650 miles, taking six hours and fifty minutes to make the trip at an estimated speed of about one hundred miles per hour, while the Cierva machine had been hauled on a truck.[14] It was then the longest Autogiro cross-country flight and revealed areas that needed improvement, such as adjustments in the cabling system that suspended the blades and kept them from drooping at low speeds and striking the aircraft superstructure. These cables were found prone to fatigue after a few hours. That problem was remedied with elastic links at the rotor hub cable attachments to control slack while in flight. Additionally, the PCA-2 prototype was modified with the addition of hydraulic dampers on the interblade bracing cables and redesigned oil hydraulic dampers by Houdaille-Hershey.

The PCA-1B, like its developmental predecessors, was a test platform for various modifications, chiefly experimentation with different angles of wing incidence to determine the most efficient configuration. Additionally, as its flights were noted by the press and public to acclaim and amazement, Pitcairn was gauging marketing potential in preparation for introducing the first certified Autogiro intended for sale—the PCA-2. The PCA-2 had been produced at the new factory in Willow Grove, Pennsylvania. Pitcairn had purchased land in February 1926 and started construction in May 1929, with the factory becoming his major production facility after the fire in November that had destroyed the Bryn Athyn facility. Although the PCA-2 was the first model specifically designed for Autogiro flight, it drew heavily on Pitcairn's previous airmail plane design. Unlike the three PCA-1 developmental prototypes, the PCA-2 had a fuselage of welded steel tubes rather than duralumin because, in consideration of the commercial intent for the model, steel was easier to repair and modify. It was a three-passenger model with two seats in the side-by-side front cockpit, and it differed from its American predecessors in that the front cockpit was close to the aircraft center of gravity, a significant factor in creating flight stability.

First flown in March of 1930, the prototype (X760W) had a gross weight of 2,750 pounds (increased in actual production to 3,000 pounds), featured a familiar 225-horsepower Wright R-760-E Whirlwind J-6–7 engine,[15] and had thirty-foot wings that were upturned at the tips. The tilting box tail had completely disappeared, and the simplified tail of the PCA-1B was used, except that it now lost its fin extension to the back of the rear cockpit that had looked so dashing on the research aircraft. The PCA-2 also featured a significant advance in Autogiro technology, the mechanical prerotator. The four-bladed rotor was positioned on a pylon located above the front cockpit on a metal tripod. The front of the forward tripod leg extended to a point midway between the cockpit and the engine mount, with the remaining legs extending to points on the right and left of the fuselage between the front and rear cockpits. But slightly in front of the forward rotor pylon spar and parallel to it was a mechanic connection extending between the rotor head and the engine. This engine drive for rotor spin-up had been designed by Pitcairn associate Agnew E. Larsen with the help of Heraclio Alfaro and Jean Nicol of Jos. S. Pecker's office (later the Machine and Tool Designing Company and Autogiro Specialties Company) which manufactured the forty-eight-pound clutch and engine drive in late 1929 to early 1930. It was initially capable of spinning up the rotor blades to eighty to ninety rpm in thirty to forty seconds, enabling the aircraft to take off

Pitcairn PCA-2 (X760W) during certification flight in fall 1930, with the PCA-1B in the background, flying past the Statue of Liberty on Bedloe's Island. (Courtesy of Stephen Pitcairn, from the Pitcairn Archives.)

with only a short taxi run to accelerate the rotor to takeoff speed. This impressed Cierva, who would promptly incorporate it into the next English Autogiro. The performance of this device was later improved with a more efficient gearbox and clutch to prerotate a larger rotor to 125 rpm, which allowed for more dramatic takeoffs with very short ground runs.

Jim Ray commented after the first flight that the PCA-2 had performed flawlessly, with a smooth rotor spin-up, a climb out free from rotor vibration, and extremely stable flight. He expressed the belief that a fixed-wing pilot could transition to the Autogiro with less than an hour's instruction and that a novice could become a proficient pilot in half the time it took to solo in an easy fixed-wing trainer. It was exactly what Pitcairn wanted to hear, and the advertising for the PCA-2 would highlight and headline the ease of flying. It would, however, be a promise unfulfilled for many, but that was in the future—in the interim a new and potentially serious impediment emerged. Confronted with the first rotary aircraft designed for commercial sale, the United States Department of Commerce admitted that no

Pitcairn PCA-2 Autogiro (X760W) during certification flight in fall 1930, with the PCA-1B in the background, flying past the George Washington Bridge, then under construction. (Courtesy of Stephen Pitcairn, from the Pitcairn Archives.)

performance standards existed to guide the Approved Type Certification. The government policy would therefore be that the development of such standards would both proceed from and apply to the PCA-2, as it was the first Autogiro submitted for the ATC. The PCA-1 series had not had this kind of certification, as it was for research and development and not intended for commercial sales. The government specified that the certification process for the PCA-2 would require a series of flight-tests extending over the next year. In fact, the PCA-2 would be granted an ATC on April 2, 1931.

Pitcairn shrewdly used the government flight-tests to gain public notice, featuring the Autogiro, accompanied by the PCA-1B as a chase plane, flying over New York City and around the Statute of Liberty. Jim Ray had even flown 2,500 miles from Willow Grove to Miami and back in January 1931. He landed often on small, rough fields and in a variety of weather conditions, all of which was part of the certification process. He also

Pitcairn PCA-2 Autogiro during fall 1930 certification flight over West 11th St. in the Wall Street section of Manhattan, flying north. (Courtesy of Stephen Pitcairn, from the Pitcairn Archives.)

gained much public notice at the All-American Air Races and the dedication of the new United States Navy Air Field. In a staged performance worthy of P. T. Barnum, Ray even landed in a Miami public park and was photographed receiving a parking ticket from the chief of police, who promptly received a ride in the Autogiro. The public response was favorable, and the Autogiro flights sometimes had the effect of stopping traffic and always of being noted in the press.

The reception at the National Air Races in Chicago in late August had been an undeniable triumph, and Pitcairn had even arranged for the PCA-1B and PCA-2 prototype to fly over Chicago—in contrast with the Cierva model, which never left the exhibit area—to great acclaim and some traffic difficulties, as people stopped to look up at the new rotary aircraft. The public appearance of the American Autogiro had been so successful that Pitcairn announced the production version of the PCA-2, with the larger, 300-horsepower Wright R-975-E Whirl-wind J-6–9 engine, to be manufactured by the newly formed Pitcairn Autogiro Company.

The first production PCA-2 had flown on November 1, 1930 and although not yet certified, Pitcairn began taking deposits. The more publicity generated by the flights of the certification model, the more public interest was generated. The larger engine produced a maximum speed of 118 miles per hour, a cruising speed of 98 miles per hour, and rate of climb of 800 feet per minute, even though the aircraft weight had increased to three thousand pounds. Pitcairn also changed the name of the Pitcairn-Cierva Autogiro Company to the Autogiro Company of America (ACA) in January 1931, with Pitcairn as president and Agnew Larsen as chief engineer. The American acclaim had also not gone unnoticed by others, and the ACA granted manufacturing licenses to Buhl Aircraft Company of Detroit, Michigan, in March of 1931, and later to the Kellett Aircraft Corporation, a company formed by W. Wallace Kellett, his brother Rodney, C[harles] Townsend Ludington, and his brother Nicholas. W. Wallace Kellett and C. T. Ludington had previously had aircraft dealerships at the Philadelphia Pine Valley Airport and had been associated with Amelia Earhart in the New York, Philadelphia and Washington Airway Corporation (NYPWA).

Even given the advent of the Great Depression, Pitcairn had reason to be optimistic about 1931. In December of 1930 Cierva had published[16] a well-received article entitled "Uses and Possibilities of the Autogiro" in the American magazine *Aero Digest* with a representation of a PCA-2 flying over New York on the cover. Furthermore, Amelia Earhart had become interested in the Autogiro. She had, after a single fifteen-to-twenty-minute flying lesson by factory test pilot J[ohn] Paul "Skipper" Lukens,[17] soloed at the Pitcairn Aviation field at Willow Grove on December 19, 1930, thus becoming the first woman Autogiro pilot. Advertising for the Autogiro and the PCA-2 was just beginning, and Pitcairn's offices soon received deposits and advanced orders from individuals and corporations seeking the convenience, safety, and publicity that seemed to accompany almost every Autogiro flight. The public's enthusiasm for the Autogiro was further encouraged in March of 1931 when David S. Ingalls, the navy's only World War I ace and assistant secretary of the navy, published an article in *Fortune* entitled "Autogiros: Missing Link," asserting that "Inventor Cierva and Impresario Pitcairn offer the most promising new flying machine in the thirty-year history of aviation."[18] It was heady praise, bolstered by the news that Pitcairn and his associates had been awarded the prestigious Collier Trophy for the greatest achievement in American aviation for 1930.

Although Pitcairn had wanted Cierva to attend,[19] it was President Herbert Hoover's schedule that dictated the timing—which occurred on April

The PCA-2, flown by James G. "Jim" Ray, the first rotary-wing aircraft to land at the White House, taking off after the ceremony of awarding the Collier Trophy at the White House on April 22, 1931, to Harold F. Pitcairn and his associates by President Herbert Hoover. (Courtesy of Stephen Pitcairn, from the Pitcairn Archives.)

22, 1931, at the White House. Hoover, previously secretary of commerce under President Calvin Coolidge, had followed the development of the Autogiro and, keen to see an Autogiro, personally requested that ceremony be held on the back lawn of the White House so that an aircraft could land and demonstrate its unique flying capabilities. It was a publicity triumph, and Pitcairn made the most of it. As that was in many ways the most significant moment in the development of the Autogiro in America, an account of it introduces this book—but it was not the only Autogiro event to capture public attention that April. Although it is likely that only the most attentive readers noted the brief announcement that the PCA-2 had received ATC 410 on April 2, 1931, the world took notice of Amelia Earhart's altitude record on April 8, 1931.[20]

Pitcairn's intent was to fan the public fires of Autogiro interest, and he set about the task with a creative ingenuity. He arranged for journalists to

Amelia Earhart in the PCA-2 Autogiro on April 8, 1931, at Pitcairn Field, Willow Grove, Pennsylvania, where she set a world altitude record of 18,415 feet. (Courtesy of Stephen Pitcairn, from the Pitcairn Archives.)

take rides in the PCA-2 and then used their columns in advertising to tout the revolutionary nature of Autogiro flight. Ernie Pyle had become the aviation editor for the *Washington Daily News* in March 1928 and won a devoted following with his human interest stories describing World War I pilots who constantly scrambled to make a living as cargo and mail pilots and as barnstormers who "gypsied from field to field, delighting crowds with wing-walks and offering thrill seekers their first flights for fees of a dollar a minute."[21] Pitcairn arranged for Jim Ray to take Pyle for a ride, and the newsman, in turn, praised the Autogiro's performance in a column dated September 26, 1930. Pyle quoted a flying companion (the front cockpit of the PCA-2 was a two-seater) as exclaiming, "That's the kind of plane for you and me, Ernie, one that comes straight down and slow," and went on to comment, "That expresses the whole thing. It's a great piece of machinery."

Pitcairn PCA-2 Autogiro (X760W) during certification flight in fall 1930, flying with the PCA-1B in the background over New York City piers. (Courtesy of Stephen Pitcairn, from the Pitcairn Archives.)

Such journalistic attention and acclaim attracted all sorts of aerial adventurers, with proposals that ranged from the preposterous to the intriguing. Each was considered—a typical example was the proposal put forth by the well-known California author and adventurer Richard Halliburton, who would publish books entitled *Flying Carpet* and *Richard Halliburton's Book of Wonders* that would continue to engage the imaginations primarily of young boys for decades.[22] Halliburton telegraphed Pitcairn on November 1, 1930, proposing that a PCA-2 Autogiro be made available for a "vagabond flight around the world by aeroplane."[23] Halliburton intended to fly a Lockheed airplane called the *Flying Carpet,* sponsored by the Shell Oil Company, and had contracts to produce ten articles for the *Ladies Home Journal* and a book for Bobbs-Merrill Company. His appeal to Pitcairn was direct:

> The Journal has a circulation of three million and goes into three million high class American homes. Each article will be read by seven to ten million

James G. "Jim" Ray taking off from a parking lot on the east side of
the Capitol building in Washington D.C. to carry Senator Hiram
Bingham to Burning Tree Country Club to play golf. Ray had to climb
out over the Senate Office Building to make the ten-minute trip in an
effective demonstration of the aircraft's usefulness—the trip usually
took an hour by car. (Courtesy of Stephen Pitcairn, from the Pitcairn
Archives.)

people. My three previous books have been in turn read in ten other countries. As they cost $5.00, they are bought by people with money.... [the]
Autogiro ship would fix the attention on my flight, a cause a sensation wherever I landed. This flight is by no means just an ambition, but already a fact
financed, equipped, piloted, publicized, waiting only for an extra gas tank to
be installed to give me 15 hundred miles radius, but with your new feature,
I can pilot my "Flying Carpet" with far greater safety into many more out-

landish places, and enjoy the advantage of having the greatest possible public interest behind me. It would promote your new ship, and my new book to the utmost.

Pitcairn politely declined the offer, recognizing that the certification process would take considerable time and, of greater importance, that a support network did not exist for the PCA-2. The Autogiro's future could only be advanced when aviators could rely on its safety, supported by a system for service maintenance and repairs. But there is little doubt that the possibility of publicity was appealing.

Publicity stunts were designed to catch the public's fancy. In addition to Ray's parking ticket in Miami, Pitcairn had the pilot land in the parking lot on the east side of the U.S. Capitol to pick up Senator Hiram Bingham and fly him to a golf outing at the Burning Tree Country Club[24] outside Washington, D.C. Pitcairn also had the PCA-2 photographed landing on the lawns of country estates, with many images of the aircraft landing at his own Bryn Athyn home, Cairncrest, and flying off to hunting or fishing camps. His advertising agency commissioned paintings, used for magazine and sales-brochure illustration, featuring the Autogiro landing at the country estate, at the foxhunt, at the Dude Ranch, and on the country club landing field having just deposited a handsome couple heading for the tennis court.[25] But perhaps the most ambitious attempt to garner public attention was that to have Amelia Earhart make the first transcontinental flight in an Autogiro in June 1931. It did not, however, work out as Pitcairn hoped.

NOTES

1. See, for example, "Autogiro Gives Air Stability," *United States Naval Institute Proceedings* 51, no. 4 (April 1925): 852–54; "De La Cierva Autogiro Achieves More Success," *United States Naval Institute Proceedings* 52, no. 275 (January 1926): 142–43.

2.

You are correct that Rawson was first to fly the C-8 on December 18, 1928. In fact this was the main reason Cierva sent Rawson along with Pitcairn to make sure that reassembly was done correctly. I was told by one of Pitcairn's employees who was there that Rawson was willing that Harold be first to fly it, but Harold felt that in keeping with Cierva's wishes Rawson should be first to check it out, even though Harold had several hours flying it in England. The procedure was also very much in keeping with the way Harold would have proceeded had it been his invention someone else would purchase and fly.... It is fact that Harold did fly it right after Rawson's check ride on the 18th. It is easy to imagine Harold's excitement to try out the new toy.

Carl Gunther, letter to author, October 26, 2001.

3. As stated by the Autogiro Company of America itself in 1932 and again in 1944. See *The Autogiro* (Philadelphia: Autogiro Company of America, 1932), p. 15; *Some Facts of Interest about Rotating-Wing Aircraft and the Autogiro Company of America* (Philadelphia: The Autogiro Company of America, 1944), p. 12 (caption to top photo). But this *cannot* be regarded as definitive, *as the 1932 publication contradicts itself* in stating on p. 31, "It was flown for the first time over American soil by Mr. Pitcairn, at Bryn Athyn, on December 18, 1928."

There is also some confusion about the pilot in the first flight. Frank Kingston Smith apparently is not the only source for the claim that *Pitcairn* made the first flight on December 18. The 1932 and 1944 publications by the Autogiro Company of America cite the 19th and, being silent as to the pilot, neither attributes that flight to Pitcairn. Brooks and Townson date that flight on the 19th, and each relates that Cierva pilot Arthur Rawson first made a test-flight after the C.8W had been reassembled—a sensible procedure probably agreed upon by both parties—and in describing the first flight(s), Smith at first relates ambiguously that "the first Autogiro made its first flight. Harold Pitcairn was the first American pilot to fly it," which, while factually true, obscures the fact that Rawson made the first flight. The accompanying photo on the same page (149) is captioned, "An historic photograph: the first rotary-wing flight in America, with Harold Pitcairn flying the Cierva C-8 Autogiro at Willow Grove in December, 1928," which is also historically accurate—it *was* the photo of *Pitcairn's* first American flight—but clearly not *the* first American flight. That honor had been claimed by Rawson. And while *Legacy of Wings* remains an affectionate tribute to Harold F. Pitcairn with much useful information, it must always be read with caution, stemming from the author's apparent dedication to depicting his subject in the most favorable light. See Peter W. Brooks, *Cierva Autogiros: The Development of Rotary-Wing Flight* (Washington, D.C.: Smithsonian Institution Press, 1988), p. 77; George Townson, *Autogiro: The Story of "the Windmill Plane"* (Fallbrook, California: Aero Publishers, 1985), p. 15; Frank Kingston Smith, *Legacy of Wings: The Story of Harold F. Pitcairn* (New York: Jason Aronson, 1981), p. 149.

4. For military notice of the Autogiro after Pitcairn's visit in July 1926, see Juan de la Cierva, "The Autogiro: Its Future as a Service Aeroplane," *United States Naval Institute Proceedings* 54, no. 8 (August 1928): 696–701; "'Autogyro' Flies from London to Paris," *United States Naval Institute Proceedings* 54, no. 11 (November 1928): 1010; Lieutenant Commander (CC) William Nelson, "The Autogiro as a Military Craft," *United States Naval Institute Proceedings* 57, no. 8 (August 1931): pp. 1092–1095; "The Navy's Autogiro," pp. 1118–1119.

5. For a history of the NACA, see Charles Greeley Abbot, *Great Inventions,* vol. 12 (Washington, D.C.: Smithsonian Institution Press, Inc., 1932, 1934, 1938, 1943, 1944), pp. 233–38; Frank A. Tichenor, "Air: Hot and Otherwise," *Aero Digest* 17, no. 7 (December 1930): 40, 124–134.

6. For a photo of the C.8W flying over Bryn Athyn, see Townson, p. 13; Smith, *Legacy of Wings,* p. 158.

7. For a photograph of the two aircraft in the tests, see Smith, *Legacy of Wings,* pp. 152–53.

8. Kathryn E. O'Brien, *The Great and the Gracious on Millionaires' Row* (Utica, New York: North Country Books, Inc., 1978), p. 83.

9. *Aircraft of the National Air and Space Museum,* 4th ed. (Washington, D.C.: Smithsonian Institution Press, 1991), see "Pitcairn Mailwing."

10. For photographs of Harold F. Pitcairn posing with the Collier Trophy in front of the PCA-2 at the White House, see Frank Kingston Smith, "Mr. Pitcairn's Autogiros," *Airpower* 12, no. 2 (March 1983): 28–49, 36.

11. Walter J. Boyne, *The Aircraft Treasures of Silver Hill* (New York: Rawson Associates, 1982), pp. 134–35; Brooks, pp. 190–91. For a photograph of the acceptance of the C.8W by the Smithsonian, see "The Autogiro." *Autogiro Company of America.* 1930, 1932 p. 29 (1932 ed.)

12. For photographs of Cierva's demonstrations of the C.19 Mk.II (G-AAKY) at the National Air Races in 1929 in Cleveland, see Edward Jablonski, *Man with Wings* (Garden City, New York: Doubleday & Company, 1980), p. 273.

13. For a description of the Pitcairn restoration and American Helicopter Museum exhibit, see "Exhibit Spotlight: Pitcairn PCA-1A," *Vertika* 7, no, 2 (October 2000): 5.

14. Pitcairn had also hoped to have the PCA-2-30 Autogiro, constructed for Heraclio Alfaro, at the National Air Races, but it had been destroyed while in flight to Chicago.

15. Brooks, Townson, and Townson and Levy agree that this was the engine of the prototype PCA-2, which was later upgraded for production to the 300 horsepower Wright R-975-E Whirlwind J-6-9 engine. See Brooks, pp. 125–28; Townson, p. 140; George Townson and Howard Levy, "The History of the Autogiro: Part 1," *Air Classics Quarterly Review* 4, no. 2 (Summer 1977): 4–18, 15. However, Frank Kingston Smith, in describing the first flight by Pitcairn pilot Jim Ray of the PCA-2, seemingly confused the prototype with the production model, ascribing the larger engine to that first flight. See Smith, *Legacy of Wings,* pp. 169–70.

16. It is interesting to note that Cierva's article on page 35 was immediately followed on the next page by an article by Don Rose, as Cierva and Don Rose would collaborate on the 1931 book *Wings of Tomorrow: The Story of the Autogiro* published in New York by Brewer, Warren & Putnam.

17. Alternatively attributed as Pitcairn chief pilot Jim Ray.

18. David S. Ingalls, "Autogiros: Missing Link," *Fortune* March 1931, 77–83, 103–4, 106, 108, 110.

19. Cierva had, by early 1931, left England to return to Spain, then in political turmoil, to attend to the safety of his wife and six children. While Cierva did not share in the Collier Trophy, he received the British Royal Aeronautical Society Silver Medal that year. Brooks, p. 129.

20. For a picture of Amelia Earhart with the factory PCA-2 after achieving the altitude record, see Mary S. Lovell, *The Sound of Wings: The Life of Amelia Earhart* (New York: St. Martins Press, 1989), photo 30.

21. See James Tobin, *Ernie Pyle's War: America's Eyewitness to World War II* (Lawrence Kansas: University Press of Kansas, 1997), p. 19.

22. Many years after his death, however, it would be asserted by literary researchers that Halliburton had not, in fact, done the daring feats described in his many books. Such posthumous assertions aside, however, it cannot be denied that Halliburton *was* a credible and recognized adventurer of the time.

23. That telegraph is currently in the possession of Michael Manning. Thanks are due to Deane B. McKercher for making a copy available to the author.

24. See Smith, *Legacy of Wings,* p. 192; for an additional photograph of golfers with the Autogiro, see George Pynchon Jr., "Something about the Autogiro," *Town & Country* 86, no. 4062 (August 15, 1931): 46–47, 46.

25. For copies of the advertising paintings, see Warren R. Young, *The Helicopters* (Alexandria, Virginia: Time-Life Books, 1982), pp. 62–63.

Chapter 4

PITCAIRN AND AMERICAN AUTOGIRO DEVELOPMENT

One of his great charms was his modesty. He never promised more than he could fulfill and he was that *very* unusual type of inventor: the man who knew more about the theory and practice than anyone else. He had the courage of his convictions and learned to fly his own machines, not merely tolerably, but extremely well. No better way of honouring his memory could be imagined than to carry to its ultimate solution the great work which he started.

John Fay, *The Helicopter*

Harold Pitcairn and Amelia Earhart's husband, George Palmer Putnam, had seen to it that the world-record for altitude flight in April was well-covered by the news media, which was always eager to cover the achievements of the photogenic Amelia. Such acclaim met each party's needs, and they sought to capitalize further with the first transcontinental flight. Seeing a publicity bonanza, the Beech-Nut Packing Company offered Earhart the use of its previously ordered PCA-2 if she would fly it coast-to-coast with the company logo painted on its side and would engage in accompanying promotion efforts, and she promptly canceled her order in favor of the Beech-Nut Autogiro. However, as Beech-Nut was scheduled to receive the thirteenth production model, Earhart, superstitious about such things, requested that she receive a lower number and in fact received C/n B-12 (NC10780). She thus displaced United States Marine Corps Reserve Lieu-

Amelia Earhart with the Beech-Nut Pitcairn PCA-2 Autogiro in 1931.
(Courtesy of Stephen Pitcairn, from the Pitcairn Archives.)

tenant John Miller, who had been the first individual to order a PCA-2 and for whom C/n B-12 had been confirmed!

John M. "Johnny" Miller, who would become a legendary pilot with an exceptionally long career that spanned eight decades (he was still flying at ninety-seven!) had been lured to aviation by the time he was five years old, watching Glenn Hammond Curtiss on his flight down the Hudson River from Albany to New York City on May 29, 1910. The flight, the first inside NYC limits and taking just over three hours, would win for Curtiss the $10,000 prize offered by the *New York World* and inspire the young Miller for a lifetime devotion to aviation. Having viewed the pioneer American aviator when he landed on the road across from the Miller farm to refuel the famous *Hudson Flyer,* Miller would later write, "[t]hat was the day, at age four and three months, when I lost interest in becoming a steam loco-motive engineer."[1]

By the time he was ten, Miller was hanging around the Curtiss Flying School at Mineola, New York, on Long Island. In 1915 the young boy met Ruth Law, a famous early woman pilot. She would be first woman to loop a plane,[2] but on that day in 1915, when she encountered the ten-year-old Miller, she talked about aviation and let him sit in the seat of her Wright Model B. It made an indelible impression on the future pilot, who still described it almost ninety years later![3]

By 1931, with a mechanical engineering education at Pratt Institute of Technology, Class of 1927, and seven years of flying experience, Miller had become the first individual to purchase a PCA-2 for a cash price of the then sizable $15,000 plus "a little extra for an auxiliary fuel tank and emer-gency flare racks for night flying."[4] Upon ordering he had been informed that he would receive production model C/n B-12 in April 1931, a delivery date later postponed to May. At the time of his order, C/n 13 was in the production line, but no order had yet been received.[5] Upon receiving con-firmation of his PCA-2 order, Miller immediately began planning a transcontinental trip,[6] a daring undertaking, as no one had previously attempted such a long flight.[7] This flight was to be in conjunction with a series of exhibition flights, and Miller kept Pitcairn sales and production officials, including Edwin T. Asplundh, fully informed of the flight plans.

Miller was understandably surprised in early May when he read in the *New York Times* of Beech-Nut's intent to sponsor Amelia Earhart's transcontinental flight! Flying to the Pitcairn Willow Grove field, he quickly discovered that the company had inserted Beech-Nut's order ahead of his and that he would now receive C/n B-13 (NC10781). This was clearly an attempt by the company to facilitate Earhart's flight, and he later claimed that "the mechanics and the test pilots leaked the information to me that the sales manager had decided that he would rather have Earhart make the first transcontinental flight for better publicity coverage."[8] Miller knew that he was regarded merely as an "unknown professional pilot with-out such publicity as Beech-Nut could provide." He also learned from the Pitcairn company pilots that Earhart's final check ride was being delayed until her aircraft could be finished and that "she told them [Pitcairn per-sonnel] that she was not interested in all the aerodynamics and short land-ing procedures," but "she just wanted to fly it across the continent and then fly around the country for a Beech-Nut advertising campaign."[9]

Miller, resorting to subterfuge in the face of the company's manipulation, announced that "if Amelia wants to make the flight she is welcome to it" but that he had to be in Omaha for the Air Races by May 17 or he would suffer a financial loss. He took a room at a nearby tourist home and, waiting to take delivery of his Autogiro, received a check ride in the PCA-1B with factory pilot "Skip" Lukens. Lukens took Miller on a single checkout ride, with five checkout practice landings. Miller, then given use of the aircraft for practice during May 9 through 12, made 110 practice landings with a total of 5.5 hours of flying logged. This averaged out to flights of about three minutes along with practice in low cloud banks with the turn indicator. Finally, on May 14, 1931, he took delivery of his Autogiro, which he would name, presumably after the recent *Fortune* magazine article,[10] the *Missing Link*. After five short test hops, Miller promptly left and headed west in PCA-2 (NC10781).[11]

An experienced professional and aerobatic pilot, Miller had gained extensive knowledge of the aerodynamics of the Autogiro from conversations with Jim Ray, Skip Lukens, Jim Faulkner, and Pitcairn chief engineer Agnew Larsen. He would need all of his abilities for the trip west. Although the normal cruising speed of the *Missing Link* was 100 mph, Miller flew at 90 mph to conserve fuel and break in the new engine. The Wright R-975-E, 330-horsepower, air-cooled radial engine consumed eighteen gallons per hour, so Miller could fly for only three hours at a time, at which point he would have only fifteen minutes of flying time on his fuel reserve. Navigation was by magnetic compass, following landmarks such as rivers or roads, and the pilot hoped that when a landing had to be made, there would be an airfield where the Rand McNally road maps showed one—it was not always the case. Miller discovered this on the second day, during which he flew from Harrisburg, Pennsylvania to Chicago. He had flown 11.3 hours in seven hops, intending to land at Maywood Air Mail Field, but that airfield had been abandoned; its replacement, later known as Midway Airport, was not yet finished or marked on the maps. Miller arrived at the site of the older field after dark and, after a perfect landing at the old site, he located the new field, to which he immediately flew, as he would have to refuel before continuing on. He napped on a workbench and, after refueling, left for Omaha at first light. He had not even eaten. He then flew an additional seven hops, 7.2 hours flying, and after arriving at the site of the Omaha Air Races, flew an additional 2 hours and made fourteen demonstration flights.

Miller remained in Omaha from May 16 to 19 and then left for San Diego. Headwinds kept him from reaching Clovis, New Mexico, on May

26, so he landed en route and installed extra fuel tanks on the front seat during the night. The next day he reached the New Mexico town, but strong headwinds on the way to El Paso consumed extra fuel, forcing him to land eighteen miles short of his destination. On May 28 he began the last leg of the journey from Lordsburg, New Mexico, before first light and, after flying four hops for 8.9 hours, landed at North Island Naval Air Station, San Diego, California. The first Autogiro transcontinental flight had taken a total flying time of 43.8 hours and was without mechanical incident. The aircraft had performed flawlessly, with the most difficult aspects of the journey for Miller seemingly to get used to the shadows of the blades passing over his head and the severe sunburn he incurred.[12] He began the return trip on June 21 after demonstrating the Autogiro for Navy officers and other interested parties and arrived back at the Pitcairn factory at Willow Grove on June 30, 1931. The factory mechanics, interested in evaluating how the PCA-2 had performed, gave it a thorough inspection— it needed only an oil change!

Miller would go on in 1932 to fly hundreds of hours in his PCA-2, thrilling crowds with his performance of the loop and other aerobatic maneuvers.[13] At Pitcairn Field on October 13, 1931, Canadian pilot Godfrey W. Dean had made the first loop in an Autogiro, a particularly challenging and impressive maneuver given the aerodynamics of the aircraft's rotor, but Miller was the most widely known pilot to perform this maneuver. He first proposed a loop in public at the 1931 National Air Races but was prevented by the Pitcairn company, which assured the air races organizer Cliff Henderson that it would prove fatal. Miller learned from Pitcairn pilots that they had been forbidden from looping, but the 1932 National Air Races at Cleveland were a different story. Miller was confident he could do the loop—he did it before an enthusiastic crowd on August 27 and continued to do it in his daily performances for the next seven days, but then tragedy happened. On September 3 as he landed and reached for the rotor brake, his aircraft was struck by a pre–World War I Curtiss pusher flown by Al Wilson. Wilson had elected to end his performance by "buzzing" the PCA-2, unaware that since the Autogiro had made a steep descent, there was a residue column of air from its rotor. Wilson's plane hit the downdraft of air and dived into the ground, resulting in Wilson's death and doing much damage to the *Missing Link*. It took twenty-seven days before it could fly again, costing Miller appearance fees, but he knew he had gotten off lucky—his friend was dead.

Miller stated seventy years later that "the PCA-2 still had the original air in one of its tires when sold with 2000+ hrs flying time. It was a first class aircraft and the safest in history, in my considered judgment the only INHERENTLY safe aircraft."[14] He received the Sikorsky Award for his part in the evolution of the helicopter and a certificate of honor from the National Aeronautic Association for his contributions to aviation, and he was made an honorary fellow in the Society of Test Experimental Pilots for having "promoted the moral obligation of the test pilot to the safety of the aerospace world."[15] His fellow society members included General Jimmy Doolittle, Howard Hughes, Charles Lindbergh, and Igor Sikorsky. A modest man, Miller replied when questioned in 1996 as to how he felt he would be remembered: "I didn't go after records or the publicity. I just went out and did the work."[16] But Amelia Earhart and her husband George Palmer Putnam were *very* interested in the publicity, and they and the Pitcairn executives who had tried to arrange for her cross-country flight to be the first, were in for a surprise!

After much preparation and orchestrated publicity, Earhart left Newark on May 29, 1931, and headed west. Accompanied by mechanic Eddie Vaught[17] and making as many as ten landings per day, she proceeded along the northern mail route to Oakland, California. At each stop she lifted children to see the cockpit, shook hands with spectators, gave interviews, and often gave out samples of the Beech-Nut chewing gum. Arriving on June 6, 1931, in Oakland, she discovered much to her amazement and her husband's mercurial anger,[18] that Miller had arrived in San Diego on May 28. Thus deprived of the transcontinental record, Earhart and her husband decided that she would claim a record by returning to the East Coast. This was not to be, as she had the first of her three Autogiro crashes in Abilene, Kansas, on June 12, 1931. Returning by the southern route, she crashed while taking off, having failed to rise quickly enough. The PCA-2 dropped thirty feet, hit two cars, and damaged its rotor and propeller. Earhart stated, "The air just went out from under me," and added, "Spectators say a whirlwind hit me. I made for the only open space available." And ever conscious of Pitcairn Aviation, she also added, "With any other type of plane the accident would have been more serious."[19] She and the accompanying mechanic were unhurt, but her attempt at the cross-country return was ended—she returned to the east coast by train.[20]

The Aeronautic Branch of the Department of Commerce, renamed in 1934 the Bureau of Air Commerce, did not accept her version of the incident and issued her a formal reprimand for "carelessness and poor judgment," based on a report made by the local inspector R. W. Delaney. The government had intended to ground Earhart for ninety days, but her friend Senator Hiram Bingham pleaded her case and secured a lesser penalty, a

Amelia Earhart with the Beech-Nut Pitcairn PCA-2 Autogiro stopping at Rock Spring, Wyoming, in June 1931. Note the man on the right wearing a gun. (Courtesy of Stephen Pitcairn, from the Pitcairn Archives.)

formal reprimand from Clarence Young, then assistant secretary of commerce for aviation.

Amelia Earhart's second Autogiro crash is known of from a single source, a letter[21] to author Susan Butler from Helen Collins MacElwee, sister of Amelia's New York, Philadelphia and Washington Airway Corporation colleague Paul Collins. Paul Collins and his sister Helen witnessed the second accident. After a "rather erratic" Autogiro flight she made after taking off from the airfield in Camden, New Jersey, she "finally landed on a fence. Amelia stepped out frustrated and furious, and announced, 'I'll never get in one of those machines again. I couldn't handle it at all.' "

Earhart's third accident in an Autogiro occurred during her subsequent Beech-Nut tour while at the Michigan State Fair in Detroit on September 12, 1931. Attempting a slow landing in front of the grandstand, she failed to level off and dropped twenty feet to the ground. She wrote her mother, "My giro spill was a freak accident. The landing gear gave way from a defect and I ground-looped only. The rotors were smashed as usual with giros, but there wasn't even a jar."[22] Although she did additional flying for Beech-Nut, her significant contact with the Autogiro finished with the end of 1931. She was already planning the solo trans-Atlantic flight of May 20 and 21, 1932, which would win her the National Geographic Society Special Medal, the first awarded to a woman pilot.

With the perspective of over seventy years, it is readily apparent that Earhart's involvement with the Autogiro was relatively insignificant. The general consensus was that she was an "impatient" pilot and that her accidents were the product of lack of training and attention to detail. The crash in Kansas appears to have resulted from forcing takeoff before the rotors had achieved high enough rotation, while the one in Detroit was the result of not having spent enough time practicing landings. To be sure, the Autogiro, despite Pitcairn's public claims of ease of operation touted in virtually every advertisement and public pronouncement, was a difficult aircraft. Amelia's friend, pilot Blanche Noyes, hired to fly a PCA-2 for an oil company, ridiculed Pitcairn's claim that "a ten-year-old boy" could fly an Autogiro. She related in her Oral History (which is part of a collection at Columbia University)[23] that the factory PCA-1B was called the Black Maria because so many pilots had accidents. So the report of Earhart's declaration after her second accident rings true, supported by an observation in *Fortune* in 1932, from an article assessing a year's Autogiro progress: "It is reported that Amelia Earhart, since her two crashes, opines that it is as hard to make a perfect landing with an autogiro as it is to make a perfect drive on the golf course."[24]

In many ways Earhart's lasting and most serious contribution to the Autogiro may have been the article published in *Cosmopolitan* magazine in August 1931.[25] It predicted that the day was fast approaching when "country houses would have wind cones flying from their roofs to guide guests to the front lawn landing area" and Autogiro hunting and fishing trips for the weekend would be common, as well as quick sorties to golf and aviation country clubs and a new convenient way to commute to work. This article almost exactly mirrored the images conveyed in Pitcairn advertising,[26] and Carl R. Gunther, Pitcairn Aircraft Association archivist and historian, has suggested that the *Cosmopolitan* article was probably

Pitcairn PCA-2 Autogiro (X760W) during certification flight in fall 1930, flying north over the Hudson River. (Courtesy of Stephen Pitcairn, from the Pitcairn Archives.)

written not by Earhart, but by either Pitcairn Aircraft or its advertising agency. That agency also authored many dramatic advertisements for American magazines, such as *Town and Country,* with spectacular Autogiro photographs and copy, and promotional brochures designed to inform and intrigue the affluent.[27] The result was a public-relations bonanza!

Based on the successful publicity and the seeming public acceptance of the PCA-2, Harold Pitcairn had cause for optimism at the end of 1932. Even if Amelia Earhart's Autogiro reputation had dimmed, Captain Lewis A. "Lew" Yancey, flying the PCA-2 owned by the Champion Spark Plug Company, was having a stellar year. He completed the first Autogiro flight from Florida to Cuba in late January 1932 and then proceeded via San Julian to Merida, Mexico, a distance of 385 miles, 135 over water, still the longest over-water flight of an Autogiro.[28] Yancey then continued on to Chichén-Itzá, where he participated in exploration of Mayan ruins. This had previously been proposed by Professor C. W. Grace of the Municipal University of Wichita, Kansas, who had seen Johnny Miller when he was in

Pitcairn PCA-2 Autogiro flying over the Florida Everglades. (Courtesy of Stephen Pitcairn, from the Pitcairn Archives.)

Wichita during May 22–24 (Miller was having a gas tank leak repaired during the transcontinental flight, and it was during this stopover that he officially christened the PCA-2 *Missing Link*).[29] Grace had written the Autogiro Company of America (ACA) on June 6, 1931, informing company officials of the Grace Mayan Expedition and requesting "all the available information concerning your plane" and inquiring as to "[w]hat kind of deal would you be willing to make on a plane of this kind for exploration purposes? Quote your lowest possible price first."[30] R.W.T. Ricker, replying for the company on June 11, 1931, referred Professor Grace to the American licensees, Pitcairn, Kellett, and Buhl (noting that "Buhl Aircraft has not yet begun its development work") and suggested that the Professor "write them direct for the information."[31] While Pitcairn was unable to quote any terms, as he was back ordered for the $15,000 aircraft, and it appears that Grace did not pursue the inquiry, Yancey and *Miss Champion* did take part in Mayan exploration under the supervision of Dr. Sylvanus G. Morley, director of scientific work for the Carnegie Foundation.[32]

Captain Lewis A. "Lew" Yancey flying the Champion Spark Plug Company PCA-2 Autogiro (NC11609) *Miss Champion* over the Temple of the Soldiers at Chichén-Itzá in the Yucatán in Mayan exploration under the supervision of Dr. Sylvanus G. Morley, Director of Scientific Work for the Carnegie Foundation, in February 1932. (Courtesy of Stephen Pitcairn, from the Pitcairn Archives.)

This resulted in spectacular publicity for both the Autogiro and its sponsor, Champion Spark Plugs. The Associated News Service "Latest World Events In Pictures" had carried a photograph of Yancey's January 21 landing in Cuba,[33] and the photographs of his low flight past the great Maya pyramid of Chichén-Itzá[34] and before the ancient Temple of Tigers[35] received wide acclaim. To top it off, Yancey broke Amelia Earhart's altitude record, with a flight reaching 21,500 feet over Boston, Massachusetts, on September 25. In a lighter manner the Autogiro had been the subject of a popular series for teenagers describing the adventures of Andy Lane[36] and was incorporated into "Uncle Don's Radio Club," where host Don Carney arrived nightly to the studios of WOR in New York in his Autogiro.

Cierva returned at the end of 1931 for his third visit to America to consult with Pitcairn, and the occasion became yet another publicity triumph for the ACA. Disembarking from the SS *Aquitania,* Cierva was met by Jim Ray landing a PCA-2 on a near-by New York pier, a well-orchestrated and

Pitcairn PCA-2 Autogiro (X760W) during certification flight in fall 1930, flying over the Battery in Manhattan. (Courtesy of Stephen Pitcairn, from the Pitcairn Archives.)

"thoroughly photographed event."[37] After Cierva cleared U.S. customs, Ray placed the inventor's bags beside him in the front cockpit and flew steeply away from the end of the pier. It was an astonishing aviation feat that could not be duplicated by any other American aircraft. Less than forty-five minutes later, Pitcairn was offering his guest a class of wine in Bryn Athyn. Cierva had come to discuss advances in Autogiro development, as it was now evident that the PCA-2 had effectively solved the problem of prerotation with its mechanical coupling of the engine and rotor, but there were other serious issues that led Cierva to leave his family during the Christmas season and come to Pennsylvania. Harold and Clara Pitcairn were gracious hosts and made a point of including Cierva in all of the family celebrations, shifting the focus of the first days of the visit from business to socializing and effectively moving the relationship between the two men from entrepreneurial colleagues to fast friends.

In the last days of 1931 Pitcairn and Cierva were in Saint Clair, Michigan to fly the first "pusher" Autogiro created by ACA licensee Buhl Air-

The Buhl Autogiro, built under license from the Autogiro Company of America, had a "pusher" propeller. Buhl's chief engineer Etienne Dormoy stands, while James Johnson is the aircraft's pilot. (Courtesy of Stephen Pitcairn, from the Pitcairn Archives.)

craft. The Buhl Aircraft Company[38] had became the third licensee of the ACA in March 1931 (the first was Pitcairn Aviation, and the Kellett brothers of Philadelphia the second). Located in Marysville, Michigan, Buhl was a successful fixed-wing manufacturer and part of a large financial, manufacturing, and real estate organization[39] headquartered in nearby Detroit. The company, founded as the Buhl-Verville Aircraft Company in 1925 with Lawrence D. Buhl as president and A. H. Buhl as vice president, was well-financed and described itself in a letter to Geoffrey S. Childs of the ACA on March 23, 1931, as consisting of "members of the Buhl family [who] all are extremely wealthy individuals in addition to their holdings in the above mentioned [family] companies."[40]

The Buhl Autogiro design team consisted of chief engineer Etienne Dormoy,[41] assistant chief engineer R. V. Doorn and nine other engineers.

The ACA also claimed that Dormoy "worked in conjunction with Agnew E. Larsen, chief engineer of the Autogiro Company, and Joseph S. Pecker, chief engineer of the Autogiro Specialties Company, to complete the necessary preliminary studies and special parts design" but acknowledged that "[t]he final design was altogether the work of the Buhl group."[42] The goal of the Buhl engineering team was to adapt a "pusher" configuration, with the engine located to the rear of the Autogiro. The company's goals for the new model were centered on visibility, accessibility, comfort, and safety,[43] a design clearly similar to Cierva's unbuilt patented design study for the C.21 of 1930.[44] The Buhl was initially powered by a 165-horsepower Continental radial air-cooled engine and specifically intended to "get the younger generation flying."[45] The "pusher" engine placement offered unparalleled visibility,[46] eliminated the propeller blast, provided for convenient conversation between pilot and passenger, and minimized exposure to motor heat and the odor of gasoline.

The design team adopted the Pitcairn rotor hub that ACA would use on its smaller PAA-1 Autogiro then in development, and forty-foot-diameter rotor blades,[47] often mistakenly reported as either forty-two or forty-eight feet.[48] The aircraft initially had a loaded weight of 1,850 pounds, later increased to 2,000 pounds (compared to the 3,000-pound loaded weight of the PCA-2), and employed the now-standard Cierva-type four-blade rotor mounted on top of a tripod pylon just in back of the rear cockpit and incorporated a Pitcairn PCA-2-type engine-powered spin-up drive. The top of the pylon also attached to the upper steel-tube boom that supported the tail. Two other booms extended from the fuselage below the engine mount from either side to secure the tail, creating a "cage" for the propeller and additional safety. The pilot and passenger sat in a tandem arrangement in a nacelle created by a steel-tube framework covered by fabric, from which fabric-covered wings extended on either side, with upturned wingtips as in the PCA-2.[49]

The Buhl was first flown by the company test pilot James "Jimmy" Johnson on December 15, 1931, and it was immediately noted that the aircraft seemed underpowered. Thirteen days later, on December 28, Cierva himself confirmed that observation when he flew the Buhl. Pitcairn and Cierva had been welcomed in Grosse Point by Henry Ford and General William "Billy" Mitchell, and Cierva is reported to have given flights[50] to Mitchell and Edsel Ford in the Buhl. The prototype remained the only one ever built, as Buhl experienced financial difficulties in 1932 and ceased aircraft production. The financial difficulties experienced by the company precluded the costly development and testing necessary for certification.

The single Buhl model was confirmed in possession of the Hiller Aviation Museum in California in 2003, awaiting restoration.

Pitcairn and Cierva shared the prestigious John Scott Award on January 15, 1932, for the invention and development of the Autogiro.[51] That award was given by the board of directors of City Trusts of the City of Philadelphia in accord with a trust created in 1816 by John Scott, a chemist working in Edinburgh, Scotland. The award was "to be distributed among ingenious men and women who made useful inventions," and the citation was "for the invention of the Autogiro, its improvement and development as a propelling and stabilizing force for heavier-than-air craft, and its introduction into America."[52] Additionally, Pitcairn and Cierva were invited to meet with President Hoover in Washington to discuss the Autogiro.[53] But it was not all honors and socializing—Cierva had come to confer on the state and future directions of Autogiro development. ACA had constructed fifty-one Autogiros by the end of 1931. Twenty-one of these were commercial PCA-2s, with an additional three in an experimental military version designated by the navy as the XOP-1[54] (denoting Experimental Observation–Pitcairn), one of which was field-tested by the United States Marine Corps (USMC) in Nicaragua.[55]

An additional aircraft, dubbed the Pitcairn-Alfaro, had been constructed by Heraclio Alfaro, the only such Pitcairn Autogiro constructed outside the factory since the original C.8W. Alfaro had apparently hoped to become Pitcairn's chief engineer, but failing in that goal,[56] he had proposed to construct an innovative variant of the PCA-2, which was originally dubbed the PC-2-30 but is more generally referred to as the PCA-2-30.[57] The contract for its construction had been signed in 1929, and the aircraft was delivered to Pitcairn Field in July of 1930. It featured considerably novel construction, with an advanced rotor design mounted on the now-standard tripod pylon with friction dampers at the blade roots. But its most innovative characteristic was clearly the extensive use of Bakelite, an early plastic compound. The rotor blades had been manufactured by the Formica Company and were covered in Micarta, as were the stabilizer and rudder, while the wings and ailerons were covered in Formica. It was first flown as an experimental prototype (X759W) on July 18, 1930,[58] and observers noted that the aircraft, even at its minimum flying weight of 1,385 pounds, lacked commercially viable flight performance. The PCA-2-30 was both underpowered and overweight, an unfortunate combination that was to prove almost fatal for pilot Skip Lukens. The rotor had been increased from thirty-four feet to thirty-eight feet in an attempt to increase lift, and plans were made to fly it at the National Air Races in Cleveland. The aircraft never made it, crashing on takeoff on August 21 at Butler, Pennsylvania. It was theorized that moisture had condensed inside the Micarta-covered blades and that the added weight from the accumulated moisture prevented the rotor from getting enough speed. In any event, the company abandoned the experiment and although Alfaro wanted to rebuild, Pitcairn did not feel there was enough promise to continue.

ACA had continued through the 1931–32 period to develop new models. Even as the PCA-2 worked its way through the certification tests and was marketed to the public, Pitcairn had overseen the development of a smaller model, the PAA-1. The designation reflected the change of the company's name—the first two models were PCA, indicating the Pitcairn-Cierva origins of the company, but after the licensing company changed the name to Autogiro Company of America, the models were designated PAA, for Pitcairn Aircraft [Company] Autogiro, the manufacturing company. When, in 1933, Pitcairn Aircraft became the Pitcairn Autogiro Company, the model designation became PA. Often characterized as "a scaled-down PCA-2,"[59] as it was two-thirds the size of the larger craft, the prototype PAA-1 (X10770) had been seen by the public at the March 1931 Detroit Aircraft Show and had gained Department of Commerce ATC No. 443 on August 7, 1931.[60] With its initial 125-horsepower Kinner B5 engine, the flight-testing revealed the aircraft to be underpowered, but the flaw was soon remedied, and the PAA-1 proved popular. It was cheaper than the PCA-2 and, even though it only cruised at a relatively modest seventy-six mph (as opposed to the PCA-2's eighty-seven- to eighty-nine-mph cruising speed), garnered many orders and became Harold Pitcairn's personal aircraft. It had a range of 250 miles and allowed the pilot to land at otherwise inaccessible locations.

Pitcairn sold forty-six Autogiros in 1931 and could not help but be optimistic—even though the country was in the grips of the Depression. Purchasers included the *Detroit News,* the first newspaper to acquire a rotary aircraft. It made 730 flights before being presented to the Henry Ford Museum[61] in 1934. Other business purchasers included the Beech-Nut Packing Company (two aircraft), Coca-Cola, Standard Oil Company of New York (SOCONY) and Ohio, the Horizon Company, Curtiss-Wright, Puget Sound Airways, Tri-State Airways (Gilbert Flying Service, Valley Stream, New York), Champion Spark Plug, and Johnson & Johnson. The Horizon Company PCA-2 alone flew 550 hours in seven and a half months in 1931, and the PCA-2s operated in all forty-eight states. The National Advisory Committee for Aeronautics purchased one and began a five-year investigation of its aerodynamic characteristics. Of greatest import, the

First production PCA-2 to receive Aircraft Type Certificate (ATC #410) on April 2, 1931, was purchased by *Detroit News,* which owned radio station WWJ. As William E. Scripps already owned a fixed-wing aircraft, the Autogiro was marked No. 2. The PCA-2 is on display in the Henry Ford Museum, Dearborn, Michigan. (Courtesy of Stephen Pitcairn, from the Pitcairn Archives.)

United States Navy had purchased three PCA-2s and assigned each to a specific role. One was sent to Anacostia Field near Washington, D.C., in early 1931, where it was flown by David Ingalls, assistant secretary of the navy for aeronautics. Ingalls was so impressed that he would write the article for *Fortune* in late March dubbing the aircraft the "Missing Link." Ingalls's flight of the XOP-1 was at the specific request of President Hoover, and Ingalls not only demonstrated the XOP-1 at the presidential retreat but he also flew with the president's son and namesake, Herbert Hoover Jr., as passenger.[62]

That same XOP-1, flown by Lieutenant Alfred M. Pride USN, made three landings and takeoffs from the aircraft carrier USS *Langley* in an aviation first off the coast of Norfolk, Virginia, on September 23, 1931.[63] The second XOP-1 was tested with pontoon floats as a seaplane, a natural use by the navy. The third was sent to the United States Marine Corps, which planned to test it in Nicaragua in the Spring of 1932 for combat effectiveness. All of this civilian success and military acceptance must have been encouraging to Pitcairn at the end of 1931, who could not know that the Autogiro would consistently fail to find a military mission in America.

The smaller PAA-1, called by *Fortune* the "Pitcairn sport model,"[64] was purchased primarily by individuals, although some businesses, such as the *Des Moines Register and Tribune,* Atlantic Seaboard Airways, Autogiro Specialties Company, and New England Giro, also bought it. And the good news did not end there—the Pitcairn organization had also developed a third prototype model in 1931, the PA-18. It was positioned between the large and expensive PCA-2 and the smaller PAA-1, with a slightly larger size but a much more powerful 165-horsepower Kinner engine, 36 percent more powerful than the PAA-1. This increase of 40 horsepower made the PA-18 an outstanding performer, with a top speed of 100 mph and a cruising speed of 80 mph, and it would prove an almost immediate success. This "improved and enlarged version of the PAA-1" initially flew on March 1, 1932, and received ATC No. 478 on April 7 of the following year. Nineteen PA-18s were sold between April 1932 and July 1933. The price was reduced from May 1932 onward to $4,940 due to slow sales in light of the deepening Depression. One of the first models (NC12678) became the personal aircraft of Ann Strawbridge[65] of the Philadelphia Strawbridge and Clothier department store family. Originally preserved by F. Sewerka, it is currently owned by Kate and Jack Tiffany of Spring Valley, Ohio, and is being restored to flying condition by Leading Edge Aircraft. Also, by the end of 1931 almost a hundred pilots had flown the Autogiro in America, and two, E. E. Law and Harold's nephew Nathan Pitcairn, had received the first Autogiro-only pilot licenses.

So when Pitcairn finally sat down with Juan de la Cierva at Cairncrest after the holiday celebrations at the end of December 1931, they shared the good news that seemingly came from all quarters. And Cierva could not help but be impressed. Not only had Pitcairn won the Collier Trophy, but the range of American Autogiros seem well-positioned for economic success, and Cierva took note of the manufacturing acumen and marketing savvy of this brash American. He was particularly impressed by the way in which Pitcairn and his associates had solved the problem of rotor prerotation with the PCA-2's mechanical coupling of the engine to the rotor by means of a shaft and gearbox just in front of the forward rotor pylon. Cierva had himself attempted unsuccessfully to develop such a mechanical prerotation device. The C.10, a small experimental single-seat Autogiro, had been ordered from builder George Parnell[66] in late 1926 but never flew—piloted by Flight Lieutenant H. A. Hamersley, it overturned and was damaged while attempting to takeoff on April 26 of that year. Slightly less than two years later, in February of 1926, Cierva had crashed the C.11, a larger experimental two-seat aircraft, known as the Parnall Gyroplane after its builder. It would be rebuilt with a smaller rotor, the wings moved forward, and an inverted V-strut bracing of each wing to the fuselage, and in that configuration it would successfully fly in October 1929. A year later it would be rebuilt again, this time at Hamble with a rotor spin-up drive from the engine. That drive, supposedly designed by engine designer Major Frank Bernard Halford,[67] did not prove successful. Demonstrated on January 8, 1930, it proved too heavy at 165 pounds, and its failure left Cierva in a quandary. Ever since the 1924 C.6bis, the rotor had been started by men pulling on ropes attached to knobs and wound around the rotor hub, much like a child's spinning top.[68] That solution was inefficient, and the aircraft was forced to taxi up and down the airfield until the airflow through the rotor had speeded up the blades to the 120 to 130 rpm sufficient for takeoff. This was spectacularly ungainly and the source of constant complaint, so Frank Courtney had even attempted to achieve sufficient rotation by winding a cable around the C.6C's rotor hub, staking the other end to the ground, and taxiing down the runway. While the blades did spin up as the staked cable rotated the rotor hub, the cable snapped off at the end and, just barely missing the pilot's head, cut the rudder in two. Needless to say, that method was abandoned after a single trial! Cierva had also patented, in January 1929, a rotor-starting device based on compressed air, steam, or water pressure, which would accelerate the blades from nozzles attached to some or all of the blade tips. The idea had been briefly tested and rejected as impractical given the current state of technology.

But Cierva's thoughts were not on prerotation, for as the Spaniard sat with the Pitcairn team, he had a dramatic announcement. The English company had begun working on the next evolutionary stage of the Autogiro, which would remove the wings and create an aircraft that bore almost no resemblance to its airplane ancestor—direct control!

NOTES

1. Letter from John M. Miller to the author dated February 28, 2001.

2. Edward Jablonski, *Man with Wings* (Garden City, New York: Doubleday & Company, 1980), p. 103.

3. Miller, letter to author.

4. John M. Miller, "The First Transcontinental Flights with a Rotary-Wing Aircraft 1931," *Popular Rotorcraft Flying* (August 1992): 11–19, 11–12.

5. Miller, letter to author.

6. Thus Frank Kingston Smith's suggestion that "[w]hen he [Johnny Miller] learned that Earhart had been advanced ahead of him on the production and delivery line, he took off for the West Coast without fanfare and beat her by two weeks" is, in its implication, incorrect. Frank Kingston Smith, *Legacy of Wings: The Story of Harold F. Pitcarin* (New York: Jason Aronson, 1981), p. 183. Miller had long planned his trip and had, in fact, contracted for air show performances at the Omaha Air Races on May 17, 1931. His sudden departure for the West was not occasioned by the announcement of Earhart's flight but the need to fulfill a previous commitment. He then, of course, continued on to the West Coast.

7. See also John M. Miller, "The First Transcontinental Rotary-Wing Flight: Part 3," *Vertika: The Newsletter of the American Helicopter Museum and Education Center* 8, no. 1 (February 2001); Ibid., "The First Transcontinental Rotary-Wing Flight" *Vertika: The Newsletter of the American Helicopter Museum and Education Center* 7, no. 2 (October 2000).

8. Miller, "First Transcontinental Flights, p. 12; Miller, letter to author.

9. Miller, letter to author; Pitcairn mechanic and pilot George Townson also claimed to have "had words" with Earhart the day of her altitude-record flights and that "she was an impatient pilot." George Townson, conversation with author, 27 March 2001.

10. David S. Ingalls, "Autogiros: Missing Link," *Fortune* (March 1931): 77–83, 103–4, 106, 108, 110.

11. Frank Kingston Smith incorrectly asserts that Miller flew the "Silverbrook Coal PCA-2 a week before [Earhart]." Smith, p. 188. It is difficult to know how this attribution could be made, as the pictures of Fred W. "Slim" Soule flying the Silverbrook Coal Company PCA-2 (NC10786) and Johnny Miller flying the *Missing Link* (NC10781) are on facing pages (182–83).

12. For a picture of Miller with a severe sunburn at the completion of the first transcontinental flight, see Miller, "First Transcontinental Flights," p. 16.

13. A copy of the poster advertising the "First Trans-Continental Autogiro" exhibition flights by Lieutenant Johnny Miller for a dollar a flight in the "Windmill Safety Plane" can be found in Henry Serrano Villard and Willis M. Allen Jr., *Looping the Loop: Posters of Flight* (Hong Kong: Palace Press, International, n.d.), plate 91. It is there erroneously attributed to have been in 1930 but obviously originates after Miller's transcontinental flight of May–June 1931. The original is in the San Diego Aerospace Museum.

14. Miller, letter to author; see also Miller, "First Transcontinental Flights, p. 19.

15. Bud Walker, "People and Planes: Captain John Miller," *Aviation History* 7, no. 2 (November 1996): 14, 16, 18.

16. Ibid.

17. Alternative, reported as Eddie Gorski. Susan Butler, *East to the Dawn: The Life of Amelia Earhart* (New York: Da Capo Press, 1999), p. 258.

18. "Amelia was disappointed and George was furious. His overreactions were well known but could be alarming to anyone witnessing them for the first time." Mary S. Lovell, *The Sound of Wings: The Life of Amelia Earhart* (New York: St. Martins Press, 1989), p. 170.

19. *New York Times,* June 20, 1921.

20. Lovell is incorrect in stating, "A replacement autogiro was hurriedly shipped to her [after the Abilene, Texas, crash] and Amelia continued her trip to Newark without further incident." Lovell, p. 171. See Smith, *Legacy of Wings,* p. 189: "Unfortunately, she [Amelia Earhart] had to complete her transcontinental trip by rail."

21. Butler, p. 260.

22. Jean L. Bakus, *Letters from Amelia: An Intimate Portrait of Amelia Earhart* (Boston, Massachusetts: Beacon Press, 1982), p. 117.

23. Oral History Collection, Columbia University, vol. 1, pt. 3, p. 17.

24. "Autogiros of 1931–1932," *Fortune* 3, no. 3 (March 1932): 48–52, 50.

25. Amelia Earhart, "Your Next Garage May House an Autogiro," *Hearst's International combined with Cosmopolitan* 91, no. 2 (August 1931): 58–59, 160–61.

26. See Warren R. Young, *The Helicopters* (Alexandria, Virginia: Time-Life Books, 1982) pp. 62–63.

27. For examples of the ACA advertisements, see George Townson, *Autogiro: The Story of "the Windmill Plane,"* (Fallbrook, California: Aero Publishers; reprint, Trenton, New Jersey: Townson, 1985) p. 151, 154–55.

28. "Captain Yancey Explores Mayan Ruins by Autogiro," *Autogiro News,* January 1932, p. 3.

29. The Wichita christening of the *Missing Link* related in conversation with author on November 12, 2001.

30. June 6, 1931, letter from Professor C. W. Grace to Autogiro Company of America (author's collection).

31. Autogiro Company of America reply to Professor C. W. Grace of June 11, 1931 (author's collection).

32. "Captain Yancey," p. 1.

33. Associated News Service, "Latest World Events in Pictures" poster, vol. 19, no. 13 of January 29, 1932.

34. Peter W. Brooks, *Cierva Autogiros: The Development of Rotary-Wing Flight* (Washington, D.C.: Smithsonian Institution Press, 1988) p. 130; for a photograph of Yancey in flight before the great Maya pyramid, see Smith, *Legacy of Wings,* p. 208.

35. "Captain Yancey," p. 2.

36. Eustace L. Adams, *The Flying Windmill* (New York: Grosset & Dunlap, 1930).

37. Smith, *Legacy of Wings,* pp. 194–95 (photo caption on 195).

38. See Brooks, pp. 145–47; Townson, pp. 17, 84, 86–87, 96; Smith, *Legacy of Wings,* 196–97.

39. Buhl letter of March 23, 1931, to Geoffrey S. Childs of the ACA.

40. Buhl, letter to Childs.

41. Buhl, letter to Childs, p. 2; see also *Autogiro News,* published by the Autogiro Company of America August, 1931, p. 1; Townson, p. 17 (but who mistakenly reports the designer's first name as André). Brooks, whose research sets the standard, here mistakenly reports this as "Dormay" (p. 146).

42. *Autogiro News,* Autogiro Company of America, October 1931.

43. "Preliminary Announcement: The Buhl Autogiro" (hereinafter "Preliminary Announcement"); Buhl letter of October 21, 1931, "Supplementing Our Recent Letter Which Answered Your Inquiry about the Buhl Autogiro" (hereinafter Buhl Supplement).

44. Cierva patented this side-by-side two-seat "pusher" Autogiro on February 11, 1930, but the design was never produced. Brooks, p. 160.

45. Buhl, letter to Childs, p. 2.

46. The "pusher" configuration would be adopted only one more time, the PA-44, two of which were constructed by Pitcairn for the United States Air Force. Generally known by its military designation, the YO-61, it looked much like the developing helicopter, with a fuselage nose of transparent plastic, which was also utilized for the doors and roof, a design calculated to yield maximum visibility. The tail, as with the Buhl a decade before, was mounted on out-riggers. Townson, p. 84.

47. See, for example, Brooks, p. 146; Townson, p. 84 (citing a "letter from Roger Ward, production manager for Autogiro Specialties, acknowledging an order for a set of 40-foot diameter blades").

48. While Townson mentions the forty-eight-foot rotor diameter without further citation, the Buhl company originally cited a forty-two-foot rotor diameter. See Buhl Supplement. This forty-two-foot figure was repeated by the ACA in its *Autogiro News* of October 1931.

49. For a picture of the Buhl, see Brooks, p. 147; George Townson, and Howard Levy, "The History of the Autogiro: Part 1," *Air Classics Quarterly Review* 4, no. 2 (Summer 1977): 4–18, 14; Smith, *Legacy of Wings,* p. 197 (picture with Etienne Dormoy [there misspelled as "Dorman"] and James "Jimmy" Johnson).

50. Brooks, p. 147.

51. For a photograph of Cierva and Pitcairn receiving the John Scott Medal, see The Autogiro, Autogiro Company of America. 1930, 1932 p. 37 of 1932 ed.

52. Ibid.

53. Smith, *Legacy of Wings,* p. 197. For a photograph of that meeting on January 21, 1932, with President Hoover, which Cierva, Pitcairn, Edwin T. Asplundh, Geoffrey S. Childs, Colonel Clarence M. Young, and Luis M. de Irujo attended,

see *Autogiro News.* Autogiro Company of America. February 1932 p. 2 (photograph caption).

54. For a picture of the XOP-1, see Townson, p. 33.

55. At least one author, George Townson, maintains that the USMC tests of the XOP-1 in Nicaragua were not fair. See Townson, p. 30. For an opposing analysis, see Charles Gablehouse, *Helicopters and Autogiros: A Chronicle of Rotating-Wing Aircraft* (Philadelphia: Lippincott, 1967), pp. 51–54.

56. Alfaro would later join the fourth ACA American licensee, L. W. Steere Engineering Company of White Plains. Steere bought a Pitcairn PAA-1 Autogiro (NC11626) but failed in subsequent Autogiro development. Townson, p. 18.

57. For a photo of the PCA-2-30, see George Townson and Howard Levy, "The History of the Autogiro: Part 2," *Air Classics Quarterly Review* 4, no. 3 (Fall 1977): 4–19, 110–14, 11.

58. Townson, p. 17. Brooks's assertion that the first flight took place on August 18, 1930, is obviously incorrect given the extensive experimentation with blades subsequently made in an attempt to increase the power and improve flight characteristics, and particularly given that all authorities agree that the aircraft was destroyed in a crash on August 21, 1930. Brooks, p. 139.

59. Townson, p. 17.

60. The best source of information on the PAA-1 is Brooks, pp. 142–44. Note that Frank Kingston Smith incorrectly asserts that the PAA-1 gained the ATC in July. Smith, *Legacy of Wings,* p. 201.

61. For a photograph of the *Detroit News* PCA-2 with its distinctive red-and-yellow marking, see Bob Ogden, *Great Aircraft Collections of the World* (New York: Gallery Books, 1988), p. 154.

62. For a picture of Assistant Secretary of the Navy for Aviation David S. Ingalls setting out for the presidential retreat on the Rapidan, see George Pynchon Jr., "Something about the Autogiro," *Town & Country* 86, no. 4062 (August 15, 1931): 46–47 (photo on p. 47).

63. For a rare photograph of the XOP-1 landing on the USS *Langley* in September 1931, see Frank Kingston Smith, "Mr. Pitcairn's Autogiros," *Airpower* 12, no. 2 (March 1983): 28–49, 32.

64. "Autogiros of 1931–1932," *Fortune* (March 1932): 48–52.

65. Although not identified in the photo, Ann Strawbridge can be seen in the front cockpit of NC11678 with a passenger in the rear cockpit. As this is identified at Wings Field, near Ambler, Pennsylvania, it is likely that she is actually flying the aircraft. Unlike the PCA-2, which was flown from the rear, the PA-18 featured dual controls. But see Townson, p. 46, where he identifies Strawbridge but claims she is the passenger, with Paul "Skip" Lukens as the owner and pilot of NC12678. As Brooks identified the Pitcairn Aeronautical Corporation as the owner, it is possible that Strawbridge had not yet acquired the aircraft and was merely getting a flying lesson.

66. David Mondey, ed., *The Complete Illustrated Encyclopedia of the World's Aircraft* (Secaucus, New Jersey: Chartwell Books, Inc., 1978) p. 396.

67. Brooks, p. 83.

68. See Peter Almond, *Aviation: The Early Years (The Hulton Getty Picture Collection)* (Köln Germany: Könemann Verlagssgesellschaft mbH, 1997), p. 343 (photograph of Cireva Autogiro in 1925 clearly shows the line of four men holding the rope on the left side of the image, preparing to spin up the rotor).

Chapter 5

DEVELOPMENT OF THE DIRECT CONTROL AUTOGIRO

Juan de la Cierva will be known to enduring fame as the outstanding pioneer in the field of rotary wing aircraft.... All helicopters and similar types of craft that have shown promise of practical performance incorporate some of the principles and inventions developed by Cierva.

Harold F. Pitcairn, "Juan de la Cierva: In Memoriam"

Cierva's brilliant insight had been to see the airplane wing differently than those who had developed the airplane. Airplane wings stalled when the air passing over the wing failed to generate enough lift at slow speed, and Cierva reasoned that stall could be effectively dealt with if the wing itself moved. His rotor was a moving wing, and indeed his Autogiro was effectively stall-proof. But although the Autogiro could remain airworthy at slow speeds, the traditional wing-based ailerons, elevators, and rudder controls became ineffective, and although professional, experienced pilots could maintain control even while landing, inexperienced fliers were encountering conditions in which the lack of flight control spelled disaster. And although there were few fatal accidents, gusts of wind often led to the Autogiro tipping over, an expensive proposition for the pilot whose rotor pounded itself into wood chips.

Cierva reasoned that the solution to this problem was to remove flight control from the wings and tail and to place it directly with the rotor hub itself. He called this *direct control* and claimed that he had been inspired with the solution while attending the London Opera House one evening when he noted the umbrellas carried by pedestrians, their black, shiny domes glistening under the streetlights and theater marquee. During the performance he visualized the rotating rotor disc as an umbrella and, in a moment of insight, realized that if the rotor disc was seen as attached to a hub that could move, moving that hub would tilt the rotor just as moving the umbrella handle tilted the umbrella domes. Development of this mechanical system would allow the pilot to achieve direct control by moving the rotor head by means of a control rod attached to the hub, effectively a hanging control-stick.

Rushing to his London quarters after the opera, Cierva had by dawn sketched a universal joint mounting for a flexible rotor hub and the attached control-stick. In one evening he reasoned that the control surfaces of the wings were now irrelevant, as direct-control would allow the pilot to directly change the rotor lifting force. His application of *cyclic pitch* control, is today found in almost every helicopter, and it eliminated the need for wings.

But Cierva had an even grander vision, which he now shared with his American colleagues. Having conceived of flexible rotor disc control, Cierva also proposed *collective control* to simultaneously vary the previously fixed pitch, or angle of incidence, of each individual rotor blade. This was known since at least 1924 with the work of the Marquis Raul de Pateras Pescara, an Argentinian of Italian descent. The Autogiro had impressed all with its ability to land in a small space, but that impressive ability was not mirrored in the takeoff. Although the mechanical prerotator significantly shortened the takeoff run, it did not allow for true vertical ascent. Cierva now proposed to combine prerotation with cyclical pitch control to achieve true vertical, or *jump,* takeoff. Always the theoretical scientist, he had calculated that a strong set of rotor blades in a flat pitch (i.e., with no incidence to the horizon) could be accelerated to at least 125 percent of the rpm necessary for liftoff, a condition known as *over-spinning.* Collective pitch would then simultaneously allow each individual blade to be angled to a lift-generating pitch, and the kinetic energy stored in the blades from the over-spinning would be expended in lift, causing the Autogiro to "jump" into the air, with the propeller driving the craft forward before the lift was fully expended. Given the PCA-2's ability to achieve the prerotation, it was obvious that if the formidable engineering challenges could be overcome to develop and integrate the collective and cyclic pitch control systems, the Autogiro would enter a new and dramatic phase. It was perhaps the finest moment of collaboration between the two aviators—Pitcairn's practical achievements coming together with Cierva's theoretical visions.

But as company owner and president, Pitcairn also realized that what was being proposed had profound implications for the Autogiro business. A functional-direct control, jump takeoff Autogiro would make each of his previous models obsolete. But it was obvious that once conceived, nothing could stop the development of the more advanced Autogiro configuration, and that even the rumor of such a development might impact the market for his existing aircraft. (*Fortune* had already stated that "[t]he autogiro is still using airplane type controls. A new type of control, better fitted for the auto-giro, may eventually make the autogiro easier to handle than the airplane or its present self.") As Cierva embarked for the return voyage to England on February 14, 1932, both engineers were committed to the same development direction and parted as colleagues and good friends. It was only later that Pitcairn learned that Cierva had informed his London associates of his ideas prior to coming to the United States and that the English company had been working on direct control and the mechanics of the jump takeoff even as he and Cierva enjoyed the Christmas holidays and a Bryn Athyn January. In the midst of the real friendship between two aviation pioneers and avowed collaborators, the seeds of distrust had been sown.

Pitcairn's optimism about the Autogiro's future was to prove unjustified—there were dark clouds on the horizon, as very real safety concerns were beginning to emerge. Cierva's article "The Uses and Possibilities of the Autogiro" in the December 1930 *Aero Digest,* with its dramatic cover showing the PCA-2 flying over a metropolitan city, excited the public's imagination, as did David Ingalls's "Missing Link" article in the March 1931 issue of *Fortune.* These articles and the spate of Pitcairn Autogiro advertisements called attention to the power and potential of the Autogiro, touting its safety and ease of flight-training, promising a virtual revolution in American aviation. Furthermore, an accompanying article in the December 1930 *Aero Digest* trumpeted the safety of the Pitcairn Autogiro for the neophyte, claiming that "a novice...is placed in much the same position as when learning to drive a motor car. Mistakes are not necessarily dangerous. If he becomes confused, he can stop and let the ship land itself."[1] But the March 1932 *Fortune* article "Autogiros of 1931–1932" took a far darker tone than that of a year earlier.

Obviously autogiros are not flying every corner of the sky. Obviously, too, consciousness of the autogiro has come up over America's horizon. The autogiro has hopped on the lawns of the White House, the Capitol, and the

Smithsonian Institution, where the first autogiro to fly in this country delivered itself to the very portals within which it is now immortalized. It has alighted on golf courses, on the piers of ocean liners, and once, when it ran out of gas, it settled at night into the back yard of a farmer. And some sixty-one commercial (as opposed to experimental) autogiros have been sold today as compared to one a year ago.[2]

And the left-hand black-bordered column of that article was entitled "Worst Autogiro Accidents" and listed ten different accidents, including Amelia Earhart's Abilene crash, a brief account of Blanche Noyes's withdrawal from flying the Standard Oil Company of Ohio's PCA-2 after two hard landings, and accidents by private and military pilots. Although the end of that accident report stated that " [i]n all cases the occupants of the machines were able to walk away from the accidents," it could not have been of much comfort to Pitcairn. And the center of the last page of the article featured, in a heavy black-bordered box, a riveting photo captioned "Behold the first commercial autogiro to be totally destroyed—but not in a flying accident. This big Pitcairn machine, belonging to United Aircraft, was idling on the ground...when a backfire started a blaze." The news photograph showed the PCA-3, which was the first (NC11671) of two PCA-2s fitted with 300-horsepower Pratt & Whitney R 985 Wasp Junior engines in a special order for United Airports, a subsidiary of the United Aircraft and Transport Corporation. As it was merely a version of the PCA-2, it was speedily granted an ATC on August 25, 1931, but was destroyed in a fire less than a month later. The photo, showing a fuselage on fire and smoke billowing from the cockpit, viscerally reinforced the theme of the article—while Autogiros accidents were not fatal, they were costly, and the aircraft had not lived up to its promise of safe aviation.

The perception of the unsafe nature of Autogiro aviation was reinforced the following year with the most deadly Autogiro accident in history. There is no mention by Brooks of the pilot or passengers, but sixty-three years later[3] John Miller relates that after the 1933 International Air Races at Chicago

> Vincent Bendix invited the participants to gather at his estate at South Bend for a dinner and party, offering free fuel to those who flew in.... As I took off and turned south to go around the south end of the lake [Michigan] I saw Charlie [Otto] take off in his PCA-2 and head straight across the lake. The wind was practically calm there.... Due to the unexpected headwind I did not have enough fuel to make it to South Bend, so landed at the Department of Commerce emergency field at McCool, Ill. where I knew that I could get

fuel out of a barrel in the little airway beacon shed. It was a very hot and humid day and we sweated at the job of getting about 15 gallons out of the barrel with the help of the man in charge of the field. Then we continued on to South Bend.... On arrival over the field Charlie's PCA-2 was not there. Flying across the lake with minimum fuel he could not be aware of his low ground speed. He ran out of fuel and went down in the lake. He and his two passengers were lost. One passenger was a well-known free-fall parachutist, Spud Manning, whom I had taken up over the air show to 15,0000 ft several times and knew well. (Free-fall jumps were new and spectacular at the time and Spud Manning was the pioneer of that sport).

The March 1932 *Fortune* observed that for all of its sales, the Autogiro had failed to achieve economic success. It pointed out that the PCA-2 then sold for the hefty price of $15,000, "whereas an airplane of similar size and power might cost $11,000...and the Pitcairn sport model for $6,750—and here is the real difficulty. An airplane to yield the same service might cost only $1,500."[4]

Fortune also noted that the Kellett Aircraft Corporation Autogiros were experiencing accidents. Rather than follow the Cierva or Pitcairn designs, Kellett initially had attempted an original design, the K-1X. Fred Seiler, Kellett chief engineer, created the final design, a small one-seat gyroplane initially powered by a forty-horsepower Szekely three-cylinder engine, later replaced by a five-cylinder, air-cooled Velie M-5 radial engine yielding sixty-five horsepower and a unique one-piece thirty-two-foot, six-inch laminated spruce rotor. The aircraft had no fixed wings but a large horizontal tail with control surfaces combining elevators and ailerons. The fuselage, constructed by the Budd Company of Philadelphia, was of spot-welded strip steel covered with fabric. The aircraft was extremely light, variously reported at between 775 and 900 pounds. The rotor head was also unique, having bearings that allowed the one-piece rotor to teeter like a seesaw.[5] Testing had begun on October 14, 1930, but was discontinued on December 3 of that year because, although the gyroplane could be taxied at sixty mph, it never left the ground. The company then abandoned its own designs, and in 1931, Kellett purchased a license from Autogiro Company of America.

The first product was the Kellett K-2 (sometimes known as the KA-1), and although it was based on the established Cierva and Pitcairn designs, the company had not completely given up on its attempts to innovate. Kellett chief engineer W. Laurence LePage, who had previously participated in the design of the Pitcairn PCA-1 experimental Autogiros, produced an open-cockpit, side-by-side two-seater with a larger rotor blade area (the chord of the blade was twenty-three inches rather than the eighteen inches utilized by the Pitcairn machines) and simplified landing gear. The fuselage was gas-welded steel tubing with a slightly faired streamlined shape, a light wood tail, and a cabin wrapped with doped airplane fabric, which could be fitted with a coupé top. It was lifted by a standard four-blade Cierva rotor on top of what became the characteristic Kellett rotor pylon—one forward strut in front of the cockpit and two lighter rearward struts attached just behind the cockpit.[6] The K-2 utilized a forty-five-pound pre-rotator clutch and gearbox developed by the Autogiro Specialties Company of Philadelphia, which had achieved a market niche supplying hubs and parts to Pitcairn, Kellett, and Buhl. The K-2 first flew on April 24, 1931, with Pitcairn pilot Jim Ray. Cierva subsequently flew the K-2 at Philadelphia on December 30, 1931. Its design was of such a proven nature that it received an initial ATC 2-431 in Group 2 on May 27, 1931, certification that was upgraded to ATC 437 on July 17, on July 17, 1931.[7] The K-2 had been flown under the original ATC by Gilbert Budwig, the chief of the department of commerce, who had noted that the Kellett model lost aileron control at low speeds and in a slow rate of climb. Approval was given for ATC 437 on the company's promise that it would take corrective action. Immediately upon receipt of an ATC, the K-2 entered into production with several advanced sales, as well as being demonstrated to the United States Army in the autumn of 1931. Twelve of the 1,556-pound K-2s would eventually sell for $7,885. Initially powered by a 165-horsepower Continental A-70-2 engine, the models sold after July 1932 would be modified with a 210 Continental R-670 engine and sold as a K-2-A. It was quickly followed by the K-3, an improved version that featured the Kinner C-5 210-horsepower engine and a more rounded shape on the elevators and stabilizer.

The K-3[8] received ATC 471 on March 26, 1932, after having been exhibited at the 1932 Detroit Air Show. Optional equipment included a coupé top and a safety nose skid to protect the propeller. It could take off with a run of only 165 feet and land in 5 to 35 feet. The K-3 was the first American Autogiro to gain international exposure, as a K-3 delivered to Argentina in that year became the first Autogiro in South America. Overseen by Mr. Leigh Wade, Kellett representative, and flown by Philadelphia pilot Edward E. Denniston at the Punta India Military Airport,[9] Buenos Aires, Argentina, the K-3 created much public and military interest, and it was not lost on the audience that it had initially been the product of a Spanish inventor. That aircraft was soon followed by three additional K-3s. And although the U.S. military had rejected the K-3 for insufficient performance after demonstrations at Wright Field, Dayton, Ohio, two were ordered by the Japanese war ministry through Okura & Company for army evaluation. They were paid for by public contributions and named at their dedication on April 16, 1933, by Major General Rensuke Isotani, representing the war minister, as *Aikkoku* (Patriotic) *81* and *82*. These overseas sales lead to increasing disagreement between the Kellett company and the Autogiro Company of America, as the latter maintained that international sales to foreign governments violated the terms and conditions of its licensing agreement.[10]

The Japanese navy was also then testing two Cierva C.19 Mk.IV aircraft (G-ABXE and G-ABXD). A third Cierva model (G-ABXF) was operated by the Tokyo newspaper *Asahi Shimbun,* which had acquired the Cierva for aerial photography and to get news stories from otherwise inaccessible locales. Although it proved inadequate, it flew almost 250 hours by 1940, mostly for advertising purposes, and a few times each year thereafter for demonstrations and annual certification tests, until it ceased flying with the end of World War II in August 1945.[11]

The Cierva C.19 series had been developed as a production aircraft intended for commercial sales in 1928 to capitalize on the boom in private flying then materializing in the UK after the C.17 had proven unsuccessful. The C.19, built at Cierva's order by Avro with the design team under the direction of Charles Saunders and Reg Calvert as chief designer and Henry Dyer as chief draftsman, was not initially successful. Being an original Autogiro design not based on preexisting World War I aircraft designs, immediate problems emerged, as the only prerotating device then available was the box "scorpion" tail. Intended as a small, low-powered aircraft with an 80-horsepower Armstrong Siddeley Genet II engine, the heavy deflector tail made it seriously underpowered. The power issue was resolved by substituting a more powerful 105-horsepower Genet Major five-cylinder engine, the model then being denoted the C.19MII, and it went on to become the first widely used Autogiro in its MkIII configuration. It would become the first Autogiro (G-AALA) to successfully compete in a race, with Reginald A.C. "Reggie" Brie taking second place, at ninety-three mph, in the Skegness Air Races on May 14, 1932.[12]

The two-seat C.19MkII had a welded steel tube fuselage, a fixed spindle rotor, and conventional airplane controls with strut-braced small stub wings with upturned tips, utilizing the now-standard Cierva four-blade rotor. Its loaded weight was 1,400 pounds, with a maximum speed of ninety-five mph, and it was on exhibit at the Seventh International Aero Exhibition at Olympia between July 16 and 27, 1929, arriving a mere eight

Juan de la Cierva, inventor of the Autogiro, making a tail-first landing of G-AAKY in August–September 1929 at Pitcairn Field, Willow Grove, Pennsylvania. (Courtesy of Stephen Pitcairn, from the Pitcairn Archives.)

weeks after the start of its design. The MkI had gained its Certificate of Airworthiness, the equivalent of the American ATC on August 2, 1929. Cierva had brought a C.19MkII to American and demonstrated it to Pitcairn on August 20 at Willow Grove, where it had been reassembled. That Autogiro, G-AAKY,[13] briefly became the most famous rotary-wing aircraft in America when Cierva flew it before the a hundred thousand people who came daily to the National Air Races in Cleveland, Ohio, from August 24 through September 2, where Pitcairn was exhibiting the prototype PCA-2 and the PCA-1B. During that period, with the technical assistance of Pitcairn engineer Paul H. Stanley (who had joined the Pitcairn company the same day as the first American Autogiro flight in December 1928), Cierva had also done the final editing of his "Engineering Theory of the Autogiro." Cierva would later, in 1934–35, author "Theory of Stresses in Autogiro Rotor Blades." Neither of his essays would be published, but they were circulated and read by rotary-wing aircraft designers and were of considerable influence for future designs.

The impact of the introduction to the C.19MkII was that the Americans became thoroughly familiar with its capabilities and had briefly considered the C.19 when deciding to develop a lighter "sport Autogiro." But, critical of the deflector tail and understandably preferring a mechanical solution, the Pitcairn engineering team had elected to pursue the development of the PAA-1. Cierva, upon his return to England in early 1930, had gone on to develop the C.19 in various configurations that were then widely sold, including models to Singapore, Germany, Sweden, Australia, and Spain in addition to those in Japan in 1932. The French Lioré-et-Olivier also acquired a license to build the C.19 but apparently never did so.[14] Cierva pilot Reggie Brie also demonstrated a C.19Mk.IV in Denmark but no sales resulted.[15]

Cierva had started an Autogiro flying school at Hanworth in May 1932 under the management of Reggie Brie and with Alan H. Marsh as chief flying instructor. Marsh had previously been instructor at the Hampshire Aeroplane Club and trained Cierva to fly in 1927. He would formally join Cierva Autogiro Company as assistant to Brie on April 5, 1932, and remain for eighteen years, until his untimely death on June 13, 1950, in the crash of the Cierva Air Horse helicopter. As Cierva intended to sell the C.19 to a wider audience, the school was intended to furnish pilot training and did so initially with a C.19MkIV (G-ABUD)[16] and then with a second aircraft (G-ABUF). The first of hundreds to learn the skills of flying the Autogiro was sixty-eight-year-old J. A. McMullen. By the end of 1932 six men and five women (including the wife of James Weir) had soloed.

The German involvement with the C.19 came in early 1931 when Professor Heinrich K. J. Focke of Focke-Wulf Flugzeugbau A. G. of Bremen inquired of a license to build the C.19MkIV through the German Cierva Company, formed by O. J. Merkel in July 1931 for such licensing purposes. The actual license agreement was been signed in December of that year and the first German model completed in May 1932. That model was first flown by Cierva pilot Arthur Rawson and Cierva in June 1932. Rawson would then leave the company, being replaced as chief pilot by Reggie Brie. After a disagreement with his board of directors in 1933 in light of the changing political circumstance in Germany, Focke left the company and established Focke-Achgelis G.mbH to specialize in the development of rotary-wing aircraft. His experience with the Cierva model would lead to the first successful helicopter in 1936 and, of greatest import for autorotational aircraft, the FA-330 rotary-kite almost a decade later.

In addition to the exported Kellet K-3s, one would briefly become the most famous Autogiro in the world when its corporate owner, the Pep

Boys chain of stores, lent it to national hero Rear Admiral Robert E. Byrd for his second Antarctic Expedition in 1933–35. The "Pep Boys *Snowman*" K-3 Autogiro (NC12615) was a converted K-2 that the company had used for advertising purposes, and it is certain that Pep Boys had been impressed by the publicity realized by Champion Spark Plugs from the involvement of Lew Yancy and *Miss Champion* in the Mayan explorations in the Yucatán the previous year. The company made the most of the publicity, even featuring a photo of Admiral Byrd christening its Autogiro at Camden, New Jersey.[17] The *Snowman* left Boston in October 1933, securely placed aboard Byrd's supply ship *Ruppert*. It was unloaded in the Bay of Whales after a largely uneventful journey on January 28, 1934, and flown to the expedition's forward base at Little America by pilot W. S. McCormick. By the end of January the K-3 had been employed for reconnaissance of sea ice by McCormick with Byrd as passenger—both were impressed with the performance. That Autogiro, unlike the Cierva and Pitcairn models, was ideally suited to such exploration, as the side-by-side seating allowed for easy communication and the optional coupé top facilitated flying in the cold climate. Byrd was reported as observing: "I was greatly impressed with the virtues of the autogiro. With its singular hovering instincts and its nearly vertical landings, it is the perfect instrument for short-range reconnaissance in the polar regions."[18] On March 24 the K-3, flown by McCormick and Byrd, struck out in the face of strong winds to find one of the expedition's missing fixed-wing aircraft. After sighting the missing aircraft, McCormick landed to check the condition of the two marooned pilots. The K-3 then returned to base to brief the leaders of the dogsled teams who would come to rescue the stranded airmen, and the subsequent rescue gained much notoriety and public acclaim. Flying resumed in early September, and weather permitting, the K-3 was used for measuring the temperature of the upper air, but the aircraft crashed on September 28, 1934, as it took off. It fell from a height of approximately seventy-five feet and was completely destroyed. McCormick survived but was found unconscious, in shock, and with a broken arm. Investigation revealed that drifting snow had weighted down the rear of the fuselage and shifted the center of gravity.

Kellett had changed the company's name to the Kellett Autogiro Company and went on to develop one more prototype two-seater in June 1933. Modified from the K-2, and called the K-4, it was awarded ATC No. 523 on December 27, 1933, but never went into production and was the last model produced without direct control, as by mid-1933, Cierva, Pitcairn, and Kellett were turning toward direct control. The press had reported in

Pitcairn PA-19 Cabin Autogiro was the largest American Autogiro built. Five were constructed, but it failed to find a market due to the Depression. (Courtesy of Stephen Pitcairn, from the Pitcairn Archives.)

late March 1932 that Cierva had flown the "first wingless aircraft," an otherwise obscure reference, but Pitcairn and his associates immediately understood its significance—Cierva was making advances in direct control! Upset that Cierva had not shared that information with him during his visit, Pitcairn, Ray, and Larsen visited England in early 1933 to gain a firsthand understanding of recent developments. What they saw was a modified early C.19 model with a spindle rotor head that could be tilted to achieve lateral and longitudinal control by means of an upside-down (hanging) control-stick. Development had not proceeded very far, and the direct control C.19MkV had only flown a few feet off the ground, with no cross-country flights. Both Pitcairn and Ray flew the experimental Autogiro and found that significant vibration made it necessary to grip the hanging control-stick with both hands, but it was apparent that Cierva had made significant theoretical progress on direct control and it would only be a matter of time until the technology caught up. Larsen also met and discussed direct control with Georges Lepère, chief of French Cierva licensee Lioré-et-Olivier, who was even then designing the C.L.10 direct control model, in consultation with Cierva, to be based on the C.19MkIV. All of this contributed to the American dedication to embark on direct control. And although the Americans returned home sure that they were not far behind in developing direct control, Pitcairn became even more suspicious of the English company and could no longer say with certainty that it remained a collaborator. It was beginning to resemble a rival, which would cause the Americans to view European Autogiro development with a growing degree of suspicion. Even as the Pitcairn engineering team solved the complex series of problems that emerged as direct control research advanced, he was careful to heed the advice of his patent-law firm, Synnestvedt & Lechner, and file a continual stream of applications. The name of his manufacturing company had changed in January 1933 to Pitcairn Autogiro Company, and his engineering team had already assigned the designation of PA-22 to the coming direct control model, but first there was another matter to be dealt with—the PA-19 cabin Autogiro.

Even as Pitcairn continued research on direct control, he had taken note of the comments made at the end of the 1932 *Fortune* article, which stated that "[t]here are other important elements of comfort, however. First the obvious one of providing autogiros with cabins."[19] The London *Times* had reported on February 10, 1931, that Pitcairn intended to build a five-passenger cabin Autogiro and Robert B.C. Noorduyn had joined the Pitcairn company as executive engineer in February 1932 at the commencement of the PA-19 project.[20] This model was to be the largest Autogiro ever constructed, equaled only by the C.34 prototype constructed in France by Société Nationale de Constructions Aéronautiques (SNCASE) in March 1939, which never advanced beyond the testing phase, no doubt because of the coming of war, but the C.34 was observed to have poor flying characteristics. The stated goals of the PA-19 were strength, reliability, ease of maintenance, comfort, appearance, and luxury. The prototype PA-19 (X13149, later NC13149)[21] was first flown by Jim Ray in September 1932 and was awarded ATC No. 509 on June 23, 1933. When introduced to the public on October 19, 1932, the "cabin" Autogiro was received with acclaim. It "rivaled the luxurious comfort of fine automobiles," suitable for women in skirts and older passengers.[22] And even though the PA-19 rivaled the passenger airplanes in terms of comfort, a far cry from the basic open-cockpit models, only five were built and it was not an economic success.

While its sheer size was impressive, weighing with passengers and cargo a massive 4,640 pounds, it was the appointments and quality of construction that most impressed. Passengers entered the PA-19[23] by walking up a retractable stairway through a wide door into a plush five-passenger cabin. Powered by a 420-horsepower Wright R-975-E2 engine and utilizing a large fifty-foot, seven-and-a-half-inch rotor, the performance of the PA-19 was as outstanding as its appearance. With a strong fuselage of welded steel tubing, it cruised at 100 mph and could reach a top speed of 120 mph. The prototype had a fore-and-aft tilting spindle that began to achieve direct control. This "tilt-adjusting" movable rotor head was the product of the ongoing direct control research and was operated by a crank in the ceiling of the cabin. Changing the angle of the rotor disc allowed the Autogiro to adjust to a greater range of center of gravity. The PA-19 flew well and had been thoroughly soundproofed with "Dry Zero" insulation blanketing. Ventilators provided for passenger comfort, and the cockpit was outfitted with a dazzling array of instruments. Pilots found it easy to fly with an adjustable seat for the primary pilot and a partially movable seat for the copilot. The control wheel and yoke were in a "throw-over" format and could be easily adjusted, allowing either pilot to fly the aircraft. The PA-19 also innovated a complete standard electrical system.

Pitcairn was optimistic by the favorable reception and foresaw a long production run, lowering the cost of each model. To encourage immediate acceptance he ordered that five be constructed and priced the PA-19 at $14,500, but only four orders materialized. Even before certification, the Year-Round Club in Florida ordered a PA-19 in February, followed by an order for two models by the Honorable A.E. Guinness of the United Kingdom. One of those, exported in 1935, was registered G-ADAM and allegedly crashed at Newtonards in Northern Ireland the same year. The second, registered as G-ADBE, later crashed at Gatwick and was stored until 1950 and finally broken up for scrap. The forth PA-19 was sold to Colonel R.L. Montgomery, a "wealthy Philadelphia sportsman" who flew it between his Pennsylvania home and Georgetown, South Carolina retreat. It eventually ended up with the U.S. Department of Agriculture. The remaining PA-19 was retained by the factory and was featured in what, after the 1931 White House landing, must be regarded as the second most dramatic moment in Autogiro history.

In 1933 Cierva returned to America for his forth, and last, visit. Arriving on the SS *Paris* on May 16, 1933, he was met by Pitcairn and Jim Ray at Newark Airport in the PA-19 and whisked to Bryn Athyn. He came to see the PA-19 and to consult with the Americans about ongoing direct control development and to receive the Daniel Guggenheim Gold Medal "for his development of the theory and practice of the Autogiro." Although Cierva had received the 1932 Fédération Aéronautique International (FAI) Gold Medal on January 11, 1933, and would receive the Elliott Cresson Medal[24]

Pitcairn PA-19 Autogiro, flown by James G. "Jim" Ray, delivering Harold F. Pitcairn and Juan de la Cierva to Soldiers Field, Chicago, for the awarding of the Carnegie Medal to Cierva on June 28, 1933. (Courtesy of Stephen Pitcairn, from the Pitcairn Archives.)

from Philadelphia's Franklin Institute in October of that year "in consideration of the original conceptions and inventive ability which have resulted in the creation and development of the Autogiro" (along with medals for Orville Wright and Igor Sikorsky), there can be little doubt that the greatest honor he received was the Guggenheim Medal for "the World's most notable Achievement in Aviation." It had only been awarded three times previously: to Orville Wright, to Frederick W. Lanchester, who had authored the vortex theory of flight in 1894, and to Ludwig Prandtl who further developed the vortex theory during World War I. Cierva was thus joining an immortal and exclusive community of theoreticians whose work had made flight possible. The medal was awarded on June 28, 1933, during a ceremony at Soldiers Field in Chicago during Engineers' Day at the 1933 Chicago World's Fair, the Century of Progress Exposition and could not have been more dramatic. Next to the exposition, where later Meigs Field would be located, was a large sports arena called Soldiers Field. On June 28, with thousands in attendance, Jim Ray flew Harold Pitcairn and Juan de la Cierva to Soldiers Field in the PA-19. It is hard to image how a greater impression could have been made—the crowd stood and cheered, the newsreel camera rolled, and flashbulbs went off as the large Autogiro landed within ten feet of its touchdown. It was a shining moment of triumph but could not erase the darkness that was falling upon commercial Autogiro development in the United States.

The United States Army had judged the Kellett K-2 to be underpowered for military use, and the Pitcairn PCA-2, dubbed the XOP-1 in its Navy markings, had not faired well either. While the XOP-1 had successfully landed on the USS *Langley,* the tests by the United States Marine Corps in Nicaragua in the Spring of 1932 had not gone well at all. While the local population cheered the XOP-1 when it appeared at Zacharias Field near Managua on June 28, 1932, seeing it as a triumph of Spanish invention, the military evaluation board was less than impressed. The original intent had apparently been to evaluate the Autogiro in combat operations against guerrilla chief Augusto Cesaer Sandino, but the board of review decided to evaluate the XOP-1 in comparative trials against a USMC Vought O2U-1 fixed-wing airplane. Tests revealed the obvious—the Autogiro could climb at a much steeper angle and fly at a considerably slower speed, but could neither carry the same load nor match the speed or climb of the fixed-wing aircraft. The board concluded that the advantages offered by the XOP-1 did not sufficiently offset its disadvantages[25] and no further military orders materialized. And the Depression effectively doomed the PA-19: Joe Jupiter, writing in volume 6 of "U.S. Civil Aviation," stated that:

> With such credentials the Pitcairn PA-19 took its place on the market of 1933; based on its ability and outstanding utility the PA-19 should have found instant favor, but being confronted with the depth of a national depression was more than a craft of this type could bargain for. There was a token interest, of sorts, but financial difficulty at the Pitcairn plant finally halted its production and further development. Actually, the cabin-type PA-19 was an aircraft too far ahead of its time.[26]

Hollywood discovered the Autogiro but it had nor resulted in a bonanza of sales, as it had been a mixed bag. Actor Edmund Lowe, the romantic adventurer, abducted social butterfly Claudette Colbert, flying her in a PCA-2 (NR784W) to a remote hunting lodge in the 1932 Paramount feature *The Misleading Lady.* While Jim Ray flew and the Autogiro performed, the aerial footage, much of it at night, did little to show off the aircraft and even less to save what was otherwise regarded by the trade publication *Variety* as "lightweight stuff of conventional pattern."[27] But if that movie failed to capture the audience, the 1933 Paramount film *International House* pandered to growing fascination with the Autogiro in a decidedly different and even less effective fashion. In that generally forgettable film, alternatively remembered as a "wacky Dada-esque Hollywood farce about an incredible array of travelers quarantined in a Shanghai hotel where a mad doctor has perfected television," the leading actor was vaudeville juggler William Claude Dukenfield, acting under the name W. C. Fields. He played Professor Henry R. Quail, pilot, con-man and alcoholic aviator. Although the movie is usually remembered, if at all, for the Cab Calloway performance of the marijuana classic "Reefer Man," it featured footage of the Kellett K-3 (NC12691) with the optional coupé top. And while the flying scenes were probably impressive to the Depression-era film-goer, any such impact was undoubtedly undercut not only by the drunken pilot meandering all over a world map in search of Kansas but also by his final arrival in the fictional Wu Hu, China, in a large simulated "prop" Autogiro (sporting the real Kellett K-3 registration NC12691), named the *Spirit of Brooklyn,* from which emerges an automobile! The movie also reflected the general lack of knowledge about rotary-wing aircraft in the dialogue between "Dr." George Burns and "Nurse" Gracie Allen. "Doctor," she asks, "what is the difference between a helicopter and Autogiro?" If the audience hoped to learn more from this answer, they were disappointed, as Burns replied, "You can't play a helicopter."

But the third Depression-era movie to feature an Autogiro was one of the most famous movies ever made, and, even though it has less than a minute of the same Kellett K-3 (NC12691) that had appeared in *International House* a year earlier, probably came closest to the Pitcairn *visual* ideal of amateur flying to the country club. *It Happened One Night,* the first film to win the four major Oscars (Picture, Director—Frank Capra, Actor—Clark Gable, Actress—Claudette Colbert, and a fifth to writer Robert Riskin), however, presents a particularly unappealing image of the Autogiro pilot. The Clark Gabel hero is an "ordinary man suffering from unemployment" and it is "only such a man . . . who can offer a woman an exciting, real, vital relationship."[28] A reporter, he meets fleeing heiress Colbert and accompanies her on a cross-country journey to marry "King Westley, the autogyro ace,"[29] who is portrayed as "an effete money-hungry playboy, without a muscle, or, it seems, an ounce of blood in his veins."[30] The Autogiro footage comes towards the very end of the film with a K-3 slowly banking into a descent and gentle landing on the broad field in back of the country club, which has been decorated for the wedding. As King

Westley exits from the aircraft in formal dress with his silk top hat and cane, the view pans slowly around the Autogiro, showing the unidentified Kellett pilot crouching in the cabin. At the moment of truth in the wedding ceremony, Colbert jilts King Westley to flee to the arms of Gable and all ends well, but the final Autogiro impression is that this is a toy of the idle rich. It must have been an unsatisfying moment for the American rotary-aircraft manufacturers. As the Depression deepened and orders for the Pitcairn and Cierva aircraft dried up, it was apparent that the industry was in deep trouble. The fourth American movie of that era to use Autogiro footage was eminently forgettable 1935 *Ladies Crave Excitement.* The Autogiro footage is not relevant to the movie, but is notable—about a minute's footage of Johnny Miller doing two loops!

Both companies knew that the future lay in direct control that would more fully realize the Autogiro's unique flying abilities and effectively answer the aviator's complaint that control was lost at slow speeds with the decline in the airflow over the traditional airplane ailerons and rudder. The Cierva C.19MkV, a single-seat direct control Autogiro had first flown in March of 1932. Registered as G-ABXP, it was the newspaper reports in America of this strange "wingless" aircraft that had so excited Pitcairn. It would be flown until scraped in 1935, and much would be learned from its various experimental arrangements. Cierva applied for a direct control patent on December 16, 1933, and eventually received British Patent No. 393,976 for a rotor disk that could be tilted in all directions.

And it is interesting to note that others were also pursuing direct control at this time, but from different directions. Most technologically significant but ultimately not historically so, was the work of Scotsman David Kay,[31] who had patented a mechanism for varying the incidence (angle) of the rotor blades of a gyroplane, called collective pitch control. He had met with Cierva on April 27, 1927, with a proposal, based on his patent, for an Autogiro that featured his control mechanism, a sideways tilting rotor, but Cierva was rejecting attempts to complicate the rotor system. In his later patent, however, it is obvious that Cierva had accepted the tiltable rotor as Pitcairn had partially done in the PA-19. In England in August 1932 Kay successfully flew his small Kay 32/1 gyroplane prototype and his Kay 33/1 gyroplane (G-ACVA) on February 18, 1935. As they were of original design, they were *gyroplanes* and featured collective pitch control and tilting rotors but retained normal elevators for longitudinal control. Although he founded Kay Gyroplanes Limited in November 1933, and his prototypes were extensively tested by the Royal Aircraft Establishment at Farnborough during September 1935 through February 1936, nothing more came of his efforts. Sixty years later Ron Herron would base his "Little Wing Autogyro" on the work of David Kay.[32]

Raoul Hafner, an Austrian helicopter designer, had by 1932 moved to England, was introduced to Cierva, learned to fly the C.19 and C.30 Autogiros and created an original gyroplane design that incorporated helicopter pitch control systems. It has also been asserted that Cierva gave permission to make use of several of his patents. This gyroplane, the A.R.III, flew in September 1935 at Heston near London.[33] It made use of collective and cyclic pitch control, developed independently from Kay's earlier work, of which Hafner was seemingly unaware. Hafner's work improved the control achieved by the combination of cyclic and collective pitch control, but he clearly used the gyroplane to advance his helicopter research. Briefly interned at the start of World War II as an enemy alien, he was released when he asked for English nationality and joined the wartime rotary-wing development efforts. The results of his labors would lead to the Rotachute, which would be instrumental in the survival of autorotational flight in the 1950s on the part of Igor Bensen.

A third unsuccessful effort to achieve direct control was that of wealthy Philadelphian E. Burke Wilford.[34] Wilford purchased the patent rights to the work of German inventor and aircraft designers Walter Rieseler and Walter Kreiser in 1925, thus predating Pitcairn's involvement with Cierva. The rights to Rieseler and Kreiser's rigid-blade gyroplane would be assigned to Wilford in U.S. Patent 1,777,678.[35] He called his craft the WRK Gyroplane (X794W),[36] and it first flew on August 5, 1931. It differed from the Cierva and Pitcairn models of the time in that utilized a rigid rotor capable of cyclic pitch variation. The pitch of the rotor blades changed as they rotated, a mechanism to equalize lift in place of the Cierva flexible blades and "flapping hinges." This use of cyclic pitch also affords a measure of control, but the Wilford model retained wing and tail control surfaces as well. Although the navy would eventually evaluate his second XOZ-1 Gyroplane in 1935–36, and it would be tested by the NACA, Wilford was never a serious contender to either Pitcairn or Kellett. This was, in part, because his test pilot Joseph McCormick, brother of William McCormick who had been with RADM Byrd in the Antarctic, died in a 1934 crash of a Wilford prototype.[37]

Pitcairn was developing the PA-22, the first American Autogiro without wings and incorporating a lateral and attitude control system into the rotor system. It "spelled the end of the fixed-spindle Autogiro"[38] but in doing so, presented the company with a terrible dilemma—the direct control Autogiro would, if not actually obsolete the previous models, make them undesirable. Pitcairn's solution would be to cease production of pre–direct control technology in January of 1934 until the development of direct control had been perfected to a commercially acceptable level. That was not immediate in coming as the PA-22 prototype (X13198) crashed from a height of fifty feet on its first flight in early March 1933 with Jim Ray as pilot. The cause of the crash remained unknown as the engineers examined the wreckage and reread the test reports. They were hard pressed to understand how Ray, perhaps the most skilled and experienced American Autogiro pilot, could lose control, but his casual observation provided the clue that finally solved the mystery. Ray described the feedback on the control system and pointed out that even though there were no wings or ailerons, *it felt* as if the upside-down, "hanging" control-stick was "rigged in reverse." The Pitcairn engineers concluded that this was, in fact, the case and that control movements had to be made in reverse of the conventional floor-mounted stick, and from that standpoint, Ray's natural impulses and previous experience had failed him. Subsequent models corrected the reversal of the control system, but there was an additional explanation. Ray also commented that as with the Cierva prototype the previous year, the hanging stick shook and was extremely difficult to manipulate due to excessive rotor feedback. The Pitcairn engineers knew that achieving direct control was not merely a measure of reversing the control system. Not only would the PA-22 prototype have to be rebuilt, the direct control system would have to be completely redesigned and refined, a costly and time-consuming endeavor. Direct control would not come easily, quickly, or cheaply. And the price would not only be collected in currency. Although it is doubtful that the Americans were aware of it, a similar crash of an experimental direct control French C.L. 10 Autogiro on December 19, 1932, at Villacoublay, had already been ascribed to the new control system. The French model, designed by Georges Lepère in consultation with Cierva, featured an overhead control, and it was felt that the cause of the crash was "the over-sensitive hanging-stick control, to which some pilots had difficulty in adapting."[39] Upon take off the aircraft rose to two hundred feet and then dove into the ground, killing pilot Pierre Martin, the first recorded Autogiro fatality.

NOTES

1. Thomas Carroll, "Relative Flight Safety of the Autogiro," *Aero Digest* 17, no. 7 (December 1930): 72.

2. "Autogiros of 1931–1932," *Fortune* (March 1932): 48–52, 48.

3. John M. Miller, "UFO Recollections: The Death of Charlie Otto," *American Helicopter Museum & Education Center Newsletter* 8, no. 1 (Spring 1996).

4. "Autogiros of 1931–1932," p. 52.

5. George Townson, *Autogiro: The Story of "the Windmill Plane,"* (Fallbrook, California: Aero Publishers; reprint, Trenton, New Jersey: Townson, 1985), pp. 88–91; Peter W. Brooks, *Cierva Autogiros: The Development of Rotary-Wing Flight* (Washington, D.C.: Smithsonian Institution Press, 1988), p. 134.

6. For pictures of the Kellett K-2 prototype (NC10766), see Brooks, p. 136; Townson, p. 94. For a photo of the second K-2 produced (NC10767), which would be converted to a K-3 model and receive ATC 471 on March 26, 1932, see "Autogiros of 1931–1932," p. 51.

7. Townson, p. 93.

8. Pictures and statistics of the K-3 can be found in Brooks, p. 137; Townson, p. 99.

9. For photographs of the K-3 in Buenos Aires, being viewed by Captains Marco Zarr and Mermos Hermosa, see Townson, p. 100.

10. Frank Kingston Smith, *Legacy of Wings: The Story of Harold F. Pitcairn* (New York: Jason Aronson, 1981), pp. 258–59.

11. Brooks, p. 274.

12. Brooks, p. 104.

13. For photos of the G-AAKY at Pitcairn Field, see Smith, *Legacy of Wings,* p. 165; Brooks, p. 101.

14. The French company would build a C.27 model that first flew in December, 1932 (later converted to a C.L.10A) a pair of C.L.10s in 1932, and a C.L.10A and a C.L.10B in 1935.

15. Brooks, p. 111.

16. Brooks, p. 114.

17. For a photo of Admiral Byrd and the Pep Boys Kellett K-3, see Townson, p. 102 (bottom); George Townson and Howard Levy, "The History of the Autogiro: Part 1," *Air Classics Quarterly Review* 4, no. 2 (Summer 1977): 18.

18. Brooks, p. 137.

19. "Autogiros of 1931–1932," p. 50.

20. Robert B. C. Noorduyn, "Pitcairn PA-19 Cabin Autogiro," *Aero Digest* 22, no. 2 (February 1933): 48–50, 48.

21. For photographs of the PA-19 in certification flights, see Frank Kingston Smith, "Mr. Pitcairn's Autogiros," *Airpower* 12, no. 2 (March 1983): 28–49, 39–40; *Aero Digest* 22, no. 2 (February 1933): cover.

22. See Smith, *Legacy of Wings,* p. 209; Noorduyn, p. 48.

23. Townson, pp. 52, 54–58; Smith, *Legacy of Wings,* p. 212; Smith, "Mr. Pitcairn's Autogiros," pp. 39–40.

24. Brooks, p. 358 n. 19.

25. Brooks, pp. 131–32.

26. Townson, p. 56.

27. John Walker, ed., *Halliwell's Film and Video Guide: 2001,* 16th ed. (Great Britain: HarperCollinsPublishers, 2000), p. 541.

28. Joan Mellen, *Big Bad Wolves: Masculinity in the American Film* (New York: Pantheon Books, 1977), p. 103.

29. Ray Carney, *American Vision: The Films of Frank Capra* (Hanover, New Hampshire: Wesleyan University Press, 1986), p. 233.

30. Mellen, p. 103.

31. For a description of David Kay's gyroplanes, see Brooks, pp. 318–19; George Townson and Howard Levy, "The History of the Autogiro: Part 2," *Air Classics Quarterly Review* 4, no. 3 (Fall 1977): 4–19, 110–14, 9 (photograph).

32. Ron Herron, "Bringing Back the Autogiro," *Rotorcraft* 33, no. 1 (February–March 1995): 12–13.

33. For a rare photograph of the Hafner A.R.III Gyroplane in flight, see Air Commodore A.E. Clouston, *The Dangerous Skies* (London, England: Cassell & Company Limited, 1954), p. 39.

34. See "The Wilford Gyroplane" in *Aero Digest* (February 1932): 56–57; Charles Gablehouse, *Helicopters and Autogiros* (Philadelphia: Lippincott, 1967), pp. 62–64; Earl Devon Francis, *The Story of the Helicopter* (New York: Coward-McCann, Inc., 1946), p. 103. (Francis claims somewhat fancifully that it was Wilford who suggested to Dr. Henrich Focke that he acquire the German manufacturing rights to the Cierva Autogiros in 1928, but although the meeting between the two is documented, neither the suggestion nor the imputed subsequent action is. See, for example, J.R. Smith and Antony L. Kay, *German Aircraft of the Second World War* [London: Putnam, 1972]).

35. *Aero Digest* 17, no. 7 (December 1930): 64.

36. For a photograph of the Wilford Gyroplane, see S. Paul Johnston, *Horizons Unlimited: A Graphic History of Aviation* (New York: Duell, Sloan and Pearce, 1941), p. 179.

37. Brooks claims that the test pilot who died was J.S. McCormac (p. 23).

38. Smith, *Legacy of Wings,* p. 213.

39. Brooks, p. 165, but the author also states that at the time of the C.L.10 accident "Cierva believed that Martin had attempted to take off with the control column locked."

Chapter 6

AMERICAN ADVANCES, THE C.30A AUTOGIRO, AND CIERVA'S DEATH

> It appears fitting, when we meet to discuss one or the other of many aspects of rotary wing flight, that we should direct our thoughts to the one man (alas no longer with us) whose creative ability and genius not only made possible the Autogiro; but whose foresight and tenacity of purpose so well and truly laid the foundations upon which the helicopter now so surely stands.
>
> Wing Commander R. A. C. Brie, "Some Problems of Helicopter Operation and Their Influence on Design"

The Depression's effect had effectively killed the PA-19, and now even the sales of the smaller PA-18 were impacted. Pitcairn lowered the price to $4,940 but even that did not help. Buhl, so optimistic with its "pusher" Autogiro the year before, had gone bankrupt. The ACA was confronted with a major new research and development effort to achieve direct control, an undertaking made more difficult by the fact that Kellett Autogiro had embarked on its own direct control efforts. Each of the three major companies, Cierva in England and Pitcairn and Kellett in America, clearly saw the prize and were determined to be, if not the first, then the most commercially successful in achieving a direct control model. As a consequence, each became more of a rival and less a collaborator, even though the interlocking series of licensing agreements and contractual obligations to share developments and patents should have made them colleagues advancing rotary flight. This had a negative impact on Autogiro development, as each became reluctant to share new developments, and there can be little doubt that had these companies cooperated, each would have benefited from the work of others and would not have been forced to spend scarce resources in duplicate research and costly testing.

The PA-22 was rebuilt, and Ray began to cautiously familiarize himself with the intricacies of direct control flight in the face of continuing potentially dangerous control-stick vibration. Cierva participated in the American research program during the summer of 1933. He made suggestions regarding the rigging and adjustment of the rotor blades, and Pitcairn perhaps wondered why Cierva's efforts were directed that way while the Americans were concentrating on dampening the overhead control-stick vibrations. The stunning answer, presented in a seminar Cierva gave to the engineering team, was "a completely new control system for rotary-wing aircraft that was far superior to anything then in existence."[1] Cierva proposed a control system based on the ability to exercise cyclic control, that is, to change the pitch of the rotor blades rather than achieving control by tilting the entire rotor disc. This was a dramatically different approach from a titling rotor head, which Cierva had previously attempted in the C.19MkIV and which the Pitcairn team was then in the early stages of developing in the PA-22. Pitcairn, however, did intend to use collective control to achieve the jump takeoff *as he had previously discussed with Cierva.* But Cierva's cyclic pitch control based on the pilot's ability to vary the pitch of the rotors as they circled the rotor disc not only provided for control, it also incorporated a jump takeoff capability. What Pitcairn was attempting to do with two systems, tilting and collective, Cierva proposed to build into one! And as Cierva proceeded to spend days lecturing on the mathematical theory that justified his approach, it was apparent to the Americans, who were talented, experienced, and as theoretically knowledgeable as anyone then involved in the development of rotary-wing flight, that they were being exposed to the cutting edge and sharing a vision of the future of the Autogiro. It was a heady, inspirational, and disheartening experience, for although the experimental work then under way with the PA-22 and its tilting rotor–collective control system was clearly and convincingly vindicated in Cierva's theory, it was equally apparent that the such an approach would be eventually supplanted by the Cierva individual cyclic control.

The Pitcairn team faced a dilemma. It was obvious that the PA-22 required a costly redesign at the very time that Autogiro sales were slowing and Cierva was going in an ultimately more productive direction.

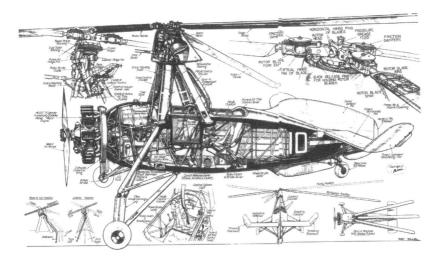

Cutaway of Cierva C.30A direct-control Autogiro. (Courtesy of *Flight International*.)

Cierva proposed, and Pitcairn readily accepted,[2] that the Americans would continue the tilting rotor–collective control system development, while the English company would concentrate on developing practical cyclic and collective pitch control. And although Brooks asserts that Cierva was proposing this division of efforts "presumably to avoid an overlap in the two companies' activities,"[3] there is another, possibly darker, interpretation. Prior to departing for America, Cierva had publicly flown the C.30A prototype at Hanworth on April 27, 1933, with an improved, working version of the control system that Pitcairn, Ray, and Larsen had seen the year before. He would receive the FAI Gold Medal in 1933, celebrating the development of direct control, and the Wakefield Gold Medal the next year for the same achievement. Subsequently, the C.30A would enter commercial production in early 1934, with deliveries commencing in July. That Cierva was suggesting that the American team continue with development of a control system he had already created was probably not obvious to the Pitcairn team. In any event, the decision to continue with the PA-22 certainly appeared the logical choice given Pitcairn's declining financial circumstances and his team's knowledge and investment of effort in and commitment to the PA-22. That decision was, however, with the perspective of history, to prove an unmitigated disaster for Pitcairn and the future of American Autogiro development, for within a decade his company would effectively be out of the Autogiro business and the Kellett organization would soon follow. But the one bright light to come out of Cierva's fourth visit, other than the honors heaped on Pitcairn and Cierva, was the sharing with the American company, as called for in their business agreement, of Cierva's s technical specifications of the cyclic and collective control system. It almost didn't happen.

After a six-week visit filled with honors, test-flights, consultations regarding the PA-22 and its problems, his private aeronautics seminar, and many social engagements, Cierva was preparing for the return to England in July 1933. He had allocated the final day of his visit for a meeting with Pitcairn's patent attorney Raymond Synnestvedt to present and explain the technical specifications and drawings of the new control system so that American patents could be sought. One can only imagine the horror of all when it was inadvertently discovered by Clara Pitcairn[4] that Cierva's ship departed at noon and not the midnight hour originally projected. Demonstrating once again the potential for the Autogiro, Pitcairn called Jim Ray at Pitcairn Field and arranged for the PA-19, a PCA-2, and PA-18 to fly to his home as soon as possible to pick up the parties and fly to Newark from which the SS *Paris* would shortly depart. The aircraft arrived and all piled aboard, with the luggage allocated among the Autogiros and Pitcairn, Cierva, and Synnestvedt in the PA-19. During the brief but successful ride to Newark, Cierva hurriedly spread his diagrams and explained the technical specification. It was presumably viewed as a humorous, frantic dash that left Cierva boarding his ship with only minutes to spare, but it is equally likely that no one present that day knew that the patents that came from those drawings would "be the keystone for the future of the entire vertical-lift industry or that they would have a critical impact on the fortunes of Harold Pitcairn."[5] Pitcairn applied for an American patent on November 16, 1933, after his firm had prepared the necessary papers, but Patent No. 2,380,585, covering the fixed-spindle rotor with collective and cyclic pitch control, was not granted until July 1, 1945. Pitcairn also received Patent No. 2,380,580 for his tilting rotor disk and jump takeoff collective pitch control.

Development on the reconstructed PA-22 proceeded at an increasingly successful pace after Cierva's departure, as the engineering team solved the presented problems and Jim Ray learned to control the hanging-stick control. The tilting rotor was developed before the more complex cyclical pitch controls that would allow vertical jump takeoffs. But by August 1933 Cierva, having already perfected the tilting rotor mechanism in the C.19MkIV and V prototypes, had achieved jump takeoffs in the C.30, the first commercial direct control Autogiro. Given that the Americans had agreed to concentrate on a tilting rotor cyclical pitch control, it is understandable that Cierva would immediately invite Pitcairn to come to England to see the latest developments, and it is equally understandable that the Pitcairn company would send Jim Ray, the most experienced pilot with direct control. Ray came and flew and then summoned Pitcairn, as the engineering team needed to view the English progress. After viewing the C.30, Larsen and Stanley concluded that it was not significantly advanced beyond the PA-22. That opinion had some justification, as the Americans had observed and flown the C.30P prototype model (G-ACFI)—sometimes erroneously[6] identified as a C.30MkIII, as the prototype was also known as the C.30P MkII—which was capable of making jump takeoffs of only a few feet, as its tilting mechanism still suffered from vibration problems. Ray and Pitcairn, both pilots, were of a similar opinion that any control achieved would soon exhaust the pilot. And it was evident that the C.30P was not ready for production or commercial sales.

If the apparent lack of English progress was not surprising, the apparent lack of cooperation was. Even though he had originally extended the invitation, Cierva seemed to have little time for his visitors. But if Cierva was excusably unavailable, and there is evidence that he was then preoccupied with political unrest in Spain, there is no rationale for the evident lack of cooperation on the part of Cierva's English associates, most notably engineer Dr. J[ames] A[llen] J[amieson] Bennett.[7] The perceived coldness and hesitation at best, and unwillingness at worst, on the part of the English engineering team to share direct control information convinced Pitcairn to take all but Jim Ray home when Cierva left for Spain. Ray, as an experienced pilot, had established a friendly relationship with the Cierva staff, and Pitcairn felt that Ray could gain more information on the direct control and jump takeoff developments than the engineers. When the Americans returned to America in late 1933, they found a dire circumstance at Willow Grove. The factory had no orders and the cash-flow possibilities were nil—the Depression finally and perhaps fatally caught up with Harold F. Pitcairn. He was a visionary, an aviation dreamer, but also a businessman from a prominent business family. He reached a necessary but unpleasant business solution. The production facility was closed, workers let go, and the remaining staff moved into a building located next to the main hanger. The only personnel retained were engineers Stanley and Larsen, pilot Jim Ray, two Autogiro mechanics, designer Harris Campbell, a bookkeeper, and a telephone operator/secretary. All other employees were notified Christmas Eve that they would be unemployed as of January 1, 1934. That year would be a bleak time for the ACA, but developmental work on the direct control PA-22 continued, each advancement leading to a patent application. Other than the technological advancement, the major product of the company during that year had to be the twenty-six patents received, four of which were personally awarded to Pitcairn, and many others were then in the application process.

But if 1934–35 was a developmental time for Pitcairn, it also saw significant Autogiro growth in Europe, with direct control models, and on the part of Pitcairn's competitor Kellett Autogiro Company. The French licensee Lioré-et-Olivier negotiated an extension to their C.19 agreement to include the manufacture and further development of the C.30, and the English company began delivery of the C.30A in July. In late 1933 and early January of 1934 Kellett chief engineer Richard H. Prewitt visited the English company to view the C.30A and study direct control. These consultations resulted in a January 1934 Kellett developmental effort that would result in the KD-1 (for Kellet Direct Control–1), a direct control version of the K-2, which was first flown on December 9, 1934, by Kellett test pilot Lou[8] Levy (who later changed his name to Leavitt).[9] It would receive ATC 712 on January 1, 1935, and eventually about ninety-seven derivatives would be built in Japan, flown against its foes in World War II.

Brie began 1934 flying a C.30P from London to Paris in two hours and twenty minutes on January 5, an impressive demonstration for the French Air Ministry, which had acquired it for testing. Cierva then flew a C.30P (G-ACIO) to France and Spain and toured his native country in February and March, a 2,500-mile direct control Autogiro tour that may also have been motivated by a desire to promote his political interests.[10] Brie also demonstrated the direct control model before the Belgian Air Force at Evère, near Brussels, in February and returned to France for the Sixth Fête Aérienne de Vincennes near Paris in February 1935. And while Cierva had tested the direct control Autogiro's ability to landing on ship on a 52-by-170-foot platform erected on the Spanish navy seaplane tender *Dédalo* off Valencia on March 7, 1934, Brie had renewed navy interest in the Autogiro with a series of landings on a smaller, 49-by-115-foot platform on the Italian cruiser *Fiume* on January 5, 1935. While the previous Cierva landings on the Spanish ship had been made while at anchor, the January 1935 flights were accomplished while the ship was steaming at twelve, fifteen, twenty-one, and twenty-four knots off Le Spezia, an impressive demonstration of the C.30 under operational conditions which led to the British Royal Navy arranging for landing trials with a C.30A Rota, as the military models were called, during the 1935 Summer Cruise. Brie duplicated his earlier success with landings on the HMS *Furious.* So even as the American Autogiro industry awaited a successful, commercially viable direct control model, the Cierva C.30 in its various configurations was achieving a civilian and potential military success.

There was also a parallel direct control effort on the part of G. and J. Weir Ltd. While Lord William and James Weir had been instrumental in bringing Cierva to England, it was not until the summer of 1932 that James proposed that the family company secure a Cierva license to build a direct control Autogiro. Although this may have been due to the advent of practical direct control, it is intriguing to suggest that it may also have been influenced by Mrs. James Weir becoming the first woman in England to get an Autogiro endorsement on her "A" license.[11] Cierva undoubtedly granted the license to his old friend because the development of the direct control C.19MkV had become an expensive undertaking, but also because Weir had proposed to develop a small, single-seat aircraft, a direction that Cierva could not but feel would prove productive as Autogiro aviation evolved into an affordable and safe form of *personal* transportation. The W.1, designed by Fred L. Hodgess and R. F. Bower in Scotland with extensive consultation with Cierva in England, was built at Cathcart, near Glasgow. The small aircraft, completed in the spring of 1933, was powered by a custom-designed 40-horsepower Douglas (Motors Ltd.) Dryad two-cylinder horizontally opposed air-cooled engine and utilized a two-blade folding rotor. The rotor disk was relatively small, with a twenty-eight-foot diameter, and the engine was relatively light, consistent with the plan to produce a personal flying machine. The prototype, featuring a light plywood monocoque fuselage, was taken to Hanworth in England, where Cierva made an initial flight in May. It was test-flown throughout the summer by Marsh and Cierva again flew it in September, possibly before Harold Pitcairn and his associates who were then visiting the English company. Marsh would remain the main test pilot for the Weir series, as both Cierva and Brie, large men, were not comfortable in the narrow confines of the small Autogiros.

Although Cierva may have considered the Weir effort as a parallel development effort to the C.19MkV–C.30 experimental direct control aircraft, the W.1 did not perform well. Due to its small size and necessarily light construction, even with a small engine, the W.1 experienced great vibration. The design also proved unacceptable as there was a pronounced lack of lateral control and inadequate provision for prerotation. The W.1 was returned to Scotland for redesign and then was returned to Hanworth in December for testing, but it soon came to an inglorious end—while landing on December 21, pilot Alan March overturned the aircraft.

The next model, the W.2, also featured a light plywood monocoque fuselage and a geared 45-horsepower Weir air-cooled engine that had been specifically designed in mid-1933 by a team under the direction of Fred Hodges and now including Ken Watson and, of greatest importance Dr. J. A. J. Bennett, who would later succeed Cierva as chief designer of the Cierva Autogiro Company. Although much more stable and controllable, its engine only produced 45 horsepower, the performance of the W.2 was anemic, and engine-induced vibration was pronounced. Although the Weir company took steps to market the W.2, issuing a sale brochure in August of 1934, the W.2 prototype remained the only one ever built, currently to be found in the aviation collection of the Royal Scottish Museum at East Fortune near Edinburgh.[12]

Even though the Weir company decided on February 5, 1935, to discontinue Autogiro development, two additional small, single-seat Weir Autogiros were built for specific testing and developmental purposes. The W.3, built by a small design team of six people, was to test the new Cierva jump takeoff rotor head with its cyclic control. The W.3 was a flying platform to test and refine the jump takeoff, and Alan Marsh flew it for the first time at Abbotsinch, near Glasgow, on July 9, 1936. Although it performed well, engine reliability was a real problem, and it never reached the commercial market. The Weir W.4 was a refined version of the W.3, with a reliable 55-horsepower Weir Pixie engine and an aerodynamically streamlined rotor pylon. The W.4 never flew, overturning during taxi tests conducted by Alan Marsh in late 1937, and was never rebuilt. The Weir company had had enough; its board met on December 28, 1937, and decided to cancel its Autogiro program, preferring instead to proceed with helicopter development,[13] as the world was then taking note of German developments, and the Weir company reasoned that their hard-won capabilities and experience in rotary flight would more profitably be transferred to the emerging helicopter.

In January 1934 Sir John Siddeley, chairman of the Armstrong Siddeley Development Company Ltd., which controlled A. V. Roe, decided that the company would commence commercial production of the C.30A direct control Autogiro. A license was obtained from Cierva Autogiro Company Ltd. early in 1934, and production commenced and deliveries started in July. Eventually seventy-eight would be produced before production ceased in June 1938. It was evident that the £31,250 price, around $50,000 today, did not deter the desire for direct control. In addition to supplying flying schools at Heston and Hamble and the Cierva Autogiro Flying School at Hanworth, the company even tested a C.30A, flown by S. J. Chamberlin, fitted with a two-way radio for the metropolitan police department for traffic control in mid-August of that year, becoming the first rotary-wing traffic reporter. In October a C.30A was exported to western Australia, intended for use in the gold fields in New Guinea, but Brooks notes that "the C.30 never got to New Guinea.... [I]ts fate is unknown." But a clue may be furnished in the first day cover of the First Australian Autogiro Flight dated November 16, 1934. That cover, from the time of the Melbourne to Portland Flight, shows an aborigine gazing up at a C.30P (G-ACIN) which is listed ambiguously by Brooks as having been scrapped in 1938. An additional C.30A was exported to Australia in June 1935, and it may be that model that was reported still in existence in November 1978.[14]

C.30 models were sold to France, Belgium, Sweden, Hong Kong, Holland, Brazil, Poland, India, Italy, Denmark, Czechoslovakia, Spain, Lithuania, Soviet Union, China, Yugoslavia, and Argentina. That it had caught the public eye surely motivated the Hon. Mrs. Victor Bruce to attempt to fly a C.30A from England to Capetown on November 25, 1934. Although her attempt started well, with a 370-mile flight to Dijon, France, it ended three days later near Nîmes, 240 miles further south in France, when she suffered slight injury in a crash landing. Brie also experimented with flights between Hanworth and the London Mount Pleasant Post Office in central London, although he did not actually land. It was the first investigation of a potential role for the Autogiro in providing airmail service between central city post offices and outlying airfields and may have served as an inspiration for the construction of a landing area on the roof of the 30th Street Post Office in Philadelphia. Pitcairn had in 1933 "advocated that it be built with a roof that would withstand landings by autogiros," and in 1935, when Kellett and Pitcairn had been flying direct control models, they would make test landings on that roof.[15] A C.30A was flown by the Spanish navy in military operations mounted in connection with a rebellion in Asturias in October of 1934, first actual combat use of the Autogiro—it would not be the last. The RAF, its interest rekindled with the advent of direct-control, ordered two new designations, each a configuration of the C.30. The Rota I, of which ten were ordered, was intended for army use, while the Rota II was equipped with floats for naval deployment. RAF Flight Lieutenants W. Humble and R. H. Haworth-Booth were trained at the Cierva school at Hanworth in September 1934 and ordered to the RAF School of Army Cooperation at Old Sarum to provide pilot training, and by December 24 the British War Office had decided to replace observation balloons with Autogiros. Although Rota service was to carve out a military role for the Autogiro, it would not prove that which was originally anticipated, and it led to the first fatal accident in England. On January 21,

1935, Flying Officer L. W. Oliver perished when, losing control in a cloud, he entered into a high-speed dive from which he was unable to recover.[16] The Rota II was tested for naval uses, but found too slow because of the weight and drag of the attached floats. And by 1935 the Americans were flying direct control models.

The Kellett Autogiro Company had begun developing a direct control aircraft shortly after company officials met with Cierva in late 1933. This direct contact with Cierva had the effect of chilling the relationship between the Kellett and Pitcairn companies, who were rapidly emerging as rivals. The KD-1 received ATC No. 712 in early January, 1935, less than thirty days after its first flight.[17] Unlike the previous Kellett Autogiros, the KD-1 was an open-cockpit, tandem model, with the streamlined rotor pylon forward of the front cockpit and the landing gear almost directly below the pylon, and the tilting direct control rotor head was controlled by means of a conventional floor-mounted stick. The model, viewed as a handsome alternative to the Cierva C.30, was aggressively promoted in a variety of configurations: KD-1, KD-1A, and KD-1B civilian variants and the YG-1, YG-1A, YG-1B, and YO-60 Army Air Corps models. It had a three-blade cantilevered rotor, and the blades could be folded straight back for easier storage. The KD-1 immediately created much interest, and the company actively sought opportunities to demonstrate its abilities. On May 25, 1935, perhaps inspired by Brie's English airmail experimental flights, Lou Levy landed a KD-1 on the roof of the 30th Street Post Office in Philadelphia,[18] and Jim Ray similarly demonstrated the PA-22.[19] But even though the new Philadelphia post office had been constructed with a reinforced roof free from projections and obstacles, it would take an additional four years before an experimental air route would be initiated with Autogiro service. There were more immediate and potentially valuable opportunities for the American manufacturers, however. The military had become interested.

The advent of direct control and the publicity being generated in Europe and the interest shown by the military authorities of several countries resulted in a 1934 government request for bids. Both Kellett and Pitcairn responded, as by then the PA-22 was flying well. With the seventh version, in which an effective direct control three-blade rotor had been installed after significant testing and a more effective tail unit with a raised stabilizer and two outboard fins was also installed, the Pitcairn engineering team felt it had finally solved the problems associated with direct control. Ray demonstrated the improved PA-22 to army, navy, and civilian officials at the Naval War College in Washington. It went through an eighth modification with the addition of outward-tilted fins on the stabilizer, resulting in increased stability, which would become characteristic of all future designs. In response to the Army Air Corps request, Pitcairn prepared two different designs for consideration. Because a Kellet KD-1 that had been acquired by the army for testing had been dubbed the YG-1, the Pitcairn PA-33 was designated the YG-2,[20] the Y denoting an army model undergoing field testing and the G for Autogiro (A already had been allocated to attack planes). It was a direct control tandem two-seater configuration with forty-six-foot rotor, later increased to fifty-foot-diameter three-blade cantilever rotor[21] mounted on a reworked PCA-2 fuselage. Powered by a Wright R-975-9 Whirlwind 420-horsepower engine, the aircraft achieved a maximum speed of 144 mph flying at 3,150 feet, even though it weighed 3,300 pounds. And as the PA-34, named the XOP-2, it was supplied to the navy with slightly different landing gear.

Military testing did not initially go well, as the stability achieved by the PA-22 did not automatically scale up to the larger military models. The YG-2 sent to the NACA experimental flight center at Langley Field, Virginia, for aerodynamic testing suffered a crash on March 30, 1936, when a rotor blade failed while flying at 120 mph. Both pilot Bill McAvoy and passenger John Wheatley safely parachuted, but the aircraft was completely destroyed by fire. The army observer was a young First Lieutenant named H[ollingsworth] Franklin Gregory, who was to become one of the most influential contributors to the failure of the Autogiro in America even as he became incredibly instrumental in making the helicopter a practical reality.

The XOP-2 ordered by the Navy Bureau of Aeronautics in 1936, but not produced until 1937, faired little better than its army counterpart to which it was almost identical, differing in having a more open tubular truss landing gear and, understandably, fittings for flotation equipment. Based on the crash of the army model, the XOP-2 rotor blades were redesigned, but the military experiences with the Pitcairn models did not produce any further orders.

The Kellett military version of the KD-1/KD-1A, the YG-1 series (YG-1/YG-1A/YG-1B), was first delivered to the Army Air Corps in October 1936 and performed well enough that subsequent models were ordered for the Air Corps Autogiro School at Patterson Field in Springfield near Dayton, Ohio, in 1938. The army had, however, first examined a KD-1 at Wright Field in 1935, considering it for military missions under the designation YG-1. The first YG-1 went to NACA in early 1936 for evaluation, where the assigned pilots were army Lieutenants H. Franklin Gregory and Erickson Snowden Nichols, brother of Ruth Nichols, the famous woman aviator. While at NACA, the army pilots were soon visited by company president Wallace W. Kellett, chief engineer Richard H. Prewitt, and Lou Leavitt, company test pilot. Gregory and Nichols found flying lessons with Leavitt to be instructive, as he was by then, with Jim Ray, one of the most experienced Autogiro pilots in America, but they also enjoyed the conversations with Prewitt. He described the aircraft's design and construction, wonderfully informative for Gregory, who would later make the decisions that would doom the Autogiro.

The pilots[22] called Prewitt, who had actually designed the YG-1, "Daddy of the Whirligig." Subsequently, the KD-1 flew in a testing program for the Field Artillery Board at Fort Bragg, North Carolina, consisting of directing artillery fire, reconnaissance, and landings in otherwise inaccessible areas. That testing, including the first use of a telephone, with a 1,500-foot cord to the ground, is recounted in H. Franklin Gregory's *Anything a Horse Can Do.* The second Kellet aircraft, a YG-1A, which had been modified with addition of military H.F. Radio, was delivered in October 1936. Although the testing showed a reconnaissance role in artillery spotting, each of the test models would suffer multiple crashes and eventually be destroyed in the testing process, in the field or in the NACA wind tunnel, but the army would open its Autogiro Training School. But as then–First Lieutenant (later Brigadier General) H. Franklin Gregory makes clear in his autobiographical book *Anything a Horse Can Do,* he and his army colleagues were even then viewing the Autogiro as an intermediate step to the helicopter. And some of the data derived from the NACA studies of the Kellett military models were to play an important role in helicopter rotor development even as the Pitcairn rotary-wing patents would prove vital to such achievements. Gregory had concluded his evaluation of the Kellett direct control (but not jump takeoff) Autogiros that they were "in reality little more than a high-lift device on a conventional airplane... that offered no more—and in many cases less—performance that the performance of many ultra-light fixed-wing aircraft."[23]

In March of 1936 *Fortune* revisited the Autogiro with an article entitled "Autogiro in 1936" and expressed a decidedly different view than the previous article by Ingalls in March 1931. In evaluating the Autogiro, the magazine viewed the coming of direct control as portending a "rebirth"— and, in evaluating the previous models praised five years earlier, stated plainly that the Autogiro had "turned out to be a lemon... for all practical purposes." While the unidentified writer recognized that the Autogiro was "still the only flying machine that could rise from a narrow lawn, loaf through the air as slowly as twenty-five miles an hour, and, if its engine died, settle to earth as gently as a parachute," went on to assert that "the trouble was...it would do those things generally only in the hands of experts; and it would *not* do, even for the experts, certain other desirable things, like flying fast and carrying a decent load. ('Half the speed for twice the horsepower' was the contemptuous jibe of airplane pilots and engineers.)" Harold F. Pitcairn, previously described as an "Impresario" in 1931, now was characterized as "a rich, scholarly Pennsylvania socialite of somewhat ascetic tendencies and mathematical bent," who with "his brothers Raymond, a lawyer, and Theodore, philosopher, artist, and Swedenborgian minister...shares the wealth of the Pitcairn Co., which has notable holdings of Pittsburgh Plate Glass." Being portrayed not as an aviation visionary but as a rich dabbler was no doubt a painful denunciation, and Pitcairn could not have helped but be further dismayed at the magazine's assertion that Cierva "had regarded the whole Pitcairn venture in the U.S. as a large testing ground on which the giro would be given a thoroughgoing workout under all sorts of conditions, while he perfected the design for market in Europe."[24] However, in the same article containing those words was a powerful *visual* testimony of the coming of age of the Autogiro, for in eleven sequential photos that started horizontally from the lower left corner of page 88 and continued vertically up the left side of the facing page and then horizontally across that page the article showed the first pictorial record of an Autogiro making a jump takeoff, a C.30 (G-ACFI) piloted by Juan de la Cierva. And in an illustration spanning the center section of the two following pages, the magazine introduced the reader to the latest Pitcairn development, the "roadable" Autogiro, capable of achieving speeds of 110 mph in the air and then, upon landing and folding its blades backward, driving along the highway at 25 mph. In final evaluation of the achievement of direct control, the jump takeoff, and the American roadable Autogiro, *Fortune* concluded that "after sixteen years the autogiro has only now become an autogiro." Sadly, however, that observation was to prove untrue, and the very developments cited for the coming Autogiro rebirth would either prove a misdirection (the roadable) or merely an essential building block of the coming helicopter.

In 1930 Amelia Earhart was working for the New York, Philadelphia and Washington Airway Corporation, which had been started by aviation-minded Philadelphians Nicholas and Charles Townsend Ludington, later partners of the Kellett brothers. Earhart was one of three vice presidents, the other two being Paul Collins and Eugene Vidal, father of writer Gore Vidal. Earhart worked seriously at the job, being in charge of publicity and complaints. She also flew over the line at least every few days, once chap-

eroning a bird and another time dealing with a woman who had announced
that she would be traveling with a lapdog but showed up with what wit-
nesses characterized as a 'heifer.'[25] She became close to her colleague
"Gene" Vidal. He was, by all accounts, handsome, bright, and an athletic
star in several sports at the University of South Dakota who graduated first
in his engineering class. While in the army after graduation, Vidal served
in the air corps and became the first flying instructor at the United States
Military Academy at West Point. He was a talented pilot and committed to
aviation, and his 1922 marriage to the daughter of the senior senator from
Tennessee, Thomas Gore, was considered a brilliant match. But by 1930,
the marriage was in name only, with a growing gap between the party-
loving socialite Nina Gore Vidal and her aviator husband, leading each to
go their separate ways. She would divorce Gene in 1935 and marry fellow
party-lover and bon vivant Hugh D. Auchincloss, future stepfather to
Jacqueline Bouvier Kennedy. But already by 1930, Gene had become
entranced with Amelia, and there is indication that the affection was
returned and that by 1933, if not before, she and Gene had become roman-
tically involved,[26] a fact apparently accepted by her husband George
Palmer Putnam.[27] Earhart wore a man's boxer shorts while flying, which
were more comfortable and suited to long flights, and while her husband
related that they were his, Eugene Vidal related that they were his. Years
later, after her disappearance, he was able to debunk reported sightings of
a woman pilot on a pacific island wearing jockey shorts.[28]

    Amelia had met Eleanor Roosevelt on November 20, 1932, and formed
an immediate friendship for the wife of the newly elected president of the
United States and admired woman of distinctive achievement. They met at
the Roosevelt home at Hyde Park, but then Mrs. Roosevelt accompanied
Amelia to her lecture at the local public high school in Poughkeepsie, New
York. During that public presentation of films and personal remarks, during which she recounted her cross-country Autogiro flight, she "diplomat-
ically praised Poughkeepsie hometown boy Lieutenant John Miller, who
had beaten her across the continent in his autogiro."[29] Earhart used her
friendship with the Roosevelts, particularly the first lady, to lobby for
Vidal's appointment as director of the aeronautics branch of the department of commerce, an appointment announced by the president on September 20, 1933. The following year Vidal championed government
support for the aviation industry, and thought that one way to accomplish
this was to create cheap, readily available airplanes. Manufacturers were,
therefore, invited to bid on the construction of this "poor man's airplane."
The prospectus requirements were formidable: a speed of 100 mph, take-

off and landing in a space thirty feet square, and a "roadable" capability. Seven experimental aircraft were developed for the program, but only one met all the requirements—the AC-35, also known as the PA-35, was produced by the Autogiro Company of America.

The "roadable" Autogiro had been a favorite idea of Jim Ray, who unsuccessfully championed roadability in 1933 for inclusion in the PA-22. The ACA received a contract from the Bureau of Air Commerce to produce a prototype AC-35 in late 1934 and commenced development that would continue into 1936. It was to be a PA-22-like direct control design, with a hanging-stick control and a side-by-side seat arrangement in an enclosed cabin. But the roadability requirement led to a novel placement of the engine, buried in the cabin to the rear of the passengers. The 90-horsepower Pobjoy Cascade engine had a shaft passing through the front cabin and connecting to the propeller, but also a crankshaft through a clutch and gear to the rear wheel that would propel the vehicle along a highway at 25 mph.[30] Additionally, there was an extension of the front crankshaft that ran up to the rotor, with a selector mechanism that allowed the pilot to prerotate the blades. It was a complex but workable solution to all the requirements of the government contract, and the AC-35, designed by a team headed by Agnew E. Larsen, featured a welded steel tube fuselage construction with wooden fairing strips and metal and fabric covering. Its original version had a boxlike tail and double-contrarotating propellers, but these were replaced by more conventional designs in final configuration.[31] The rotors could be folded back and the aircraft stored in a seven-by-twenty-four-foot space.[32] It was scheduled for delivery to the government in the late summer of 1935 but was not delivered until the fall of the following year, having first flown on March 26, 1936. Jim Ray had flown the test flights and found it to be a stabile, reliable aircraft, and fun besides. Ray would often fly to a small Pennsylvania town, land on a deserted road, fold the blades, and drive around town. He always drew a crowd when he extended the rotor blades and took off from a road outside town.

It is likely that Gene Vidal smiled when, in response to Pitcairn's inquiry as to where the AC-35 (NX70) was to be delivered, he replied that since it was, after all, "roadable," it should be delivered to the front door of the Department of Commerce building on Fourteenth Street in Washington, D.C.[33] It is equally likely that both Pitcairn and Ray readily took the challenge, for it promised terrific publicity for the AC-35. On October 26 Pennsylvania Avenue was blocked off between Thirteenth and Fifteenth, and Jim Ray landed in front of the Occidental Restaurant and the Willard Hotel, folded the rotor blades, and drove to the Department of Commerce building, where it was accepted by Secretary Roper.[34] It was a publicity triumph, and the film of that landing is still shown by the Smithsonian National Air and Space Museum.

The government subsequently lent the AC-35 back to Pitcairn for further development of a powered rotor and several modifications were made, but the project was cancelled in 1940. Pitcairn then restored the AC-35 to its 1936 form, and it was returned to the government in 1942 and accepted into the Smithsonian National Air and Space Museum in 1950. It was driven to the museum through the streets of Washington, D.C. Pitcairn had even offered the licensing rights to the PA-35 to Cierva Autogiro Company Ltd., but the English company was not interested—it was moving in different directions.[35] There is a curious postscript to the AC-35, and it represents one of the two unsuccessful attempts in the late 1950s through the early 1960s to revive the Autogiro. In 1961 Skyway Engineering Company, Inc., located in Carmel, Indiana, acquired a license from the ACA for the design and patent rights, blueprints, analyses and flight test results to the AC-35, with the intent to produce and market a modernized version of the roadable Autogiro. One model was built and successfully flown at Terry Field near Indianapolis. The Skyway prototype (N35133)[36] even used some components from the original aircraft, but the company experienced internal problems and the project failed to go forward beyond the single prototype.

At the successful conclusion of the AC-35 project in 1936, and with the PA-22 having been developed into an effective flying-test platform, Pitcairn decided to take his engineering team and their families to England for an extended working vacation to consult with Cierva and his colleagues. But soon after Pitcairn, Stanley, Ray, and Larsen had arrived, they learned that Cierva was absent in Spain and that cooperation was not readily forthcoming from the engineers and officials of the English company. Cierva, preoccupied with the deteriorating conditions, had good reason to be afraid, as Spain was then in civil war and his brother Ricardo was being held by the Communist forces. He had earlier that year, with Louis Antonio Bolin Bidwell, London correspondent of the Royalist Madrid newspaper *A B C*, secretly arranged for a clandestine flight by a twin-engine English airplane piloted by Cecil W. H. Bebb, which altered the course of history. The flight departed Croydon, London's airport on July 11 supposedly to take retired British Army officer Major Hugh Pollard, his daughter, and her friend on vacation to Casablanca. After dropping the passengers off, the flight continued to Las Palmas to pick up General Francisco Franco and his aide and fly them to Tetunan, in Spanish Morocco, on July 19. Franco was in a strategic position from where he would assume command of the Spanish army. The Spanish government had exiled General Franco to quell his popularity with the people and the army, and shortly after his return, he would lead the army in revolt against the Communist government, initiating the Spanish Civil War.[37] So Pitcairn and his colleagues were understanding of the delays in meeting with Cierva, but they grew increasingly concerned about the coldness of the English team. Even though Pitcairn had moments of doubt in the past concerning the English company, of which he was a director, he trusted that upon Cierva's return a productive relationship could continue. And although, in light of the past, that was probably true, it was not to be, as Cierva perished in the crash of the KLM D.C.2 (PH-AKL) flight, bound for Amsterdam from the airport at Croydon Aerodrome, London, as recounted in the introduction to this book.

Many paid tribute to Cierva in published obituaries, and the Royal Aeronautical Society posthumously awarded him its prestigious Gold Medal. Harold Pitcairn paid tribute to his friend of almost a decade in writing

> Juan de la Cierva will be known to enduring fame as the outstanding pioneer in the field of rotary wing aircraft.... All helicopters and similar types of craft that have shown promise of practical performance incorporate some of the principles and inventions developed by Cierva.[38]

NOTES

1. Frank Kingston Smith, *Legacy of Wings: The Story of Harold F. Pitcairn* (New York: Jason Aronson, 1981), p. 222.

2. Peter W. Brooks, *Cierva Autogiros: The Development of Rotary-Wing Flight* (Washington, D.C.: Smithsonian Institution Press, 1988), p. 152; but see Smith, *Legacy of Wings,* p. 223, where the decision to continue with a tilting rotor–collective pitch control system is solely a Pitcairn decision.

3. Brooks, p. 152.

4. Smith, *Legacy of Wings,* p. 225. Brooks fails to mention this episode.

5. Smith, *Legacy of Wings,* p. 225.

6. Ibid., p. 226; but see Brooks, p. 182.

7. D.Sc., Ph.D., D.I.C., F.R.Ae.Sc., and Professor of Aerodynamics and Deputy Principal, College of Aeronautics in 1961.

8. Gablehouse reports this as "Lew" but is probably confusing him with Lewis "Lew" A. Yancey. Gablehouse, Charles. *Helicopters and Autogiros: A Chronicle of Rotating-Wing Aircraft.* Philadelphia: Lippincott, 1967, p. 198.

9. For Lou Levy, see "Autogiro in 1936," *Fortune* 13, no. 3 (March 1936): 88–93, 130–31, 134, 137, 89; John M. Miller, "Test Flying for Kellett Autogiro Corporation," *Rotorcraft* 30, no. 7 (October–November 1992): 22–28, 22; Ibid.,

"The First Scheduled Rooftop Flying Operation in Aviation (Autogiro Air Mail Service at Philadelphia, 1939–40)," *Rotorcraft* 30, no. 6 (September 1992): 24–33, 24; Hollingsworth Franklin Gregory, *Anything a Horse Can Do: The Story of the Helicopter,* Introduction by Igor Sikorsky (New York: Reynal & Hitchcock, 1944), p. 55 ("and Lou Levy (later his name was changed to Leavitt), test pilot for Kellett").

10. Brooks, p. 358 n. 24.

11. Ibid., p. 112.

12. Ibid., p. 170; Bob Ogden, *British Aviation Museums and Collections,* 2nd ed. (Stamford, Lincolnshire, England: Key Publishing Ltd., 1986), pp. 91–92; Ibid., *British Aviation Museums* (Stamford, Lincolnshire, England: Key Publishing Ltd., 1983), p. 64–65.

13. Brooks, p. 174.

14. Martin Hollmann, "One of the Last C.30A Autogiros Founds in Australia," *Gyroplane World,* no. 26 (November 1978): 2–3.

15. Miller, "First Scheduled Rooftop Flying," p. 24.

16. That accident and other European fatal crashes are recounted by Brooks, including reference to a fatal accident in August 1935 at Willow Grove in a PA-18, where John Miller's passenger Robert Swenson died. See Brooks, pp. 192–93. But neither George Townson nor Frank Kingston Smith make any mention of this accident, nor was it mentioned in "Autogiro in 1936."

17. For pictures of the KD-1, see Brooks, p. 231; George Townson, *Autogiro: The Story of "the Windmill Plane,"* pp. 107, 111; Richard Howe, "Kellett KD-1/ YG-1 Autogyro (Photo Essay)," *American Aviation Historical Society Journal* 23, no. 1 (first quarter 1978): 49–50; George Townson and Howard Levy, "The History of the Autogiro: Part 2," *Air Classics Quarterly Review* 4, no. 3 (Fall 1977): 4–19, 110–114, 14 (Kellett KD-1 in TWA markings).

18. For a photograph of Levy making a test delivery of mail on the roof of the 30th Street Post Office, see Townson, p. 116.

19. For photographs of the PA-22 landing on the post office roof, see Smith, *Legacy of Wings,* p. 239.

20. For a picture of the YG-2, see Brooks, p. 235; Townson, p. 65.

21. Brooks, pp. 214–17.

22. Gregory, p. 55.

23. As cited in Smith, *Legacy of Wings,* p. 252.

24. "Autogiro in 1936," p. 93.

25. Susan Butler, *East to the Dawn: The Life of Amelia Earhart* (New York: Da Capo Press, Inc., 1999), p. 239.

26. There had been rumors that when Earhart returned from her ill-fated round-the-world flight, she would divorce George Palmer Putnam and marry Paul Mantz, a well-known pilot, but even as these rumors had been widely circulated, Mantz had recently married and was seemingly happy. Walter Winchell, in fact, told the American radio audience that such rumors were untrue but that Earhart

would divorce and then marry "an aviation inventor." Lovell, however, argues that Earhart and Vidal were not lovers, citing a lack of mention of the relationship in Earhart's letters, but such reasoning is not convincing, because Earhart was generally circumspect in her letter writing. Butler states unequivocally, "Gene and Amelia were undoubtedly lovers," and cites the opinions of Katherine (Kit) Vidal, Gene's widow, whom he married in 1939, and his son Gore Vidal. See Mary S. Lovell, *The Sound of Wings: The Life of Amelia Earhart* (New York: St. Martins Press, 1989), p. 265; Butler, pp. 291–95.

27. Butler, pp. 277–79.

28. Ibid., pp. 291–92.

29. Ibid., p. 281.

30. *Aircraft of the National Air and Space Museum,* 4th ed. (Washington, D.C.: Smithsonian Institution Press, 1991; see "Pitcairn AC-35"); but see Brooks, p. 219, citing a top speed for the AC-35 of twenty-five to thirty mph.

31. For pictures of the various configurations of the AC-35, see Smith, *Legacy of Wings,* pp. 243–45.

32. *Aircraft of the National Air and Space Museum.*

33. For a contemporary account of the delivery of the "roadable" Autogiro, see "Roadable Autogiro," *Aviation* 35, no. 11 (November 1936): 33–34.

34. For photographs of the AC-35, rotors folded back, being accepted by the department of commerce, see Walter J. Boyne, *The Aircraft Treasures of Silver Hill* (New York: Rawson Associates, 1982), pp. 137–38.

35. Smith, *Legacy of Wings,* p. 259.

36. For a picture and diagrams of the Skyway AC-35, see Townson, pp. 71–72.

37. Brooks, p. 359 n. 28.

38. Harold F. Pitcairn, "Juan de la Cierva: In Memoriam," Autogiro Company of America, Philadelphia, January 9, 1939.

Chapter 7

PITCAIRN AFTER CIERVA

Cierva's flair was for the elegance of mathematics and the romance of aerodynamics. He cared little for mechanical engineering; as far as he was concerned, that was a rather crude relative of the plumbing business. The Autogiro was his, and his alone, and he could not bear to contemplate submerging its beautiful simplicity in a welter of shafts, clutches, and gears—which would not be his.

Pioneering English pilot Frank T. Courtney, *Flight Path*

Cierva's death had a great personal impact on Harold Pitcairn. Even though the American had sensed an underlying rivalry in terms of the two companies, he remained close friends with the inventor. As a member of the board of directors of the English company Pitcairn had participated in its affairs (fifteen trips to Europe between 1928 and 1936, and four consultations with Cierva in America), but after Cierva's death, a distancing had taken place. The English engineers seem to be very hesitant about including Pitcairn's colleagues Agnew Larsen and Paul Stanley, who were then resident in England—a hesitation that soon descended into resentment and then outright hostility. Additionally, Dr. J. A. J. Bennett, Cierva's successor, seemed unwilling to share the results of current research, effectively refusing to respond to Pitcairn's legitimate inquiries. Pitcairn had a vital interest in this work, as the Autogiro Company of America had rights to the work of the English company, which influenced and was then incorporated into and its

Agnew E. Larsen and Harold F. Pitcairn with the aerodynamic cyclic/collective pitch rotor head of the PA-36 Autogiro, the patented control system used in later helicopter development. (Courtesy of Stephen Pitcairn, from the Pitcairn Archives.)

own work. It was now evident that the collaborative spirit and practices that existed during Cierva's lifetime were a thing of the past.

By February 1937 Pitcairn, concluding that the relationship between the English company and his associates then in England had deteriorated to such an extent that it no longer made any sense to maintain a presence, recalled Larsen and Stanley. However, he asked Jim Ray to remain, in part because of Ray's cordial personal and professional relationship with chief test pilot Reggie Brie. In this manner Pitcairn hoped to gather information on what was happening in Europe. The news was not encouraging; Ray, observing that the Bennett engineering group seemed to have little to do after the departure of the American observers, complained that he had little to occupy his time and thought it appropriate to return home. Pitcairn, however, asked him to gather information on the French and German helicopter programs that were already fueling an active rumor mill. In this case the German rumors proved to be founded and ominous. Ray transmitted the information he had gathered on the achievements of Anton Flettner and, after spending several weeks in Bremen, a confirmation of the seminal work of Focke-Achgelis & Company's Fa-61 helicopter.[1] The Germans had been most anxious to show off the latest aviation development and had willingly demonstrated the Fa-61 from a distance. But in the Fall of 1937 the Fa-61 had been flown before the world's most famous living aviator, Colonel Charles A. Lindbergh, then a personal guest of Hermann Göring, who wanted to proclaim the advances in German air power. Lindbergh's reports of the vertical flight achievements of the Fa-61 were insightful and were perceived as accurate, authoritative, and alarming.

To Pitcairn, however, whatever the comfort derived from the Lindbergh observation that this was a prototype capable of lifting only its pilot, the most problematic and troubling report was that the Fa-61 had used some of Cierva's supposedly closely guarded patents to achieve effective collective pitch control. It had become immediately and painfully obvious to Pitcairn that the company on whose board he served had, at best, not kept him informed of licensing agreements with Focke (and, as it turned out, with Flettner) and, at worst, actively betrayed Autogiro development. (Indeed, on October 23, 1964, Professor Focke delivered the Fifth Cierva Memorial Lecture, in which he paid tribute to Cierva's work as being *the* significant contribution to the eventual development of the helicopter. Igor Sikorsky would later state that Cierva had shortened the development of the helicopter by ten years.) Pitcairn immediately departed for England and a confrontation with his fellow board members and with Dr. Bennett.

The board of directors of Cierva Autogiro Company, Ltd. directly informed its American member that the Fa-61 had been developed under license from Cierva Autogiro of its cyclic/collective pitch control rotor hub *and that in a cross-licensing arrangement Cierva Autogiro had received a license to build Focke-Achgelis helicopters!* Pitcairn was stunned—it was obvious that, with the Cierva Autogiro Company seeking to form a consortium with Focke-Achgelis and the French Bréguet Company to build Fa-61-derived aircraft, the development of Autogiros in Europe was, if not ending, receiving a major setback. Far from being discouraged, however, he returned to America rededicated to development of the jump takeoff and convinced that its development would henceforth be an American enterprise. This, for all practical consideration, effectively ended the relationship between the American and

Pitcairn PA-36 Whirlwing Autogiro in 1939. (Courtesy of Stephen Pitcairn, from the Pitcairn Archives.)

English companies, and Pitcairn would proceed alone with the development of PA-36, considered by many to be the most beautiful Autogiro ever built. But it would be the last Autogiro designed by the original Pitcairn company.

Kellett was also active, and it had received much publicity when a KD-1A accompanied the MacGregor Arctic Expedition, sponsored by the United States Weather Bureau during the summer of 1937.[2] The Autogiro began flying in the Arctic, piloted by Navy Lieutenant Commander I. Schlossbach, in November of 1937. In May 1938 R. Johnson had his arm broken by the propeller, a reminder that Autogiro aviation could still be hazardous. The rotary-wing flights of the Kellett aircraft ended with the conclusion of the expedition on July 7, 1938. It was the first such aircraft to fly in the Arctic, beating a TsAGI Russian A-7 autogyro, which did not fly in that part of the world until later in 1938.

GERALD HERRICK: HV-1, HV-2A

There had also been Autogiro research and developments in the late 1920s and throughout the 1930s in America, Germany, and Russia that, while not significant for mainstream technological progress, would prove influential thirty years later in the development of the Fairey Rotodyne, the ultimate but doomed Autogiro achievement of the early 1960s.

The first of these, the work of Pitcairn's fellow Philadelphian Gerald Herrick, was the HV-2A Convertaplane,[3] an initial attempt to combine fixed- and rotary-wing flight (he called his various iterations Vertoplane, Convertiplane, Convertoplane, and the generic Convertaplane). Assisted by Ralph Herbert McClarren, then with the Franklin Institute of Philadelphia, Pennsylvania, Herrick and his associates sought to combine the best features of fixed-wing flight and the Autogiro. He had carefully considered Cierva's developments and by early 1931 had decided that while the safety of autorotation was obvious, the problem was the relative lack of efficiency in horizontal flight. His ingenious solution, developed after much wind tunnel experimentation, was a symmetrical airfoil, mounted on a central rotor pylon that allowed aerodynamic adjustment for control. This was, in essence, a biplane with a two-bladed single cantilever upper wing—the Herrick Convertaplane took off as a conventional biplane and then converted into a gyroplane, with the upper wing rotating to provide lift on a central pylon. Herrick's initial design, the HV-1,[4] dubbed the Vertoplane (X11384), flew for the first time on November 6, 1931. It was powered by a tiny three-cylinder 48-horsepower Poyer engine. Taking off in biplane mode later in the first attempt to convert from biplane to gyroplane, the pilot released the upper wing in transition to autogyro mode to descend, but the aircraft vibrated uncontrollably and almost immediately dove to the ground in a crash that killed its pilot.

Undeterred, Herrick immediately set about designing a new model. By 1936, after much redesign and experimentation, he had constructed the Herrick HV-2A (X13515). Called the Convertaplane, this was a much more sturdy craft with significant design improvements. The upper wing/rotor was now reduced in size (to twenty-four feet) in comparison to the lower wing (twenty-eight feet), and an electric motor was employed to start the engine. Flight-testing of the biplane, with full cantilever wings but lacking struts or wires between the wings, began in October of 1936 at Boulevard Airport in northeast Philadelphia with a pilot more distinguished by his drinking and carousing than dedication to the project. Herrick soon replaced him with George Townson, who had been active with Pitcairn Aviation. The gross weight of the aircraft was 1,700 pounds, and the engine was the 125-horsepower, air-cooled, five-cylinder Kinner. After the HV-2A flew satisfactorily as a fixed-wing craft, the flight-testing turned to the autogyro (rotating wing) mode.

The HV-2A was not able to spin up the upper wing (rotor) mechanically, as no connection had been made to the engine. In the air, the upper wing would be released to rotate, initially powered by several five-eighths-inch rubber bungee cords inside each upper wing half and running through an aluminum tube. Each of these wing cords was connected to a cable that wound around a spool. Prior to takeoff, two people would grasp each wingtip and walk twice around the central pylon in the opposite direction to autorotation. The upper wing was then locked in the biplane cantilever wing position. When the pilot released the lock, the bungee cords would cause the wing to rotate for two turns at 60 rpm. The now-spinning wing would then rotate freely and the flow of air through the disk would increase its speed to 220 revolutions per minute in autorotation.[5]

Understandably, the first tests of the HV-2A's autorotation abilities were not in conversion from horizontal flight but in takeoff as an autogyro. There, bungee cords could not provide sufficient rotation to achieve flight, so the HV-2V was taxied around the perimeter of the airport with the flow of air slowly increasing the rpm of the rotor. Takeoff was achieved with 180 rpm, and once airborne, autorotation increased to 220 rpm. The rotor-craft flights proved the abilities of the HV-2A in autorotation flight, so the decision was made to attempt a mid-air conversion. However, in both auto-gyro and biplane modes, higher-than-expected drag was noted as well as a tendency to veer to one side. Although the former was never solved, the latter was controlled by pilot technique.

For safety the first conversions from biplane to autogyro were at low level—and were successful. A public demonstration on July 30, 1937, gained national publicity[6] when the media photographed the air-to-air Convertaplane conversion at 1,500 feet, and the inventor was heartened by expressions of interest by United States Navy, but no funds were provided for further development. With the public demonstrations in Germany of the helicopter and the coming of World War II, interest dimmed in Herrick's vision although he continued to design (but not construct) more sophisticated machines, earning him the title of "dean of convertible aircraft designers." The HV-2A made more than 100 air conversions prior to retirement to the Smithsonian Institution's Silver Hill Restoration Facility in 1954. While he was the first, and for many years the only, designer of a successful Convertaplane, such was the power of this conventional/rotating wing combination that other designers and innovators were also attracted to it at the time. While Herrick remains the most well-known, notice should also be taken of the Russian I. P. Bratukhin and the German Anton Flettner.

IVAN PAVAL BRATUKHIN: TSAGI 11-EA

By the mid-to-late 1930s Russia, aviation designer Professor Ivan Pavel Bratukhin already had several years of rotorcraft design experience. The Soviet Union had made a major commitment to the study and development of aircraft with the creation of the Tsentralnyi Aero-gidrodinamicheskii Institut (Central Aero-Hydrodynamics Institute), the TsAGI. It was the first time a scientific institution combined basic studies, applied research, structural design, pilot production, and testing of aircraft.

Bratukhin, born in the village of Yaschera (in the modern-day Kirov region) on February 25, 1903, had become a member of the Communist party at the age of seventeen, when the struggle for control of Russia was still in progress. When TsAGI set up a helicopter research section under Boris Yuriev, Bratukhin briefly joined it in 1926, but he soon left for additional studies at the Bauman Technical School in Moscow, from which he graduated in 1930. Returning to TsAGI, he was placed in charge of a brigade that developed the 11-EA between 1936 and 1938. This model sought to combine the capacities of helicopter, autogyro, and fixed-wing aircraft, but it failed to accomplish its task and its development fell victim to politics, a fate later suffered by the Fairey Rotodyne.

Under conceptual development since 1933, the Bratukhin-designed 11-EA (for *Experimentalnyi Autozhir,* or "experimental autogyro")[7] was, under conceptual development since 1933, constructed in 1936. The 11-EA had the appearance of a conventional two-passenger aircraft with an Autogiro pylon topped with a six-bladed rotor consisting of three shorter rigid blades capable of feathering (changing pitch) and three longer, articulated blades. As this craft was to take off as a helicopter, Bratukhin placed counter-torque propellers (rotating in opposite directions from each other) on the forward edge of each wing, which would push and pull in opposite directions in helicopter (hovering) mode. In forward flight, however, both propellers pulled forward even as the rotor was unloaded as an autogyro. The 11-EA had a streamlined fuselage and was powered by a 630-horsepower Curtiss Conqueror engine mounted in the forward part of the fuselage, with a large fan-equipped radiator in front. Tethered flight-testing began in 1936 and continued until the next year. These tests, only in limited helicopter mode, revealed control problems due to the complex six-bladed rotor. However, external circumstances—including Stalinist purges, forced relocations of key personnel, and fear of being accused of sabotage if tests encountered difficulties—slowed testing, and subsequent developments by 1937 doomed this compound aircraft.[8]

The reluctance to advance to full, untethered flight-testing in 1937 was dictated by the official retribution for failure increasingly exacted by Soviet dictator Joseph Stalin, who saw treachery in such lack of success. Bratukhin and his associates were constantly confronted with reports of arrests of fellow engineers who had failed to deliver the desired results. As precisely observed by Lennart Andersson in *Soviet Aircraft and Aviation 1917–1941,* "No one wanted to risk making a mistake and be accused of sabotage."[9] Indeed, Bratukhin's test pilot, Aleksei Cheremukhin was arrested and imprisoned with other bureau colleagues.[10] By late 1938 the political climate was such that it was judged improvident and unacceptably risky to continue the development of this compound aircraft. The 11-EA was rebuilt between late 1938 and December 1939 as the 11-EA-PV, a pure helicopter version, with the wings replaced by framework booms with auxiliary rotors on each side for torque control. It began flight-testing in October 1940, and no one mourned it when it last flew in 1941. Thus did the 11-EA fade, having failed to realize a potential of combining the benefits of its individual components—that would wait for the Fairey Rotodyne in 1957. But, as will be seen, Bratukhin would make a contribution in the early 1950s in the eventual design of the Kamov Ka-22, the *Vintokrulya* ("Screw Wing"), the *Russian* Rotodyne.

ANTON FLETTNER: FL 184, FL 185, AND FL 265

That the German Anton Flettner[11] is little remembered as a pioneering designer of rotary-wing aircraft is no doubt, in some measure, occasioned by the rather consistent destruction or abandonment of his models! His prototype helicopter of 1932, distinguished by the placement of a small engine and tractor propeller on each blade of a two-bladed rotor (thus effectively avoiding with the issue of torque derived from a airframe-mounted engine) was destroyed shortly after a successful tethered flight when it overturned in a storm. Flettner's next design, a two-seat Autogiro dubbed the Flettner Fl 184 (D-EDVE)[12] was constructed with a three-blade rotor and tractor-propeller-powered 140-horsepower Siemens-Halske Sh 14 radial engine. This machine was also destroyed when, in

September 1936, the prototype crashed while making a left turn in preparation for landing into the wind. It was later determined that an incorrectly set stabilizer, which could not be controlled during flight, had forced the model into a seventy-degree dive from about 330 feet.[13]

Flettner then created a combination Autogiro/helicopter, the Fl 185 (D-EFLT).[14] The machine was designed to take off as a helicopter with a rotor powered by a 140-horsepower Siemens-Halske Sh 14 engine equipped with a cowl and frontal fan for cooling. The engine transmitted power to the rotor and two variable-pitch airscrews mounted on outrigger arms extending from the fuselage by means of a gearbox located just behind the cowling. In helicopter mode, the airscrews rotated in opposite directions, thus effectively resulting in antagonistic thrust designed to counter rotor-torque. But with a free rotor in autorotation, the pitch of the airscrews could be altered to provide forward thrust as they accelerated, with full power then redirected from the rotor. Unfortunately for Flettner, this prototype was abandoned after only a few test-flights, as he bowed to official policy and concentrated on a pure helicopter (a course of action similar to that that would be taken by Bratukhin two years later).

Flettner would go on to develop the best helicopter in World War II, the "synchropter," a machine suggested by the work of Dr. J. A. J. Bennett, based on two off-set intermeshing counter-rotating blades. It was Flettner's dual-rotor design that gave rise to the term *eggbeater*.

FOCKE-WULF FW 186 JUMP TAKEOFF AUTOGYRO

In Germany, Focke-Wulf Flugzeubau GmbH of Bremen held a Cierva license to manufacture the C.30, but in 1936, the company developed a jump takeoff autogyro,[15] the Fw 186. Focke would leave the operational company in 1933, forced out by the Nazis, who questioned his political reliability,[16] and he was succeeded by Kurt Tank. Focke, who with Gerd Achgelis formed the company that bore their names, had already successfully flown the Fa-61 on June 26, 1936, but the Focke-Wulf company, in a manner similar to Pitcairn, had committed in 1936 to develop a jump takeoff autogyro as an alternative to the newly developed helicopter. This was to meet German military requirements for a short takeoff liaison, reconnaissance, and light transport aircraft. The company had already produced about thirty C.30s under its license by 1938, but development of the jump takeoff model, not under a Cierva license, proceeded in secret as part of the German rearmament program through 1937. The prototype[17] FW 186

(D-ISTQ) flew in July 1938, but this two-seat tandem open-cockpit aircraft was not successful. It featured a three-blade rotor mounted on a cantilevered Kellett KD-1-type pylon, with its jump takeoff mechanism supposedly based on principles developed by England's Dr. J. A. J. Bennett.[18] The first German jump takeoff model, its performance was found lacking. The German military authorities, then actively planning for war, abandoned development in favor of the Fa-61 helicopter, a technological shift that would occur in 1940 in America when public funds were withheld from Autogiro development in favor of an Fa-61-type helicopter proposed by an upstart company. Focke-Achgelis also developed the Fa-225, combining a glider fuselage with an Autogiro rotor system, effectively creating a rotary-wing glider, but it was never utilized.[19]

Each of these—Herrick, Bratukhin, Flettner, and Focke-Wulf—attempted significant developments in autorotational technology. Had they succeeded, each would have guided that technology down other paths, but they did not. Yet the impressive technological achievement embodied in the Fairey Rotodyne during the 1957–62 time frame vindicates their work. Although the Rotodyne merits extensive discussion later in this book, it was primarily with Pitcairn and the Autogiro Company of America and Kellett Autogiro Company that technology continued toward an inglorious end. But first Pitcairn would produce the most beautiful Autogiro ever constructed—the PA-36.

Pitcairn had already developed jump takeoff capability in the PA-22, his flying-test platform, and had returned from Europe in February 1937 committed to the development of a new Autogiro that would be the state-of-the-art. And there was the growing imperative of the changing world situation. Furthermore, of greatest importance to rotary aircraft development, the Fa-61, the German helicopter, was then publicly demonstrated. This aircraft, initially flown by test pilot Ewald Rohlfs for just twenty-eight seconds on June 26, 1936, weighed just over 2,100 pounds and was powered by a 160-horsepower Bramo radial engine mounted in the front of the fuselage. It superficially resembled an Autogiro as there was a wooden propeller in the front center, but its blades had been cut down to the size of the cylinders and its only function was to cool the engine that powered the two three-bladed rotors, which were mounted on lateral outriggers on either side of the fuselage. Its forward motion was derived from the rotors, and as such, it is often credited as the first "practical" helicopter. For although there was seemingly little public awareness of the speed records of seventy-seven mph and altitude achievement of 7,800 feet set in June of 1937, the public took sharp notice when the world's first female helicopter pilot, Hanna Reitsch (dubbed the German "Amelia Earhart" and recently given the honorary rank of Flugkapitän [Flight Captain] in recognition of her many research flights in gliders and warplanes), flew the Fa-61 inside the Berlin Deutschland-Halle,[20] a large meeting hall, before thousands of spectators in February 1938. It has been claimed that the spectators consumed so much oxygen that it reduced the Fa-61's engine power, necessitating airing of the arena during the demonstration flights.[21] Chosen in part for her photogenic and propaganda appeal and perhaps in equal part for her petite, slim stature, given the limited lifting ability of the small helicopter, she gave an inspired performance. Pictures of her controlled indoor flight stunned the world in general, and aviation designers and military leaders in particular—including American army aviator First Lieutenant H. Franklin Gregory, who would become instrumental in rotary-wing development—and insured that rotary development funds went to the helicopter and not the Autogiro.

Military leaders, stunned when Hitler had revealed the new *Luftwaffe* in 1935 as a prime image of a resurgent Germany, now saw with the Fa-61 evidence of secret technological progress with the newest aerial weapon. Years later, when the female helicopter pilots would found an international organization called the Whirly Girls, Reitsch fittingly became Whirly Girl #1, a recognition that has lasted longer than the dubious honor she achieved at the end of World War II. Toward the end of the war, Reitsch, then a military test pilot, flew with a male test pilot through Russian flak to visit Hitler in the Berlin Bunker. The legend, which then circulated, that Reitsch had subsequently flown Hitler to safety, resulted in her imprisonment and intense questioning by Allied intelligence officers after the war's end.[22] But while that was years away in early 1937, Pitcairn was certainly aware of the military buildup then beginning in Europe and its implications for America. It would lead into a legislative lobbying effort designed to produce government support for his new model. Pitcairn's older brother Raymond was well-connected in local and national politics, so it was logical to turn to an aviation-knowledgeable and sympathetic member of the Pennsylvania congressional delegation, Representative Frank J.G. Dorsey. Pitcairn requested that the congressman sponsor a bill to fund experimental Autogiro development *for the military*, and after drafting, Dorsey introduced H.R. 8143 in early 1937, thereafter known as the Dorsey Bill. The bill, reflecting the *specific* concerns of his constituents, was titled "To Authorize the Sum of $2,000,000 for the Purpose of Autogiro Research, Development and Procurement for Experimental Purposes." Congressional hearings by the House Committee on Military

Pitcairn PA-36 Whirlwing Autogiro, making a jump takeoff in 1941, flown by Frederick "Slim" Soule. (Courtesy of Stephen Pitcairn, from the Pitcairn Archives.)

Affairs were scheduled early in 1938, and Pitcairn resolved to have a flying model of the PA-36 ready to convince the legislators.

The Kellett Autogiro Company expressed great interest and it was apparent that the two would be rivals for the government funds. It had, after all, already received a contract to supply seven KD-1s, dubbed YG-1A/Bs by the military, to the Army Autogiro School, established on April 15, 1938, for pilot training and field-testing under the direction of First Lieutenant H. Franklin Gregory *at the same time that Pitcairn had closed his production line.* This rivalry had been simmering for several years as Kellett rankled over the superior position of the Autogiro Company of America as *the* Cierva licensing agent and patent repository in America, and Pitcairn retained resentment of the Kellett independent consultations with Cierva on direct-control. And there was the lingering issue of the international sales of Kellett K-3s—Pitcairn continued to maintain that this was a violation of the ACA licensing agreement. But more importantly, and of greater relevance to the pending legislation, Pitcairn felt that the army testing of K-3s was not in the best interest of the Autogiro. He had faith that the future of Autogiro technology would be better demonstrated by what he termed the "third generation"—the direct control, jump takeoff PA-36.

It was not surprising, then, when it became known that Kellett was also working on the collective pitch control system necessary for jump takeoffs under the direction of its chief engineer Richard H. Prewitt.[23] Prewitt had previously visited Cierva in late 1933 to discuss direct control, a meeting viewed with continuing suspicion by Pitcairn and his associates. Pitcairn had always taken legal steps to protect his discoveries and patents, and he now took other steps made necessary by the perceived rivalry. Engineers were admonished to exercise caution—all papers were to be accounted for; locked desks and safes were the order of the day; and all public statements had to be screened by legal counsel, approved, and then only repeated, not elaborated upon. No hints that might help the Kellett engineers were to be given, even in a casual conversation.

Agnew Larsen, who had been placed in charge of the PA-36 development program, expressed confidence that he could build the prototype for $50,000 and by the time of the Dorsey Bill hearings. He would be proven wrong on both counts. Jim Ray had asked to return to America from England, but as the company had no flying prototype, there was no piloting work and Pitcairn felt it would be more productive for Ray to continue gathering information on European rotary-flight developments. The company would require Ray's piloting skills when the prototype was ready for testing, but again events would overtake the parties and Ray would not, in fact, be the test pilot. The Dorsey Bill hearings began on April 26, 1938, but Ray remained in Europe, as the PA-36 development had slipped badly and Pitcairn did not even have a mock-up, much less a flying prototype. But that did not deter Pitcairn, who appeared as the undeniable leader of the Autogiro community in America by virtue of the ACA and the Pitcairn Autogiro Company—he had meticulously prepared a booklet illustrating Autogiro development and touting the PA-36 as the latest model. But he had misestimated the sentiments of those who appeared before the committee; the hearings would prove a disaster to the future of Autogiro development in America.

There is no doubt that those who testified during the two days of Dorsey Bill hearings were influenced by the images from a few months earlier of Hanna Reitsch flying the Fa-61, and this clearly showed from their comments. Assistant Secretary of the Navy Charles Edison flatly stated that the navy, having already evaluated the Autogiro, was not interested in rotor-craft unless it possessed ability to hover. Then Professor Alexander Klemin, dean of the Guggenheim School of Aeronautical Engineering of New York University, who had made known a favorable view of the Auto-giro, cited the recent Hanna Reitsch Fa-61 flights and ventured the seem-ingly innocent opinion that the language in the Dorsey Bill, with its use of the word *Autogiro,* might reasonably be construed to include all types of rotary-wing aircraft. And, he added, that being the case, perhaps this implicit meaning should be directly stated to include all such rotary-wing aircraft. No one present, especially Harold Pitcairn, took much notice of this redefinition of the bill's terms, but the inclusion of all types of rotary-wing aircraft would now encompass helicopters, surely relevant to First Lieutenant H. Franklin Gregory, who was of such a lowly rank that there was no possibility he would be called to testify before the committee. But he was perhaps the most experienced army officer in rotary-wing, aircraft based on his experience with the Army Kellett K-3 models, and his rela-tively junior officer status had not prevented the army from assigning him to command its Autogiro School nor would it prevent his subsequent assignment to administer the Dorsey Bill funds.

Witness after witness, citing potential uses, enthusiastically testified as to the benefits of rotary-wing flight, and by the time that Pitcairn testified it was apparent that the Committee members were favorably disposed to the Dorsey Bill. As it was near the end of the second day, Pitcairn's testi-mony was highly abbreviated, and having placed his prepared Autogiro booklet in the record, he soon left after answering a few questions, appar-ently satisfied with the results. But he failed to understand the gravity of the opinions that had emerged, namely that rotary-wing flight would be of benefit, not specifically *Autogiro* flight. He further failed to realize that two of the questions asked were seminal to a far different result than he had anticipated. He had admitted to the committee that he had spent $3,250,000 in Autogiro development since 1928 and that the PA-36, the end product of that investment, did not even yet exist in prototype. Although he was apparently satisfied with the hearings, it is clear that the final impression was that although rotary-wing flight would prove of ben-efit, those benefits were not likely to emerge from the Autogiro. It is not surprising, then, that the committee took only ten minutes after the close of the hearings to substitute in the bill's title "Rotary Wing and Other Air-craft" for "Autogiro." Although this substitution may have seemed innocuous, it would doom the Autogiro, for First Lieutenant H. Franklin Gregory would insure that no government money would flow to the Pitcairn PA-36 development or procurement from the Dorsey Bill, which was passed by the House, sponsored by Senator Logan in the Senate, and signed into law by President Roosevelt as the Dorsey-Logan Act on August 1, 1938.

Anticipating accelerated development of the PA-36, Pitcairn cabled Ray and requested that he return. It would not be a happy homecoming, and "Big Jim" would soon leave the company, to all intents and purposes let go for economic reasons, as the development of the PA-36 prototype dragged on. But a contributing fact was Ray's opinion, based on his observations of European helicopter development, regarding the future directions for Pitcairn Autogiro Company.

Ray returned to America on October 26, 1938. He had traveled on the same ship that brought Austrian designer Raoul Hafner, who was going to attend an international symposium on rotary-wing aircraft jointly sponsored by the Philadelphia chapter of the Institute of Aeronautical Science and the Franklin Institute. It had been arranged by engineers Ralph Herbert McClarren and E. Burke Wilford, inventor of the Wilford Gyroplane, which was distinguished by its use of non-Cierva technology. Such was the significance of this symposium that both Heinrich Focke and Louis Bréguet, invited but unable to attend, asked Hafner to read their scientific papers. Ray attended with associates Paul Stanley and Agnew Larsen of the Autogiro Company of America and even delivered an impromptu but well-received account of his European observations. In light of subsequent developments, it was unfortunate that Harold Pitcairn, ill and unable to attend the symposium, was forced to rely on the observations of Larsen, Stanley, and Ray, who did attend. Even if he had heard the presentations by such rotary-wing luminaries as Richard H. Prewitt of Kellett Autogiro, Havilland D. Platt, W. Laurence LePage, E. Burke Wilford and Gerald Herrick, it is unlikely that events would have worked out differently, but Pitcairn might have at least glimpsed the future.

LePage had first met Harold Pitcairn while on loan from England's National Physical Laboratory to the Massachusetts Institute of Technology to assist them in developing wind tunnel testing programs during the time that Pitcairn had contracted for testing of his fixed-wing aircraft designs. LePage had previously participated in the testing of Cierva models performed by Vickers at Weybridge in January and February of 1925, and it is likely that he discussed his experiences with the Cierva models with Pitcairn. He then decided to remain in America and had joined the Pitcairn company and worked on the PCA-1 series.[24] LePage subsequently resigned shortly after the Pitcairn Company changed its name to Autogiro

Company of America, because he failed to become chief engineer. Although he had been originally hired to assist Geoffrey Childs in administrative and technical matters, he had been active in the development of the PA-1 Fleetwing airmail plane and later gained advanced experience with rotary aircraft in the development of the first American models. He became an aviation engineering journalist after leaving Pitcairn, but soon joined the rival Kellett Autogiro Company and was active in its design efforts through the successful development of direct control. After leaving Kellett with a decline of the Autogiro market in 1937, he found himself a member of a new breed of innovators, an "independent consulting engineer" making use of his extensive theoretical and practical experience in rotary-wing aircraft. It was then that LePage encountered another independent, Havilland D. Platt, a talented mechanical engineer from New York who had gained fame as the inventor of the first successful automatic automobile transmission.[25] It was to prove a fortuitous meeting, as they would go on to form the Platt-LePage Helicopter Company—and it was to that company that the Dorsey-Logan funds would go, channeled by H. Franklin Gregory.

Gregory would later write in 1944 that "in the Autogiro we had seen a way to vertical flight and to many of us it seemed the next step toward the helicopter. Primarily this was the reason for the Army's gyro school and its exhaustive research into rotary-wing aircraft. Thus one of my main tasks in the new job at Wright Field was to look for a successful helicopter,"[26] but these sentiments were apparently not evident to Pitcairn.

Pitcairn, having recovered several weeks after the symposium, hosted a belated homecoming for Jim Ray in his Bryn Athyn home. Ray was forthright and blunt in his observation that the Fa-61 and Flettner machines clearly pointed towards helicopter development, but he went further and asserted that continued development of the Autogiro was ill-advised. Pitcairn, Stanley, and Larsen were stunned when Ray recommended that the PA-36 jump takeoff Autogiro be discontinued and the company's efforts be redirected toward helicopter development. It was unacceptable advice, both personally, as Harold Pitcairn was emotionally committed to Autogiro development and had been for over a decade, and economically, as the company had an irreversible financial investment in its Autogiro research program. Both Stanley and Larsen, who had also been at the Franklin Institute seminar, disagreed with Ray, and it was apparent the company could not accommodate both opposing views. Jim Ray had to go and in fact was terminated at the end of 1938,[27] and although Pitcairn would characterize the decision as economic belt-tightening because of the Depression, Ray justifiably "attributed his termination to this disagreement concerning the continuation of the Autogiro program."[28] Ray, as well as Lou Leavitt (formerly Levy), would later take part in the flight-testing of the Platt-LePage XR-1 helicopter, developed with the Dorsey-Logan funds.

Although Larsen had promised that the PA-36 prototype would be flying by the April Dorsey Bill hearings, it was only ready for ground testing almost six months later. But even given the delay, Pitcairn's faith did not waiver, for the PA-36, with its gleaming aluminum fuselage, was beautiful. That faith would remain, even though he was informed by the English company in late December 1938 that it had decided to abandon the Autogiro business and concentrate on helicopters, and in the face of increasing American military focus on the helicopter. Pitcairn was still a director of Cierva Autogiro Company Ltd. but had not participated in the board's decision, so he traveled to England to clarify the future directions in Europe and inadvertently missed the 1939 Franklin Institute rotary-wing gathering, at which the army's H. Franklin Gregory, now promoted to captain, met Russian inventor Igor I. Sikorsky. It was to prove a fateful meeting for the future of rotary-wing flight in America.[29]

For construction of the aluminum PA-36,[30] Pitcairn had turned to the Trenton, New Jersey, Luscombe Airplane Company,[31] which was experienced in such construction. Pitcairn first made a PA-36 wood mock-up, which was used by Luscombe to fit all aluminum fabricated parts before the actual assembly. This was a difficult period for the Trenton company, and the Pitcairn commission proved a lifesaver, as the monies paid for the PA-36 were virtually the only major earnings during this period, but it was not easy money. An aluminum Autogiro had never been built, and the Luscombe and Pitcairn engineers faced unique engineering challenges. Although the PA-36 "borrowed heavily from the PA-35,"[32] it became an aluminum, two-place, side-by-side cabin Autogiro, powered by a Warner Super Scarab 165-horsepower engine, and weighed in at 2,050 pounds, with a forty-three foot three-bladed jump takeoff rotor.[33] The engine, as in the PA-35, was buried in the fuselage in back of the cabin in order to make the aircraft "roadable" but, of the two bodies constructed by Luscombe, only one ever flew and it was never roadable.

At Luscombe the initial fabrication design work was completed by William B. Shepard. Those designs were approved by Pitcairn's Agnew Larsen, and Luscombe hired a night shift to expedite the actual construction, presumably in response to demands by Larsen, who was already falling behind his promised production schedule. The first metal was cut in late October, and almost immediately unique engineering challenges emerged. The design itself was incredibly ambitious, and it was readily evident that no other Autogiro in history had achieved such standards. The component parts of the metal seats were to join with the inner cabin fuselage in a seamless manner to form the fuel and engine oil tanks. It required production to an exceptionally precise tolerance, a task made increasingly difficult by the changes in the angle iron jigs used to fabricate the aluminum. The jigs would expand or contract depending on whether it was day or night—it proved impossible to correct the constant day/night error of one-eighth inch, and production could continue only when the parties and a department of commerce inspector agreed that such a small error was acceptable.

The fuel tank proved especially a challenge for Luscombe—due to its construction, the tank was actually part of the fuselage and the metal joints forming the tank had to be first sealed with neoprene tape and then riveted to the fuselage. But when the fabricated tank was vibration-tested by Pitcairn, it proved unsound. The tank itself fell apart as seals failed, leading to a dangerously unbalanced and almost immediately unstable structure. Larsen, then under considerable pressure from Pitcairn, declined to redesign the tank with Luscombe but arranged to have a second tank built by the Fleetwings company, located nearby Bristol, Pennsylvania. That tank passed the vibration test[34] and was successfully installed in the aircraft. Luscombe records reveal that the construction of the PA-36 was hard work—the Autogiro bulkhead panels had to be hand-formed from duraluminum over maple wood forms and then sent to a second factory to be heat-treated. It was a delicate process, for too much heat would melt the metal, but too little would result in an unacceptably brittle product.

Aware of the importance of the Pitcairn work for the company, Luscombe supervisory personnel tried to maintain cordial relations with Larsen all through the complex design and fabrication process, and for the most part it appears they succeeded. But it was a different picture with the shop personnel, who often felt the pressure during Larsen's frequent visits—they referred to the PA-36 as "Larsen's Goon" behind his back.[35] At least once the workers' attitude resulted in a practical joke that has become part of Luscombe history, but not apparently of Pitcairn's. Employee Henri D'Estout filled a workman's glove with water and attached it to bottom rear of the aluminum aircraft's fuselage. And while the employees laughed at the appearance of an "udder" on the Pitcairn, it was never determined if Larsen shared the joke as he was making an unexpected inspection visit at the time.

The PA-36 had a large central tail fin and outward-slanting fins on the tail stabilizers on either side of the fuselage. The original propeller configuration featured two single-blade counterweighted counter-rotating propellers, but this resulted in unacceptable vibration, and the two propellers were replaced by a single four-blade propeller. The final model, dubbed the Whirlwing, was unveiled for flight-testing in the early spring of 1939. As Pitcairn had let Jim Ray go, the company turned to former Kellett chief test pilot Lou Leavitt and experienced Cierva test pilot Alan Marsh, who had come from England to aid in the testing. Marsh had had experience with the English company in jump takeoffs but Leavitt did not, and that presented an immediate problem that resulted in the replacement of the Kellett pilot. The PA-36 prototype was too heavy to allow for jump takeoffs with two pilots, so Marsh could not instruct Leavitt, and the former was unwilling to make the attempt. He and Pitcairn quarreled about the readiness of the prototype for jump takeoffs; it became apparent that although the unproven prototype might be ready, the veteran pilot was not. Leavitt stormed out, leaving Pitcairn high and dry, leading to a somewhat desperate search for a pilot, as Pitcairn was counting on demonstrations before military officials to produce government orders. He first approached Jim Ray, who had experience with jump takeoffs in the PA-22, but understandably received a cool reception. Ray had just started a new job with a West Coast airline and had no intention of returning to a company that had let him go. Fortunately, there was another experienced Autogiro pilot nearby in Bloomfield, New Jersey, performing aerial applications for the treatment of Dutch elm disease for the department of agriculture—Frederick W. Soule. Known to his friends as "Slim" because of an alleged resemblance to Charles Lindbergh, Fred Soule had originally been instructed by Pitcairn factory pilots Ray and Jim Faulkner and by mid-1939 had accumulated thousands of hours flying a variety of Autogiro models. In July of that year Alan Marsh, before he returned to England, which was then facing war in Europe, briefed Soule in jump takeoff techniques, and the stage was set for flight-testing.

NOTES

1. For photographs of the Fa-61 (also called the Fw-61), see Hollingsworth Franklin Gregory, *Anything a Horse Can Do: The Story of the Helicopter*, Introduction by Igor Sikorsky (New York: Reynal & Hitchcock, 1944), p. 59; Earl Devon Francis, *The Story of the Helicopter* (New York: Coward-McCann, Inc., 1946), facing p. 70 (photo by W. Laurence LePage); Martin Hollmann, *Flying the Gyroplane* (Monterey, California: Aircraft Designs, Inc., 1986), p. 29.

2. Peter W. Brooks, *Cierva Autogiros: The Development of Rotary-Wing Flight* (Washington, D.C.: Smithsonian Institution Press, 1988), p. 231.

3. For photographs of the Herrick aircraft models, Walter J. Boyne, *The Aircraft Treasures of Silver Hill* (New York: Rawson Associates, 1982), pp. 140–42; Charles Gablehouse, *Helicopters and Autogiros: A Chronicle of Rotating-Wing Aircraft* (Philadelphia: Lippincott, 1967), p. 65; George Townson, "The Herrick Convertaplane," *American Helicopter Museum & Education Center Newsletter* 4, no. 3 (3rd Quarter 1997): 3–4; Walter J. Boyne and Donald S. Lopez, *Vertical Flight* (Washington, D.C.: Smithsonian Institution Press, 1984), p. 175.

4. Brooks, pp. 23–26 (photo on p. 25).

5. For photographs of the HV-2A in conversion from fixed-wing to gyroplane flight, see Lieutenant Victor Haugen, "Principles of Rotating Wing Aircraft," *Aeronautics* 2, no. 7 (October 16, 1940): 429; Brooks, p. 25.

6. See, for example, "Vertaplane," *Time* 30, no. 6 (August 9, 1937): 21–22.

7. Everett-Heath, in an otherwise outstanding book, comments on the 11 EA: "Why the word 'autogiro' was used to describe a helicopter is not known," seemingly oblivious to the origins of the design and ignoring that the aircraft was intended as a convertaplane, taking off and landing as a helicopter but flying as an autogyro. See John Everett-Heath, *Soviet Helicopters: Design, Development, and Tactics* (London: Jane's Publication Company, 1983), p. 5.

8. Ibid., pp. 11–12.

9. Lennart Andersson, *Soviet Aircraft and Aviation 1917–1941* (Annapolis, Maryland: Naval Institute Press, 1994), p. 335.

10. Everett-Heath, pp. 146–47.

11. For a discussion of Flettner, see Roger Ford, *Germany's Secret Weapons in World War II* (Osceloa, Wisconsin: MBI Publishing Company, 2000), pp. 57–58; J.R. Smith and Antony L. Kay, *German Aircraft of the Second World War* (London: Putnam, 1972), pp. 589–90.

12. For a photograph of the Fl 184, see Brooks, p. 27; Smith and Kay, p. 590; George Townson and Howard Levy, "The History of the Autogiro: Part 2," *Air Classics Quarterly Review* 4, no. 3 (Fall 1977): 13.

13. See Martin Hollmann, *Helicopters* (Monterey, California: Aircraft Designs, Inc., 1986), pp. 119–25.

14. For a photograph of the Fl 185, see Hollmann, *Helicopters,* pp. 73, 122; Townson and Levy, "History of the Autogiro: Part 2," p. 12.

15. As this was not built under a Cierva license, it is more properly called an *autogyro.*

16. Brooks maintains that Focke still remained a member of the board of directors of Focke-Wulfe even after he left (p. 255).

17. For photographs of the Fw 186, see Brooks, p. 254; Smith and Kay, p. 143; Tony Wood and Bill Gunston, *Hitler's Luftwaffe* (New York: Crescent Books, 1978), p. 160.

18. Brooks, p. 256.

19. See Smith and Kay, p. 603; Ford, pp. 55–56; but see Bill Gunston, *Helicopters at War* (London, England: Hamlyn, 1977), p. 30 (there incorrectly listed as the Fa-325).

20. For photographs of Hanna Reitsch flying inside the arena in Berlin, see Charles Gablehouse, *Helicopters and Autogiros: A History of Rotating-Wing and V/STOL Aviation,* rev. ed. (Philadelphia: Lippincott, 1969), p. 67 (note that the caption of the photo is misleading).

21. See Warren R. Young, *The Helicopters* (Alexandria, Virginia: Time-Life Books, 1982), p. 74 (photograph of Reitsch conferring with Focke before the exhibition flights).

22. See C. R. Roseberry, *The Challenging Skies: The Colorful Story of Aviation's Most Exciting Years 1919–39* (Garden City, New York: Doubleday & Company, 1966), p. 428.

23. He would summarize some of his ideas in Richard H. Prewitt, "Possibilities of the Jump Take-Off Autogiro," *Journal of Aeronautical Sciences* 6, no. 1 (November 1938).

24. Brooks, p. 135.

25. Jay P. Spenser, *Whirlybirds: A History of the U.S. Helicopter Pioneers* (Seattle and London: University of Washington Press, 1998), p. 100.

26. Gregory, p. 87.

27. Brooks erroneously asserts that "Ray had left the company in September 1937" p. 223).

28. Frank Kingston Smith, *Legacy of Wings: The Story of Harold F. Pitcairn* (New York: Jason Aronson, 1981), p. 267.

29. Gregory noted the meeting at p. 103 of his book.

30. For photographs of the PA-36, see George Townson, *Autogiro: The Story of "the Windmill Plane"* (Fallbrook, California: Aero Publishers; reprint, Trenton, New Jersey: Townson, 1985), pp. 73–76; Brooks, p. 223; Smith, *Legacy of Wings,* p. 269; Ibid., "Mr. Pitcairn's Autogiros" *Airpower* 12, no. 2 (March 1983): 28–49, 45–46.

31. For information on Luscombe and pictures of the construction of the PA-36, see James B. Zazas, *Visions of Luscombe: The Early Years* (Terre Haute, Indiana: SunShine House, Inc., 1993), pp. 187–89.

32. Ibid., pp. 187–88.

33. Statistics taken from Townson, pp. 141–42; see also Brooks, p. 222.

34. Zazas claims that this test was less "vigorous" than the Luscombe/Pitcairn test (p. 189).

35. Ibid., pp. 190, 204.

Chapter 8

PITCAIRN, THE KELLETT BROTHERS, AND THE COMING OF WAR

It is this fact that has made aviation pay so dearly for progress. It has cost the lives of many brave and useful men and an immense sum of money spent and gone in crashed airplanes. All progress has its price, of course, but rarely is the price so high. Every other development in transportation has paid early attention to safety; subsequent progress has depended on it and counted on it. Whenever assurance of security has been lacking, other matters have been considered comparatively unimportant until it was attained in a reasonable degree. But because the limitations of the best airplane of today are in essentials the same as those of twenty years ago, the business of aviation has borne an extraordinary burden of waste and loss, sometimes to such an extent as to shake seriously the public's confidence in its future.

Juan de la Cierva and Don Rose, Wings of Tomorrow

The flight testing of the jump takeoff was to assume even greater significance as the government was finally moving forward with the Dorsey-Logan Bill funding. The bad news was that the original funding had been reduced to $300,000 in Public Act #61, passed by the Seventy-sixth Congress, and the House Appropriations Committee had designated the army as the administrative agency, and the army had in turn appointed Captain H. Franklin Gregory as its administrator. Gregory, by now familiar with the rotary-wing developments of Pitcairn, Kellett, and Sikorsky, had become convinced, and informed his army superiors, that the future lay with the helicopter. Pitcairn continued, however, to have faith in the PA-36 and scrutinized the Circular Proposal drafted by Gregory, the first step in the allocations process. That document would contain the requirements for aircraft and was drafted in consultation with officials from various government agencies that had vested interests in rotary aircraft, including the departments of agriculture and the interior, survey agencies,[1] NACA, and the army and navy. That meeting was conducted by Colonel C.L. Tinker, in the absence of General Henry H. "Hap" Arnold, then chief of the Army Air Corps, on May 31, 1939. The assembled officials[2] agreed, prodded by the expert briefing supplied by Captain Gregory, who was the junior officer present but who had become project officer for all the army's rotary-wing aircraft, that the proposed Dorsey-Logan rotary-wing aircraft should be able to make a vertical takeoff over a fifty-foot obstacle. At the time only the German Fa-61 could achieve such performance, and to all intents and purposes this flight requirement eliminated the Autogiro. Furthermore, by the end of the summer of 1939 the PA-36 had yet to make a jump takeoff, but Pitcairn was optimistic that it could meet the government requirements. The world had been stunned with the German development of the Fa-61, but on September 1, 1939, that faded into the background as German tanks crossed into Poland—a sure and quick triumph that saw the disastrous confrontation of the old and the new as Polish troops on horseback gallantly charged Panzer tanks, with predictable results. Europe was at war, and the United States was fifteen months away from officially entering the conflict. Gregory was even more convinced that the future of military rotary-wing flight lay with the helicopter, not the Autogiro.

Slim Soule was confident that he could make jump takeoffs in the PA-36, and he took ten days to gradually increase the height of the "jump." He would prerotate the rotor blades in flat pitch beyond that which was necessary for flight-lift, a condition known as *over-spinning,* and then automatically declutch the rotor and snap the blades into a positive lifting angle. He is reputed to have joked, when one of his jumps resulted in a hop of 120 feet, that he had "caught up to the Wright brothers' first flight."[3] In the autumn of 1939 Soule made the first true jump takeoff,[4] and everyone cheered as the aircraft lifted upward and flew away, circled Pitcairn Field, and then landed. The second jump takeoff, however, almost resulted in tragedy when during flight Soule experienced severe vibration in the control stick, experienced a loss of control, and while attempting to land in a field, struck a telephone pole with one of the rotor blades. It was fortunate that the accident occurred while so close to the ground so that Soule was not hurt nor was the aircraft extensively damaged. It was also fortunate that one rotor blade survived, for examination readily revealed the source of the vibration, and it was one that Pitcairn had seen before. As with the NACA testing of a PA-33 at Langley, Virginia, three years before, there was a lack of venting at the tip of the blade. The centrifugal force of the hollow rotating blade caused a buildup of interior pressure, and the blade ribs failed. When the ribs inside the blade failed, the leading edge of the blade became detached and the blades began to vibrate uncontrollably. At this point Larsen had spent approximately $200,000[5] in developing the PA-36, almost four times the sum he had originally promised, but of greater importance to Pitcairn, the failure came at the same time that Gregory was drafting the request for bids for Dorsey-Logan funds. And while the PA-36 could not hover, Pitcairn hoped that when restored to flying condition with new and improved rotor blades, it could accomplish the jump takeoff required by the Dorsey-Logan army criteria. This clearly strained but did not break his faith, as Slim Soule had not jumped higher than four to five feet before flying away and the testing program was then already a year behind schedule.

The testing led to two other accidents, each stemming from the increased complexity of the jump takeoff mechanism. The rotor spin-up drive failed to declutch during the takeoff, and the rotor snapped into lift condition while still powered. The motor had enough power to over-spin the rotor for the jump but not to take off as a helicopter, so the Autogiro gyrated back and forth before landing hard on the ground, causing the frame to bend. In one of these failures the air-cooling vanes on the whirling propeller shaft came into contact with the engine starter gear, causing the vanes to shatter and cut through the aluminum fuselage. Soule was unhurt, as the fuselage was cut a few inches in back of the pilot's seat, but it took months to repair the serious damage to the fuselage.

Testing continued, and the entire rotary-wing aircraft industry intently scrutinized the final version of the army's Circular 40-260 which, in accord with federal regulations, set forth the bill's flight requirements. Both Pitcairn and Kellett, the only companies in the United States with real rotary-win experience, submitted bids, and because Kellett had not yet perfected the jump takeoff, its Bryn Athyn rival felt that it had the upper hand. The PA-36, then being rebuilt, *could* do a jump takeoff, and Pitcairn and his associates felt that this achievement outweighed the fact that the takeoffs did not reach the fifty feet cited in the circular. But then again, no aircraft then in America could do that!

Pitcairn PA-38 design concept submitted by Pitcairn Aviation in a bid for the Rotary Wing competition in conjunction with the funding authorized by the Dorsey Bill authorization. (Courtesy of Stephen Pitcairn, from the Pitcairn Archives.)

So Harold F. Pitcairn felt confident; he had the best possibility. He even commissioned and put forth a new jump takeoff design—the PA-38, which was a two-cockpit jump takeoff Autogiro equipped with sliding canopies, but the rear was open to allow the backward-facing observer to operate a swiveling machine gun. It was a large aircraft of 4,200 pounds loaded weight, designed for observation, cargo including mail, insecticide dispensing, and firefighting. By the time that this proposed model was presented to the public, it was advertised as a product of the Pitcairn-Larsen Autogiro Company, Inc., but it was never realized in either a civilian or military form, and neither were the Kellett proposed models successfully considered. They were even less likely candidates than either the PA-36 or 38, even though Kellett was much in the news.

Johnny Miller had joined the Kellett Autogiro Company as a test pilot in 1937, and he had extensive experience in both his own PCA-2 and the Kellett KD-1 direct control model, and its military configuration, the YG-1. In September Miller had written the cover article for *Popular Mechanics Magazine,* entitled "The Missing Link in Aviation," echoing both the

March 1931 David Ingalls article in *Fortune* and his own PCA-2 of the same name. The cover painting depicted[6] a KD-1 descending to land on top of a checkerboard landing area on top of a skyscraper surrounded by a stylized metropolitan setting. The article predicted that the Autogiro would soon find a reconnaissance role with the military and could be counted on for landing on building roofs. Miller had landed a Kellett KD-1 in 1935 on the newly completed 30th Street Post Office in Philadelphia (along with Jim Ray in a Pitcairn) and was anticipating regular Autogiro airmail service the following year. In 1938 the KD-1 had been redesigned for the mail route as a KD-1B and it would receive ATC No. 712 in December 1939, and Kellett had begun to approach major airlines to bid on the mail route, feeling that such an established aviation enterprise would have a better chance of gaining government approval. To publicize the mailing-carrying capabilities of the KD-1B, Miller had landed on the roof of the Chicago post office and made several landings on Washington, D.C., streets as part of Airmail Week celebrations in 1938, the twentieth anniversary of airmail service. The KD-1B had an enclosed sliding canopy that was fitted over the rear cockpit and a covered compartment holding 350 pounds of mail in the front cockpit. A two-way radio had been installed along with a turn-and-slip indicator and directional gyro in the instrument panel. The aircraft had emergency landing flares, but night flying was never scheduled.

Eastern Air Lines (EAL) had entered into an agreement with Kellett to apply for the airmail route, with its president Captain Edward "Eddie" Rickenbacker himself interviewing Miller, who considered the World War I ace a personal hero. Rickenbacker was impressed when Miller stated that he felt that there would be a 75 percent on-time performance, and knowing that Miller was the only qualified Kellett pilot, inquired what it would take to gain his services on the mail route. Miller replied that he expected Eastern to double his Kellett salary. Eastern Airlines bid for the government contract and received a one-year commitment at $63,000 for a 75 percent completion rate. Later Eastern agreed that Miller could remain in its employ with the rank of captain *and,* of greatest importance, seniority from the date the mail route started in 1939. The company set up a rooftop office next to the elevator shaft, with a radio station and meteorological instruments with a service manager and mail handler. At the other end of the route, Camden Airport, the company arranged for office and hanger space. At six miles, it was and still ranks as the shortest scheduled air route in history and the first with a rotary-flight aircraft.

Miller spent time preparing for the flight analyzing the effect of wind on the post office roof, for although its construction had anticipated such flights with a reinforced roof, no thought had been given to wind and turbulence. On July 1 Miller delivered the KD-1B to Eastern Air Lines at the Camden, New Jersey, airport and commenced his employment as an EAL captain, from which he would retire twenty-five years later. His career spanned from World War I "Jennys" in 1923 to jet aircraft, and he even flew himself to the Experimental Aircraft Association (EAA) "Air Venture" at Oshkosh, Wisconsin, in 2002 at the age of 96! In preparation for the start of the airmail service[7] on June 6, 1939, Miller made several test and practice flights, learning how to deal with the air currents that swirled around the rooftop, for which he partially blamed Pitcairn.[8] Miller tossed hundreds of pieces of toilet paper into the wind streams from the south edge of the roof to determine the direction and force of the turbulence, observing that the roof was subject to an updraft at the building edge but that airflow then converted to a downdraft in the center of the roof. That research explained several hard landings and takeoff difficulties, and led to techniques that made flying safe even in strong northwest winds up to sixty mph. EAL had arranged for Miller to carry noted musician André Kostelanetz[9] sitting on top of mail sacks in the forward cockpit while making a practice run, an event that Kostelanetz's agent subsequently refused to allow to be publicized. The experimental airmail run commenced[10] with great public fanfare in the first week of July and continued for an uneventful year, ending the first week of July 1940.[11] The schedule consisted of five flights each day, six days per week; Miller had seen the need for a reserve pilot, but it was not recognized by EAL management until he suffered a bout of flu and the flight had to be shut down for two weeks in the autumn of 1939. He was then given permission to secure the services of a reserve and finally hired former Pitcairn factory pilot Skip Lukens, who had checked Miller out in the original PCA-2 in 1931 before the transcontinental flight. Now Miller became the teacher, as Lukens did not have experience with direct control, but he soon became proficient with about twelve hours of dual instruction in the Kellett KD-1 prototype that EAL rented. During the year of operation, Lukens and Miller made over 2,300 takeoffs and landings from the post office with better than a 95 percent on-time flight record, but there were two accidents, and as luck would have it, both while Lukens was at the controls. In one the Autogiro overturned on the roof, pushed over by a strong wind gust. Parts of the shredded rotor blades fell to the street below, giving rise to rumors that the Kellett had crashed,[12] which circulated for many years among those who collect airmail stamps. Additionally, Lukens made a forced landing with only partial power in a vacant lot while flying to the Camden, New Jersey, airport, but was soon in the air and on his way when the engine resumed full power. (Miller was convinced that the cause was carburetor icing). The experimental airmail route had established an outstanding safety and completion record that was profitable,[13] but it ended after a year. War was coming, and the attention of the aviation world was no longer directed toward roof landings. And although Kellett Autogiro Company received much publicity from the experiment, it would not result in a successful bid for Dorsey-Logan funds.

In addition to the Pitcairn and Kellett companies, two others had submitted bids in response to the Dorsey-Logan circular, the Vought-Sikorsky Division of the United Aircraft Corporation and a new company named the Platt-LePage Helicopter Company. The former was not in the running, as its prototype helicopter, the VS-300, had been destroyed in a test-flight and Gregory could not rely on the company's theoretical submission, and this left the latter as the only alternative to Autogiro companies. Platt-LePage had already done developmental work on two prototypes by 1940, but the models had not been successful. When word had reached LePage of Focke's success with the Fa-61 with Hanna Reitsch's public flight in 1937, he had gone to Germany and developed a quick rapport with Focke, whom he found similar in age, theoretical approach, and enthusiasm. And although flow of information between the two aviation inventors had stopped with the coming of war to Europe in September 1939, LePage had observed the Fa-61 in Bremen[14] and was confident in embarking on a similar design. By the time that Gregory opened the bids on April 15, 1940, the third Platt-LePage prototype had achieved some success at hovering flight, but it had never carried a person. It is likely, then, that Pitcairn was confident, because in the PA-36, he clearly had the most technologically advanced jump takeoff Autogiro and rotary aircraft in America. But he did not remain confident for long.

The Dorsey-Logan funds were awarded to the upstart Platt-LePage Helicopter Company based on scaled-up drawings of its third model helicopter, the PL-3, which the army now called the Platt-LePage XR-1 (short for "Experimental Rotary aircraft number 1"). The contract called for delivery of a flyable helicopter by January 1, 1941, barely seven and a half months away. And even though Pitcairn was quietly told that although the PA-36 *was* the best engineered submission, and that Gregory and his army colleagues had decided that the future of military rotary-wing flight lay with the helicopter, not the Autogiro, Pitcairn resolved to undertake efforts to convince the army that his Autogiro should also be acquired. The XR-1, based on the twin-rotor, outrigger system that LePage had seen on the Fa-61, would not make its first tethered flight until May 12, 1941, piloted by

Lou Leavitt.[15] The flight was observed by Platt-LePage Company consultant, aviation pioneer Grover Loening, famous for his seaplane designs,[16] who commented that "this craft has tremendous possibilities, but there is still a long hard way ahead."[17] That view echoed the earlier opinions of those who observed the Fa-61—it flew, but it was not yet a useful machine. So it was certainly understandable that Pitcairn, convinced of the merit of the PA-36, prepared it for effective demonstrations before senior military officers. He was also convinced that the military continued to be interested in the Autogiro because Gregory authorized additional funds to upgrade two Kellett aircraft already in the army's inventory. But although the funds were authorized, Pitcairn failed to take into account that they were *not* from Dorsey-Logan, *and* he seriously misunderstood why Gregory wanted two advanced Autogiros.

By mid-1940 the world was reading grim daily reports of war in Europe, the successful withdrawal of almost 338,000 English soldiers from the beaches of Dunkirk, and the new English Prime Minister Winston Churchill's ringing declarations of continued resistance. The army benefited from increased government budget allocations to prepare for what all knew was a coming conflict, and it had allocated some of the funds to Kellett Autogiro Company to retrofit two of the YG-1Bs already at the Army Air Corps Autogiro School. The YG-1B[18] was a modification of the YG-1A, based on improvements designed by Richard H. Prewitt, primarily the addition of collective pitch control, which allowed the pilot to vary the blade pitch from zero to four degrees. This improved collective pitch control, the latest development in the 1936–39 attempt to achieve the jump takeoff, did not achieve that goal but did significantly accelerate takeoffs[19] and had first been delivered to the army on December 29, 1937. As these aircraft were already in the army inventory, no public bidding was required, and the modification added a more powerful engine, a Hamilton Standard constant-speed propeller, and a long-stroke cantilever landing gear for vertical landings. This resulted in jump takeoff capacity comparable to the direct-control YG-1Bs, with the resulting experimental model being designated the XR-2 following the Platt-LePage helicopter, which had already been designated the XR-1. The XR-2 was completely destroyed[20] in 1941 by ground resonance, uncontrollable vibration while attempting a jump takeoff in 1941. As the motor was spinning up to achieve a jump takeoff, the craft literally shook itself apart in less than five seconds, a problem that had been experienced by all Autogiro developers and which had already been solved mathematically and technologically by Paul Stanley of Pitcairn. The destruction of the XR-2 would prove to be significant, as it led to serious investigation of ground resonance, which would be solved by Bob Wagner of Kellett Autogiro and Prewitt Coleman of NACA. Their work would enable future rotary-wing aircraft to avoid the problem, a significant advancement and a necessary development for the technology to move forward. A second YG-1B was converted for the army, known as the XG-1B, and was used primarily as a test bed for improvements flowing from the analysis of the accident. A stiffer rotor pylon was added with interblade dampers to deal with vibration, but these improvements did little to eliminate severe control system and rotor vibration, problems that had contributed to the destruction of the XR-2. The army solution was to redesign the rotor hub itself to dampen vibration, a solution incorporated into the army's second Kellett experimental Autogiro, denoted the XR-3. It was also a converted army YG-1B, with a newly redesigned rotor hub featuring a fixed spindle with collective and cyclic pitch control. As such it harkened back to the comment that Cierva had made during his 1933 visit, when he suggested that the Americans should concentrate on the tilting hub/overhead stick control and the English on a fixed spindle with cyclic/collective control system. Now that system was installed in the XR-3 and jump takeoffs to fifteen feet were possible. It proved a useful experimental platform for new rotor and controls systems and useful for the development of the helicopter, which was exactly what Gregory intended it for! In fact, when sold as surplus to General Electric in 1945, it was used for helicopter rotor development. Although it played no significant role in Autogiro development, it would prove perhaps the most significant aircraft for the survival of autorotational technology, as one of the XR-3 pilots at General Electric was a Russian immigrant named Igor Bensen.

The PA-36 was ready for demonstration flights by mid-July. It had been rebuilt and made lighter by the removal of the "roadability" equipment, as that capability was not militarily required, and it was capable of making a stunning jump takeoff and dramatic flyaway. Pitcairn justifiably felt that this was the best performance ever achieved by an Autogiro, and he invited Gregory and the press to watch the shiny aluminum aircraft make its public debut, confident that it would rekindle military interest, as the PA-36 was clearly the most capable rotary-wing aircraft flying. It could meet all the Dorsey-Logan requirements but not all of the military's—it could not hover, but Pitcairn no doubt felt that with the increased efforts to prepare for war, the government would be willing to order the PA-36. By all accounts the demonstration flights at Pitcairn Field, with Slim Soule as pilot, were all that Pitcairn hoped they would be, and having invited the media, the company reaped a public relations bonanza. Even today, the films[21] of the aluminum aircraft jumping straight up and flying away over an eighteen-foot barrier amaze, just as they did theater-goers who watched the newsreels. Gregory was impressed but noncommittal, for unlike Pitcairn he was aware of developments even then unfolding with the Sikorsky VS-300 developmental helicopter that would soon doom even the success Pitcairn was enjoying with the PA-36. Pitcairn thought to capitalize on the sensation created by Soule's demonstrations by having a spectacular takeoff and landing at the 1940 World's Fair then at Flushing Meadow in New York, and officials agreed, but an accident to the PA-36 doomed the effort at the last minute. However, the subsequent government/ military demonstrations at College Park, Maryland, and Bolling Field in February 1941 went very well.[22] The most impressive part of Soule's flying was the jump takeoff from the midst of a circle of parked automobiles. Pitcairn, ever the entrepreneur, touted the advantages of the PA-36 to officers serving on the staff of Army Air Corps Chief General "Hap" Arnold, and the general that ordered both Captain Gregory take an additional look at the PA-36 and that the aircraft make a five-thousand-mile tour of military bases[23] to "investigate possible military applications."[24] The military demonstrations impressed all who watched the PA-36 make flawless jump takeoffs and pinpoint landings, but it was too late—the fate of the Autogiro had been decided at a meeting the previous December.

Pitcairn's biographer, Frank Kingston Smith, incorrectly asserts that after Captain Gregory left Pitcairn Field and the Soule demonstrations of the PA-36 in October 1941, he visited the Vought-Sikorsky factory to confer with his friend Igor Sikorsky as to the status of the VS-300 helicopter, that the Russian immigrant[25] flew a demonstration flight and then offered the controls to the army officer,[26] and that as a result of this collaboration, the orders for the PA-36 never materialized. Although it is tempting to ascribe the failure of the PA-36 to such a conspiracy, Gregory's 1944 book tells a different story.[27] He indeed had been a friend and kept in touch with Sikor-sky since their first meeting at the 1939 Franklin Institute rotary-wing gathering, but he had been officially interested in the Vought-Sikorsky developments *since* 1938, when he and Major Carl F. Green visited with Sikorsky associates Michael Gluhareff[28] and Boris P. Labensky and were briefed on the development of a mechanism to control helicopter flight. Subsequently the VS-300 in its original configuration flew successfully in free-flight on September 14, 1939, and Gregory was then invited in 1940 to make a test-flight. He likened his first helicopter experience to riding a bucking bronco, and it is from that that the title of his book *Anything a Horse Can Do* describing the helicopter was derived. But of greater importance, that flight convinced him of the ultimate success of the helicopter, which he communicated back to his army superiors. Several weeks later he met with Igor Sikorsky and Serge Gluhareff in Sikorsky's car to guarantee confidentiality and share his view that the army should fund development of the VS-300 even as it was funding the Platt-LePage XR-1. While it is not possible to know for certain if Gregory's belief was due to a conviction that a parallel research and development effort was desirable (as he states) or whether he truly was convinced that Sikorsky's single-rotor technology was superior[29] to the Fa-61 twin-rotor scheme employed in the XR-1, what is certain is that he was proposing to commit the remaining Dorsey-Logan funds. The officer and the inventor agreed that Sikorsky would seek the additional funds that would be necessary from the Vought-Sikorsky company. In the end, the government would expend considerably more than the remaining funds to develop the Sikorsky helicopter, which would be called the XR-4, and it would lead to the single-rotor configuration as the model for future helicopters and bring the United States military firmly into the world of rotary-wing flight in a manner that a decade's involvement with the Autogiro had failed to achieve. The funds were officially allocated on December 17, 1940, the thirty-seventh anniversary of the Wright brothers' first flight, by sixteen officials[30] meeting in Washington, D.C. The military had made its commitment to the helicopter even before Pitcairn had sent Soule on his tour of military bases, and it really did not matter how dramatic or successful the jump takeoff demonstrations were—there were no orders nor further expression of army interest. Pitcairn withdrew the PA-36 in mid-1941; it never received certification and was cut up for scrap the following year in response to the need for aluminum to support the American war effort. It was an inglorious end to the most beautiful Autogiro ever built, but by that time Pitcairn was out of the Autogiro business.

In 1940 the Pitcairn Autogiro Company was inactive, and for business reasons it had changed the name of the manufacturing company to Pitcairn-Larsen Autogiro Company. That business entity was created especially to fill an order from an old English friend, Reggie Brie, now an RAF Wing Commander (equal to a Lieutenant Colonel) serving in Washington, D.C., with the British Air Purchasing Commission and now considered the "United Kingdom's leading military authority on rotary-wing aircraft."[31] Brie had been Cierva chief test pilot and sought to provide convoy protection for the Royal Navy, which was then absorbing terrible losses to German submarines in the Battle of the Atlantic. Being very familiar with the capability of the Autogiro for shipboard landings, he had visited Pitcairn in late 1940 to explore the ability of the American to provide Autogiros to be flown from the decks of merchant ships for testing under combat conditions.[32] Pitcairn replied that although America had not yet entered the war, the economy anticipated the coming conflict and it would be impossible to manufacture a new model; the company could reacquire older PA-18 two-place, open-cockpit Autogiros (which in turn had been derived from the earlier PAA-1) and retrofit them with direct control, jump takeoff systems that would be suitable for British maritime service. Pitcairn's motivation was undoubtedly complex, as he was well aware that the small British order would hardly be profitable, yet it was a patriotic effort and there was certainly the possibility that British success in utilizing the Autogiro under wartime conditions might lead to a more favorable view by the United States Navy, which had consistently rejected the aircraft since the USMC evaluations in Nicaragua almost a decade before.

The British Air Purchasing Commission issued a contract on November 5, 1940, for seven of what came to be called the PA-39,[33] in reality a conversion of the older PA-18 Autogiros that had been repurchased from their owners who could not, in most cases, fly due to wartime flight and fuel restrictions.[34] These aircraft had been flying for a decade—now they would be made ready for war. Pitcairn-Larsen reworked the fuselages to include triple vertical tail surfaces, including a central fin and rudder and outward-slanted end-plates, which had been previously developed for a latter configuration of the experimental PA-22 and later incorporated in the aluminum PA-36. The models, with wings removed, were retrofitted with a three-blade cantilever rotor with collective and cyclic pitch control, resulting in an impressive jump takeoff capability powered by a 165-horsepower Warner-Scarab engine. The aircraft were scheduled for delivery in late 1941, and as each was completed, it was test-flown by Slim Soule and accepted by Wing Commander Brie. Five were to be sent to England, with two remaining in the United States—one for testing and demonstration purposes by Brie and the other retained by the Pitcairn-Larsen factory for further development and, of great importance to the principals, in anticipation of future orders once the Autogiro had demonstrated its wartime worth. In May of the following year Brie did, in fact, make extensive shipboard tests of the PA-39, landing on the British escort carrier HMS *Avenger* in Long Island Sound and in Chesapeake Bay off Newport News, Virginia, from a platform constructed on the *Empire Mersey,* a British merchant ship, taking off and landing at anchor and while the vessels were underway. And while the tests achieved success, the remaining five British PA-39s did not.[35]

In a supreme irony, of the five PA-39s crated and sent by ship to England, only two reached their destination, the others and all the spare parts for the order having been lost when the ships upon which they had been placed were sunk by German submarines![36] And the two Autogiros that did reach England did not actively participate in the war effort. Even before they were to arrive, the mission had changed from maritime to communications, and the models had been allocated to the RAF. Only one of the two that did reach England ever flew, probably flown by former Cierva pilot Alan Marsh at Duxford and later at Boscombe Down. None were recorded as having survived the war, probably damaged in accidents and as spare parts were unavailable, not repaired. And as the mission had changed, it became known that there would be no future Royal Navy orders—the PA-39 program was dead and with it Pitcairn's hopes of impressing the United States Navy. The short production run had cost a great deal of money that now could not be recouped in future orders— Harold Pitcairn was effectively out of the Autogiro business and with all the feelings of a patriotic citizen searching for an avenue to make a contribution as his country went to war. The feelings of anger and frustration were vented at Agnew Larsen, who had failed to deliver the PA-36 in time for the Dorsey Bill hearings and just supervised the ill-fated PA-39 program. And there is little doubt that Larsen, who with Ray had been there from the very first when Pitcairn Field was little more than a cow pasture, also had deep feelings about the latest efforts—both men were driven, talented, and patriotic. Pitcairn vented and Larsen responded, a predictable escalation from which there was seemingly no retreat.[37] The association between the two friends, which had lasted over twenty-four years, ended with Larsen's resignation, abruptly ending the confrontation.

Ray and Larsen were now gone, and there was seemingly no immediate future for Pitcairn's Autogiros. So a new venture was born, as America faced war on a scale never before seen. Even though Pitcairn had retained engineers Paul Stanley and Harris Campbell to design civilian versions of the PA-36 and larger passenger configurations,[38] nothing would come of these grandiose visions until the Fairey Rotodyne flew in 1957. Pitcairn responded to a more immediate need for military aircraft, and the name of the company was changed to A.G.A. Aviation Corporation (for autogiros, gliders and airplanes). Although the corporate name may have reflected the optimistic belief that it would be called upon to provide each, no Autogiros or airplanes were ever ordered by the military—only large, gawky assault gliders. That did not save the company, and it was acquired by Firestone Tire and Rubber Company in early 1942. Reflecting the lack of government or civilian interest, Firestone renamed the company G&A Aircraft and took over the lease of the Pitcairn Aircraft factory at Willow Grove, Pennsylvania. But there was one last moment for the Pitcairn Autogiro engineering team—such was the shock of America's entry into the war that the government contracted for a new kind of specialized configuration designed for reconnaissance and observation, based on the Pitcairn jump takeoff mechanisms and the earlier Buhl Autogiro "pusher" rear engine placement. This model, of which only two were ever built before the contract was cancelled, had a plastic-enclosed cabin for maximum visibility, a configuration that later influenced the Sikorsky XR-5 helicopter design. Although the initial contract for what the military designated the XO-61[39] (Experimental Observation)[40] was for a static test frame and six test models, it was later reduced to five and then cancelled altogether. Only two were ever built,[41] and certification was never sought. Development had taken its toll, as problems with ground resonance and engine cooling were encountered. The focus of the company had shifted to the development of its other product, the troop-carrying glider, and by the advent of production helicopters. It would be the last Autogiro manufactured in America. As the G&A Division of Firestone, the company belatedly attempted to get into the helicopter business. Its XR-9, delivered to Wright Field in 1946, and an enlarged version called the XR-14 were too little, too late and could not compete for military orders with those developed previously. The G&A Division then attempted to market a civilian version but was unsuccessful and finally went out of business in 1948.[42]

By 1943 Harold Pitcairn was faced with three other developments that signaled the end of his active involvement with the Autogiro. The U.S. government, responding to the need for new training facilities for pilots, let him know through the offices of the aviation department of the Fourth Naval District in Philadelphia that it was preparing to acquire his inactive airfield for military use by the navy. Rather than force a formal condemnation proceeding in accord with eminent domain, Pitcairn sold the field to the navy for its appraised value. He had been offered premium sums for the airfield, but now its purchase price was set on the basis of the surrounding farmland, a much smaller return, and Pitcairn Field became the United States Naval Air Station (NAS) Willow Grove.[43] Pitcairn had also concluded that the PA-36 would fly no longer and ordered the two models constructed by Luscombe to be cut up into aluminum scrap and donated to the war effort in 1943.[44]

The year 1943 also saw what was perhaps the most patriotic gesture of Harold Pitcairn, an extraordinary act to aid America in its greatest time of need. Pitcairn *personally* held 19 rotary-wing patents in 1941, and the ACA additionally held 145 patents, granted from the dawn of American Autogiro flight. Almost all of the subsequent helicopter development in America was based on Pitcairn patents, and the ACA had legally binding rights to royalties for each and every helicopter utilizing its patents. Igor Sikorsky readily recognized that these royalties were due, as he had gained a license to use certain of the Pitcairn patents at the insistence of H. Franklin Gregory, and he insisted that the parent company of Vought-Sikorsky, United Aircraft, make provision to pay them. But then Harold F. Pitcairn astounded all with an offer that was incredibly generous and revealed his deep patriotism. He wrote the commanding general, Army Air Forces Material Command, Wright Field, in a letter of July 22, 1943:[45]

> For some time we have been giving a lot of thought to what contribution the Autogiro Company of America might make to the country in this time of conflict, beyond the engineering assistance which we have made available to our licensees. For the good of the war effort and to conserve public funds, we have decided to reduce to a nominal rate the royalty charged for the fruits of our fifteen years of invention, development, and experience in rotary-wing aircraft. Therefore, on machines and equipment supplied to the United States Government by our licensees, we will reduce our royalty from 5% on the basis of fully-equipped machines to eighty-five one-hundredths of one percent (.85%) of the [government] contract price, to be effective for the duration of hostilities with Germany, Japan, and Italy.

The offer was accepted by the Wartime Royalty Adjustment Board after being extended to "the end of all present hostilities, plus six months," and it has been estimated that under then-current helicopter contracts the ACA lost over five million dollars in royalty payments.

The Kellett Autogiro Company did not fare much better with regard to the military, although it had provided the advanced direct control XR-2 and XR-3 experimental models. In 1941, with the country anticipating entry into war and a functional helicopter still a long way off, the army contracted for eight experimental observation Autogiros from Kellett. Called the XO-60/YO-60, this was a true direct control jump takeoff model, derived from the KD-1/XR-2 series. The first aircraft was completed in February 1943, with six more delivered by the end of the year for testing. The latter models were denoted YO-60s because, unlike the XO-60, which had XR-2-like long-stroke cantilever landing gear, the YO-60s had standard truss landing

Kellett YO-60 Autogiro. (Courtesy of Stephen Pitcairn, from the Pitcairn Archives.)

gear. The three-blade rotor achieved the jump takeoff by means of a novel collective pitch control, which was later sold to the Autogiro Company of America, further enhancing its patent protection in the field of rotary-wing flight, a fact that would become important in the early 1950s when Harold Pitcairn would sue the government to enforce his patents. The rotor could fold for easy storage, and the visibility afforded by the large transparent canopy was enhanced by the placement of windows in the floor of the cockpit.[46] The army performed extensive testing but found that they offered little advantage to the cheaper L-3 Aeronca, L-4 Piper, and L-5 Stinson liaison aircraft currently in military use[47] and that the Autogiros suffered from higher initial cost and maintenance. They were subsequently rejected for liaison duty as a result of a report by Autogiro test pilot E. Stuart Gregg, who was sent to the Sikorsky factory in late 1943. Gregg flew the YR-4B helicopter for a few hours and stated in his report that it was abundantly clear that the "primitive but functional machine could perform vertical takeoffs and landings with far greater ease and dexterity than a jump giro." He would claim in March 2001 that his report resulted in "virtually killing further military procurement of autogiros."[48] Gregg also confessed that he "often used a YO-60 to fly home for lunch, landing in my front yard. I also flew it to the local golf course, where I landed on the ninth fairway, parked behind the caddy shack, and put in a quick nine holes of therapeutic golf." It was, ironically, the perfect early 1930s version of Pitcairn's advertising, but this was wartime, and aviation procurement personnel were not concerned about rounds of golf.

None of the YO-60s were assigned to operational reconnaissance units. The last, denoted XO-60, was delivered to Wright Field on December 6, 1944, and is currently in the Smithsonian's National Aeronautical Collection, officially received on May 1, 1949.[49] With the end of the XO-60 program in 1943, Kellett reverted to its original name and became Kellett Aircraft Corporation, and it also belatedly turned to helicopter development. It built and flew the XR-8 in 1945, "looking like a flea topped by canted side-by-side rotors with intermeshing blades, it was the first U.S. helicopter to employ the 'synchropter' configuration developed by Anton Flettner in Germany shortly before the war."[50] The experimental XR-8 is today found in the collection of the National Air and Space Museum,[51] but the later Kellett XR-10 did not survive the fatal crash that killed its pilot, whose parachute became entangled in the eggbeater rotors in 1948, and Kellett, which had been in bankruptcy since 1946, did not survive as an operational company. Thus ended the two remaining American Autogiro companies and the Autogiro in America—they had failed to find either a civilian or military market and had come to helicopters too late.

World War I had seen vast advances in aviation technologies—war has that power. The aerial novelty at the outset became a deadly weapon by the time that the guns fell silent at the eleventh hour of the eleventh day of the eleventh month. World War II would see a similar progress, with jet planes and missiles, including an experimental German bomber that flew to within 12.4 miles of New York City and back to its French base.[52] Autogiros would also go to war, but they would not be American, and there would be no significant advances in technology occasioned by conflict. So although the wartime role of Cierva's technology is not well known, what is readily evident is that the Autogiro should not have survived the war and the coming of the helicopter, and it almost did not. It would come down to a surplus Kellett XR-3.

NOTES

1. Biological Survey and Coast and Geodetic Survey departments.
2. The May 31, 1939, meeting laid the foundation for American military rotary-wing development and insured that the helicopter would eventually join

America's arsenal. Those attending were J. P. Godwin, Department of Agriculture; Frederick C. Lincoln, U.S. Biological Survey; Charles M. Kieobee, Division of Air Military Service; Captain L. T. Chalker, U.S. Coast Guard; John Easton, Civil Aeronautics Authority; Lieutenant Commander C. L. Helber, Bureau of Aeronautics; Roy Knabensheue, Department of the Interior; C. S. Helds and C. W. Crowley Jr., NACA; Major W. C. Crittenberger, Cavalry; Major R. W. Beasley, Field Artillery; Lieutenant Colonel Dale D. Miniman, Coast Artillery; and Lieutenant Colonel E. W. Fales, Infantry. Hollingsworth Franklin Gregory, *Anything a Horse Can Do: The Story of the Helicopter,* Introduction by Igor Sikorsky (New York: Rynal & Hitchcock, 1944), p. 91.

3. Frank Kingston Smith, *Legacy of Wings: The Story of Harold F. Pitcairn* (New York: Jason Aronson, 1981), p. 276.

4. For a photograph of the PA-36 in jump takeoff mode, see Smith, *Legacy of Wings,* p. 276.

5. Peter W. Brooks, *Cierva Autogiros: The Development of Rotary-Wing Flight* (Washington, D.C.: Smithsonian Institution Press, 1988), p. 224; $200,799.11 cited in Smith, *Legacy of Wings,* p. 279.

6. John M. Miller, "The Missing Link in Aviation," *Popular Mechanics Magazine* 70, no. 3 (September 1938): 346–51, 134A–135A.

7. For a description of the mail service, see George Townson, *Autogiro: The Story of "the Windmill Plane"* (Fallbrook, California: Aero Publishers, 1985), pp. 122–25.

8. Miller noted that the top of the six-story post office was oriented north-south and featured twenty-foot-high structures on the east and west sides, effectively creating a channel for wind. "The north and south ends had upwards ramps about 3' high and about 30' long. They were not a good idea, but there they were. *Someone, probably at Pitcairn, had suggested them as logical in the absence of any previous experience or wind tunnel tests.*" John M. Miller, "The First Scheduled Rooftop Flying Operation in Aviation (Autogiro Airmail Service at Philadelphia, 1939–40)," *Rotorcraft* 30, no. 6 (September 1992): 24–33, 29 (emphasis added).

9. Ibid., p. 29; but see Townson, p. 122, where the erroneous claim is made that the passenger was "Jose Utubti, popular pianist," presumably a reference to José Iturbi. The debate is clearly resolved in Miller's (and Kostelanetz's) favor, as a photographic record has survived of the occasion; see John M. Miller, "Civil Uses of the Autogiro," *Aeronautics* 2, no. 10 (1940), p. 622.

10. There is some controversy regarding the commencement of the Philadelphia-Camden Airmail Service–Experimental Airmail Route 2001. Townson relates that the route began on July 5, 1939. "Autogiro Airmail," *American Helicopter Museum & Education Center Newsletter* 2, no. 3 (Spring 1996): 3. However, Brooks states that the flights began on July 6, 1939, and continued through July 5, 1940 (p. 232). Brooks is undoubtedly correct for the following reasons. Townson was not a first party to the airmail service and was also incorrect concerning the

musician passenger; Miller himself cites *test and practice* fights "through July 5" (Miller, "First Scheduled Rooftop Flying, p. 29); and the first-day post office covers clearly are dated July 6, 1939 (author's collection). But Miller himself is inconsistent—in 1940 he stated that "[s]ervice was begun...on July 7, 1939." Miller, "Civil Uses of the Autogiro," p. 617. And Miller also states, "The last scheduled flights were made on July 4, 1940. My log book shows that I made four round trips, so I assume that Lukens made one last round trip." Miller, "First Scheduled Rooftop Flying," p. 32.

11. For photographs of Miller taking off from the roof of the Philadelphia post office, see *Some Facts of Interest about Rotating-Wing Aircraft and the Autogiro Company of America* (Philadelphia: Autogiro Company of America, 1944), p. 20; Townson, pp. 123–25; George Townson and Howard Levy, "The History of the Autogiro: Part 2," *Air Classics Quarterly Review* 4, no. 2 (Summer 1977): 18; Smith, *Legacy of Wings,* p. 258; Brooks, p. 232; Miller, "Civil Uses," pp. 613–19, 622.

12. Miller, "First Scheduled Rooftop Flying, p. 32; John W.R. Taylor and Kenneth Munson, *History of Aviation* (New York: Crown Publishers, Inc., 1972), p. 214.

13. EAL was compensated at $3.86 per airplane mile and at the end of the year had made a profit of almost $23,000. According to Rickenbacker, however, writing in 1946, although EAL made a profit, the post office found the cost high and the capacity of the Autogiro limited so decided to cancel the route. "The Great Silver Fleet News," *Eastern Air Lines* 10, no. 4 (July–August 1946): 18.

14. Spenser claims that this took place in Berlin. Jay P. Spenser, *Whirlybirds: A History of the U.S. Helicopter Pioneers* (Seattle and London: University of Washington Press, 1998), p. 101.

15. In February 1936 Leavitt had taken a seventeen-year-old by the name of Frank Nicholas Piasecki for a ride in a Kellett KD-1. Piasecki would later state: "With that experience my interest exploded into the desire to build my own design. The helicopter particularly appealed to me, vertical lift being the hot topic in aviation at the time." Piasecki was a junior engineer working on the XR-1, gaining invaluable rotary-wing experience—he would go on become a premier helicopter innovator. Spenser, *Whirlybirds,* pp. 97–98. Piasecki would also gain the first helicopter pilot license in America. Smith, *Legacy of Wings,* p. 318.

16. See Edward Jablonski, *Man with Wings* (Garden City, New York: Doubleday & Company, 1980), pp. 185–88; David Donald (ed.), *The Complete Encyclopedia of World Aircraft* (New York: Barnes & Noble Books, 1997), pp. 587–88; David Mondey (ed.), *The Complete Illustrated Encyclopedia of the World's Aircraft* (Secaucus, New Jersey: Chartwell Books, Inc., 1978; updated by Michael Taylor, 2000), pp. 342–43; for a commentary on Frank Courtney's involvement with both Amelia Earhart and the Loening amphibian seaplanes, see Frank T. Courtney, *The Eighth Sea* (New York: Doubleday & Co.; also published as *Flight Path,* London: William Kimber, 1972), pp. 205–6, 263, 265.

17. As quoted in Gregory, p. 99.

18. While Brooks cites that the latter modified YG-1B is sometimes referred to as the YG-1C, this designation is not used elsewhere and is not accepted by either George Townson or Frank Kingston Smith.

19. See the description and photo in Brooks, p. 234.

20. For a photograph of the destroyed XR-2, see Brooks, p. 235.

21. "Army-Air Force Newsreels 1941," Traditions Military Videos, www. militaryvideo.com. (Accessed April 21, 2003.)

22. Brooks, p. 224. But Frank Kingston Smith asserts that the Bolling Field demonstration actually took place in October of 1941, but as he notes the subsequent tour of military installations by Soule in 1941, the February date is undoubtedly correct. Smith, *Legacy of Wings,* p. 286.

23. Including Forts Bragg, Benning, Knox, and Sill.

24. Brooks, p. 224.

25. Gregory would claim that the "proudest moment of [Sikorsky's] life...was the day he became a United States citizen" (p. 104).

26. Smith, *Legacy of Wings,* pp. 291–92.

27. See Gregory, pp. 105–13.

28. See Earl Devon Francis, *The Story of the Helicopter* (New York: Coward-McCann, Inc., 1946), pp. 117–18, where he details the secret research in 1925 by Michael and Serge Gluhareff, working for the Sikorsky Aero Engineering Corporation (then housed in a Long Island, New York, barn), into jet reaction helicopter rotors powered by compressed air.

29. In fact Sikorsky was proposing the third form of helicopter, the other two being the German twin-rotor system Fa-61 and Anton Flettner's intermeshing rotor system, first suggested by England's Dr. J.A.J. Bennett, the synchropter, which would give rise to the term *eggbeater* as a synonym for helicopter. In the end Sikorsky's single-rotor system would prove superior, but it is not known if that was Gregory's conviction. See Charles Gablehouse, *Helicopters and Autogiros: A History of Rotating-Wing and V/STOL Aviation,* rev. ed. (Philadelphia: Lippincott, 1969), p. 77.

30. The participants in the December 17 meeting were A. Gordon Calloway, Department of Agriculture, and Donald Hamilton of its Forest Service Division; Roy M. Martin, Air Mail Service Division of the Post Office Department; CDR W.J. Kossler, U.S. Coast Guard; Alan L. Morse, Technical Development Division, Civil Aeronautics Administration; LCDR J.M. Lane, Bureau of Aeronautics; R. Paul Wessner, National Park Service (Department of the Interior); F.J. Bailey Jr. and John W. Crowley, NACA; Majors B.W. Chidlaw and J.F. Phillips and Captain R.L. Montgomery, Material Division of the Office Chief of the Army Air Corps; Major Rex E. Chandler, Field Artillery; Gregory himself representing the Material Division, Wright Field; and Captain V.R. Haugen, Aircraft Laboratory of the Material Division. The participants viewed a film of the Sikorsky helicopter and agreed that it would be wise to fund the development of a second helicopter

in addition to the XR-1 and that "beneficial results would be obtained in the comparison of the Vought-Sikorsky helicopter with the one already under construction by Platt-LePage." Gregory, p. 112.

31. Willian E. Hunt, *"Heelicopter": Pioneering with Igor Sikorsky* (London: Airlife Publishing Ltd., 1998), p. 132.

32. Frank Kingston Smith maintains, without citation, that this plan was not enthusiastically embraced by the Royal Navy Admiralty, alluding to the desperation of the moment. Smith, *Legacy of Wings,* p. 294.

33. For photographs and engineering drawings of the PA-39, see Brooks, pp. 227–28; Townson, pp. 77–78; Smith, *Legacy of Wings,* pp. 296–97.

34. Townson, p. 77. However, not all those approached were willing to sell their aircraft back to Pitcairn-Larsen. This was the case with the PA-18 owned by Ann Strawbridge, currently being restored by Kate and Jack Tiffany of Spring Valley, Ohio (Leading Edge Aircraft).

35. Brooks (p. 228) notes:

The two American-held PA-39s were purchased back from the British. One was given to the Aeronautical Section of Princeton University and its fate is unrecorded. The other was used by the Firestone Tire and Rubber Company (which had taken over the G&A Aircraft Company in 1943) to test the effects of the drag and weight of rocket units at the blade tips of the rotor.

Townson (p. 80) chronicles the fate of one of the two repurchased PA-39s:

The autogiro [PA-39] was donated to Princeton University's Forrestall Research Center. Princeton University never flew it. About 1959, it was sold to one of their mechanics who assembled it and John Miller, one of Kellett's former test pilots flew it. Later it was used by Umbaugh Aircraft when they were first promoting their gyroplane. Their test pilot was Fred "Slim" Soule who had flown all the PA-39s originally. Then it was bought by Ryan Aeronautical of San Diego. It was badly damaged in a landing accident there. It has since moved through several owners and the son of Harold Pitcairn, Stephen Pitcairn acquired and restored it for static display at the EAA Antique Airfield in Oshkosh, Wisconsin.

36. But see Brooks (pp. 228–29):

None of the seven PA-39's (BW828 to BW834) were ever employed on actual operations in the intended role. Five were to have been used to by the RAF for communication duties but only two reached the United Kingdom and apparently only one (probably BW833) was flown there—initially at Duxford by Alan Marsh and probably later at Boscombe Down. Three (BW828–BW830) were damaged—according to one account, deliberately sabotaged—in January 1942 in Canada while being loaded for shipment to the United Kingdom and were scrapped together with spares in the same consignment. At the time it was stated that they had been lost at sea when their ship was torpedoed.

37. Frank Kingston Smith even maintains that Larsen's retorts eventually included some remarks "of a somewhat personal nature," with no specification.

Smith, *Legacy of Wings,* pp. 300–1. For a somewhat suspect discussion of some of the rumors that circulated about the Pitcairn family, see (with caution) Kathryn E. O'Brien, *The Great and the Gracious on Millionaires' Row* (Utica, New York: North Country Books, Inc., 1978), pp. 83–89.

38. For schematic drawings of the passenger Autogiros, see Smith, *Legacy of Wings,* pp. 364–67.

39. For pictures of the ill-fated XO-61, see Smith, *Legacy of Wings,* p. 308; Brooks, p. 240; Townson, p. 84.

40. Also designated as YO-61. See Brooks, p. 240; Townson, p. 84; but see Smith, *Legacy of Wings,* p. 308, where the model is called the "Firestone OX-61."

41. Brooks, p. 240; but see Townson, p. 84, where the author maintains that only one aircraft was completed.

42. The Autogiro Company of America continued as a licensing company into the late 1950s and attempted to license Skyway Engineering to produce the AC-35 "roadable," an attempt detailed earlier in this book, which was unsuccessful.

43. After the war NAS Willow Grove was designated a Naval Reserve Training Station. The navy subsequently enlarged the facility to its present 1,100 acres, and in 1994 the name was again changed, this time to Naval Air Station Joint Reserve Base (NAS JRB). For an official history of the military base see http://www.nasjrbwillowgrove.navy.mil/history.htm.

44. Smith relates this to the loss of the airfield and claims that this occurred in 1943. Smith, *Legacy of Wings,* p. 316. But see Brooks (p. 352), who claims that this occurred in 1942.

45. As cited in Smith, *Legacy of Wings,* p. 307.

46. For photographs of the Kellett YO-60, see Brooks, p. 239; Townson, pp. 126–31 (with chief test pilot Dave Driscoll).

47. For a depiction of the use of a light "spotter" plane for reconnaissance in World War II, see the 1965 movie *Battle of the Bulge,* starring Henry Fonda.

48. E. Stuart Gregg, "Jump Ship," *Smithsonian Air & Space* 15, no. 6 (March 2001): 14–15.

49. *Aircraft of the National Air and Space Museum,* 4th ed. (Washington, D.C.: Smithsonian Institution Press 1991); see "Kellett XO-60". Note that Townson (p. 131) maintains that the Smithsonian exhibit is one of the YO-60s that was damaged and then repaired, but his claim is not credible.

50. Spenser, p. 381.

51. *Aircraft of the National Air and Space Museum;* the XR-8 is listed as being "In storage."

52. Roger Ford, *Germany's Secret Weapons in World War II* (Osceloa, Wisconsin: MBI Publishing Company, 2000), p. 30.

Chapter 9

THE AUTOGIRO GOES TO WAR: THE ALLIES

Well, Wallis, I would rather see a man with a bit of fire in his belly who really wants to fly, than some of the perfect specimens I get. I am going to prescribe a pair of flying goggles for you with a corrected lens on one side. Don't bother to put them on, but if you get an eye shot out, put them on and bring the aeroplane home.

Air Commodore Livingstone to Ken Wallis,
allowing him to continue flying in World War II, "Profile:
Wing Commander K H Wallis," *Popular Flying*

"A honeymoon in Cairo in a brand new autogyro..."
Dick Haymes singing to Helen Forrest
in "I'll Buy That Dream" (1945)

ENGLAND

Although the American military had failed to adopt the Autogiro, it was a different story in Europe. Cierva had been encouraged to relocate to England chiefly by James G. "Jimmy" Weir, well-known Scottish industrialist who had previously been secretary of state for air. Given Weir's connections and munitions procurement/aviation sophistication, it is not surprising that the military had been intrigued with the potential of Cierva's Autogiro since 1925. But military interest was also evident in

Europe by the mid 1930s with the first commercially available Autogiro, the C.30 and C.30A were widely purchased by foreign military authorities and civilian manufacturers who supplied the military for evaluation prior to World War II. C.30s were exported to Hong Kong, Brazil, Czechoslovakia, India, Italy, the Soviet Union, Lithuania, Germany, China, Australia, Argentina, Sweden, and Austria, and were purchased by Lioré-et-Olivier in France and the air forces of England, Poland, Spain, Belgium, and Yugoslavia as well as the Danish army.[1]

Additionally, while the authorities in Japan had evaluated the C.19 Mk.IV, the Japanese military adopted a Kellett model as their basic configuration and produced more military Autogiros than any other World War II combatant. And therein perhaps lies the reason why the military record of the Autogiro is largely unknown: Autogiro development largely stopped with the coming of war in 1939, and these aircraft were primarily used by defeated military powers. The French (defeated initially) and the Japanese (defeated eventually) were the major users of the Autogiro, and England alone among the victors employed a squadron of Autogiros in a highly specialized, and today almost totally unremembered, vital role. In the United States and Soviet Union, helicopter development eclipsed the Autogiro by the start of the war, and the Autogiro played a small and relatively insignificant wartime role. But that leaves Germany, and with the success of the Fa-61 and Flettner helicopters, there would seem no impetus for Autogiro development, yet there were seemingly insignificant developments of a rotary kite that was to prove of greatest significance to the survival of Cierva's vision.

The Royal Air Force had found the C.19MkIII unacceptable, but in 1933 Cierva pilot Reggie Brie demonstrated the C.19MkIV during the annual army maneuvers at Salisbury. This included taking senior officers along as passengers, to favorable reviews, and the introduction of direct control induced the authorities to reevaluate a military role. The following year Brie would stress that the direct control C.30 was *much* easier for the average pilot to fly under all conditions, a condition important for the British military that contemplated wide adoption.[2] The military, defining two distinct roles for the C.30A, army cooperation (liaison) and naval functions, subsequently adopted the direct control C.30A. The air ministry ordered ten C.30As for the former on July 9, 1934, called Rota I (from *rota*ry aircraft) and later ordered two naval models with floats called Rota II. The first ten were built by Avro and given the number 761.[3] In September of that year Flight Lieutenants W. Humble and R. H. Haworth-Booth were trained with the C.30A and assigned to instructor duty in the RAF School of Army Cooperation at Old Sarum, where six Rotas had been accepted for service by November 22. The Rota's obvious reconnaissance and observation potential prompted the War Office to officially end the dangerous World War I practice of utilizing captive observation balloons, a Christmas Eve decision that probably occasioned much relief on the part of those assigned to such hazardous duty.

By the following September the six Rotas assumed a military role in combined RAF/Army war games but with only limited success, as the C.30A's performance under actual battlefield conditions left much to be desired. The aircraft required a ground run of 450 feet for takeoff and often suffered from ground resonance experienced in landing on rough ground, defined as "self-excited mechanical (potentially destructive) vibration on the ground of a rotary-wing aircraft involving a couple between the blade motion and that of the supporting structure or of the whole aircraft."[4] These were seen as serious handicaps and led not to additional military duties but to a serious course of research including wind tunnel testing of a model at the National Physical Laboratory and, in 1937, in the French Chalais-Meudon wind tunnel. The RAF made no further efforts to acquire additional Rota I models; they would make use of light airplanes for observation, communication, and reconnaissance functions in World War II, as did their allies and enemies, but the Rota I remained in the RAF inventory with seemingly no defined role at Old Sarum.[5] That would change with the coming of war in 1939.

Qualified pilots were quickly inducted into RAF service, and Squadron Leader R.A.C. Brie, flying a C.30A, played a unique role in calibrating the United Kingdom's new radar chain, which was soon to play such a large part in the Battle of Britain. Brie's success soon led in July 1940 to the creation of specialized units used mainly for radar calibration. Thus, now having finally found a vital role for the Autogiro, the military requisitioned civilian C.30As and, along with the remaining Rota I aircraft, assigned them to eight Radio Servicing Units as part of the No. 74 (Signals) Wing. These units were later consolidated into No. 1448 Flight, based at RAF Hendon,[6] Odiham, and Duxford. This mixed squadron of Autogiros and Bristol Blenheim Mk IVs was initially commanded by Flight Lieutenant M.J.B. Stoker and later by Brie, who was then promoted to Wing Commander. In June 1943 the Autogiros were ordered to No. 529 Squadron, the RAF's first operational rotary-wing unit, operating from Halton and Crazies Hill near Henley-on-Thames until disbanded on October 20, 1945. At its largest, 529 had seventeen C.30s in service and accumulated a total of 9,141 flying hours. Brie was not available for command, as he had been ordered to the British Purchasing Commission in Washington, D.C., but the 529 was led by another Cierva veteran, Squadron Leader Alan Marsh.

Flying Officer Norman Hill described in a 1963 magazine article the procedure by which radar calibration was achieved. He described the flight of July 14, 1943:

> Working with CHF (Chain-Height-Find), Rye3, I had to orbit about a dozen marks on land and sea. Special markers were first dropped for the sea runs, around which the smallest possible orbit had to be maintained for a period of three to six minutes, at altitudes of two, three, and four thousand feet, while the special squegger aboard transmitted signals to the radar stations.[7]

But such activities and reconnaissance missions in which the Autogiro ventured into harm's way always carried with it the possibility of confrontation with the enemy. As Lieutenants Gregory and Nichols had discovered in testing the Kellett YG-1 in 1936, the Autogiro's turning and descending abilities could outmaneuver fighter planes, which generally only had one pass, but those drills, conducted during artillery-spotting exercises, did not feature live ammunition.[8] But Flying Officer Hill[9] was about to discover what it was like to face real ammunition and enemy pilots determined to down the Autogiro.

The sun was setting late on the July afternoon as Hill was completing his final calibration exercise—he passed through some disturbed air and realized for the first time that there were other aircraft in the area. When trying to find the cause of the disturbance, he first noticed an aircraft flying below his position and became alarmed when realizing that it was a well-armed German fighter, a Focke-Wulf Fw 190 capable of flying at speeds well above 300 mph. Although the Rota theoretically had a maximum speed of 110 mph, Hill's experience was that aircraft's top speed was 85–90 mph,[10] so he knew all that stood between him and death were the unique flying characteristics of the Autogiro! Pretending that he was unaware of the Fw 190, which had looped upward from below the Rota and was then positioned for a strafing run with its cannons and machine guns, sweating profusely with his hand on the stick, Hill waited until the last moment before tilting the rotor head backward, causing the Rota to slow and flare upward and the German plane to pass harmlessly overhead. Hill then pushed the stick hard to port, causing the Rota to turn and dive toward the ground, a maneuver the German pilot declined to follow. But even as Hill struggled to regain control of the Autogiro, which was locked in a steep dive, a *second* Fw 190 appeared and closed for a kill. Again

Hill's flying ability and the aircraft's capabilities saved the day, as he deliberately turned directly toward his attacker, both presenting the smallest possible profile to the attacking aircraft and likely scaring its pilot, a technique that worked. The second aircraft broke off the attack at what seemed the last minute and passed *below* the Rota. The entire encounter had taken only three minutes and used up most of Hill's fuel, but it carried the planes considerably inland, where Hill was able to continue in steep but controlled descent to a safe landing, while the Fw 190s presumably returned to their bases across the English Channel. The Autogiro had survived, in much the same manner as established by Gregory and Nichols seven years before.

FRANCE

Cierva's early demonstrations in France had attracted the distinguished aviation engineer Captain Georges Lepère, who built the first cabin Autogiro, the C.18, in June 1929.[11] Avions Weymann-Lepère had been formed in January 1929, predating the formation of the Pitcairn-Cierva Autogiro Company of America by several weeks, and acquired a Cierva license when the English company had ordered a metal cabin Autogiro in early 1929. It was built for Loel Guinness as an entry into the November 1929 Guggenheim Safe Aircraft Competition. The aircraft had been first flown at Villacoublay on August 12, 1929, and it is thought that Cierva himself may have been the pilot, just prior to his departure for America. The C.18 was taken to the Pitcairn factory in America for reassembly and testing prior to the Guggenheim competition, but it experienced high vibration levels and never did enter the Guggenheim.

When Lepère left the company in 1930, it was renamed Éstablissements Aéronautiques Weymann, and it continued with a Lepère project for the French navy, the Weymann CTW.200, also known as the WEL.200. This was a side-by-side two-seat, with dual controls in the open cockpit. The fuselage was of a chromemolybdomen steel tube construction covered by fabric. It featured a four-blade rotor mounted on four-strut pylon attached to the fuselage just forward of the cockpit.[12] With its upturned wingtips and box deflector tail for rotor spin-up, the French Autogiro bore a strong resemblance to the Cierva C.19MkIII. The CTW.200 was exhibited in mock-up at the Twelfth Paris Solon in late November and early December 1930 and was first flown Easter week. Cierva came from England to pilot the initial flights, but it was subsequently flown by Weymann test pilot Pierre Martin. Martin also flew demonstration flights of the C.19MkII's in

May in the air show at Orly, and the French navy ordered two Weymann-Lepère Autogiros, but it "is not known whether both Autogiros ordered by the French navy were completed and delivered."[13] The second model was denoted the CTW.201, a heavier cabin model of the previous CTW.200, with a more powerful engine. It was the second European Autogiro to have an engine-powered prerotator but it was not relevant for military development in France, as such efforts had by then shifted to the Lioré-et-Olivier company, which Georges Lepère had joined. But it was clear that from 1930 onward, French military authorities were interested in the Autogiro. That interest was to result in a significant but doomed deployment of Autogiros in the early days of World War II.

After leaving his former company, Lepère joined the French aviation firm founded by Fernand Lioré and Henri Olivier in 1906. Largely due to his rotary-wing enthusiasm, Lioré-et-Olivier acquired a license from Cierva in 1931 to manufacture and sell the C.19MkIV, but they apparently did not do so, and that license acquisition may have been to establish a relationship with the English company. This led to a confusing circumstance, where Weymann held the French Autogiro design license, while Lioré had the license to build the C.19MkIV, which was resolved when Cierva himself proposed in March 1932 that the design licensee be transferred to Lioré but allowed Weymann to continue with the construction of its own models, the CTW.200 and CTW.2001. The license was officially transferred in February of the following year. The result was a series of Autogiros denoted as the C.L. series (for "Cierva Lioré"), and they would furnish the French military with the largest military Autogiro component in World War II.

The direct control Autogiro provided the impetus for the French military to seriously consider rotary-wing aircraft. Lepère had worked with Cierva in England during the last three months of 1932 to design the direct control C.L.10 Autogiro, which featured a hanging-stick control column for pitch control but which also utilized a wheel attached to the end of the control column for lateral control. This was different from that being developed in England and did not prove successful. Cierva, flying the second C.L.10 on November 24, 1932, at Orly, found it unstable, with an overly sensitive tilting rotor head control. There were several sequential Cierva Lepère models, but they did not prove successful. The first C.L.10 was modified in England and became known as the C.L.10A. The Cierva-Lepère C.L.10B, an extensive modification of the original C.L.10, was produced in France in 1933. The Cierva-Lepère C.L.20, a prototype constructed by Westlands at Yeovil in August 1934, a side-by-side direct control cabin two-seater, proved underpowered when first flown by Cierva and Alan Marsh on February 4, 1935. It never received a Certificate of Air Worthiness and was scrapped in 1938. Nothing ever came of the plan to market this aircraft, much like the fate of the PA-19 Cabin Autogiro in America, which, by all accounts, was a much more impressive aircraft.

As in England, it would be the C.30A direct control Autogiro that was embraced by the French military. Cierva had flown the C.30P (G-ACIO) to France and Spain early in 1934, and Reggie Brie had made a notable non-stop flight from London to Paris in two hours and twenty minutes on January of that year, making an impressive delivery to the French Air Ministry, which had acquired the C.30P for evaluation. On February 8 Cierva flew another C.30 to Paris, where he demonstrated its capabilities before representatives of the French army and navy at Villacoublay, along with flights by Lioré test pilot Lucien Bourdin. The enthusiastic reception prompted an expansion of the Lioré C.19MkIV license to include the C.30, and Fernand Lioré established a separate department for the production and development of Autogiros under the direction of Ingenieur Pierre Renoux and Roger Lepreux for flight-testing. Autogiros had caught the public's fancy, and a separate gyroplane license category had been established in January of 1935 with the support of the newly established Club Autogire de France—Roger Lepreux obtained one of the first rotary-wing licenses and was undoubtedly congratulated by club president Juan de la Cierva!

Lioré, under the terms of its expanded Cierva license, ordered four Avro C.30As in 1935, the first of which was flown in July from England to Paris by Lepreux. This was turned over to French officials for evaluation and extensively flown in army war games in September at Le Val d'Alion and Mourmelon. The second C.30A arrived in October and garnered extensive publicity and acclaim when Lepreux landed in front of the Grand Palais on the Champs Elysées, where the Fourteenth Salon de l'Aéronautique was being held. The third C.30A arrived by the end of 1934, and the final aircraft was in France by April of the following year. That the French military was serious in considering the C.30A was evident in the final model, which had been fitted with a locally produced 203-horsepower Salmson 9Nd engine and a Ratier propeller in anticipation of French production. The flight-testing of the first three produced a favorable evaluation, and the government authorized purchase of the four aircraft on December 28, 1934. The English Autogiros receiving their Certificats de Navigabilté in the 1935–36 period were assigned to the Flight Test Center at Villacoublay until being transferred to the Armée de l'Air in 1939 in preparation for war.

France had designated its naval air arm as the Aéronautique Navale (l'Aéronavale) in 1925, and the Aviation Militaire itself became the Armée de l'Air in 1933.[14] Lioré-et-Olivier received an order for twenty-five Autogiros on April 25, 1935, for the Armée de l'Air, which intended to use them in reconnaissance and artillery-spotting roles. This order came quickly after delivery of the fourth Avro-built aircraft, with the first five C.30As coming from England as components for French assembly in June of that year, but delivery was delayed until January of 1936 so that modifications requested by the French authorities could be made. This delay, also allegedly occasioned by French claims that the English production drawings were inadequate, reflected the growing tension between the English and the French manufacturer, clearly intent on local production. These modified aircraft, known as the LeO C.30s,[15] could almost achieve a jump takeoff and stimulated naval interest. Of the original order, four were allocated to l'Aéronavale in late 1935, and the Armée de l'Air received the last of the twenty-five in July 1936. In extended operations and testing, the army did not embrace the C.30A, as it was found to be too slow and to have a poor climb rate, occasional lack of stability, inadequate landing gear for rough field landings, and easily damaged rotor blades.

The Autogiro was seen as a replacement for the artillery observation balloons used in World War I, and had it been that kind of war, it might have achieved success. But as the Polish cavalry discovered when its gallant but doomed officers charged German tanks, this was to be a different kind of conflict, and therein is to be found the explanation for the lack of Autogiro military success and its quick exit from the European war arena. As often occurs, the French generals were planning to fight the last war, which had been distinguished by static confrontation along entrenchment lines and had created the Maginot Line of hardened defensive positions to oppose the German Siegfried Line. Had the conflict mirrored the static preparations, the Autogiro might have readily assumed a valuable reconnaissance role, as previous American experience and later British experience demonstrated that an Autogiro could successfully evade fighter aircraft. But the devastating German blitzkrieg doomed the planned Autogiro military role and quickly drove it from the field of battle.

The army had planned in 1937 to use forty-six Autogiro units, each consisting of three aircraft assigned to reconnaissance roles, but this had been reduced to thirty units by 1938. It was not anticipated that war would come before 1941, and Aircraft Plan V of 1938 called for the activation of six units in 1938 and eight in each of the next three years. This also reflected the preparation for war as early as 1936, when the French had become seriously alarmed by the resurgent German military Observation units and had begun receiving Autogiros on November 16, 1938. At the beginning of September in 1939 before the German attack on Poland, fifty-five Autogiros had been delivered to the Armée de l'Air, (sixty-four LeO C.30s would be delivered by the end of 1939) with fifty-two available, but only twenty-eight were in operational status, with an additional five then under repair. Three were in an Autogiro training unit (Center d'Instruction à l'Observation sur Avions Autogyres) and sixteen in a storage depot (Entrepots de l'Armée l'Air). Thus did the French Autogiro go to war, with the first reconnaissance missions being flown in October 1939 by Adjudant de Zimmer of Squadron (Group d'Aviation d'Observation) 1/514 over enemy territory.

The French navy was also using the C.30A to track torpedoes from submarines and surface ships and to calibrate gunnery director radars on the larger worships. A total of eight Autogiros were in service with the navy at the beginning of September 1939, but four more were acquired from the air force by March of the following year, while only forty-seven remained in air force service. By May 10, when the Germans attacked westward, the air force had only eighteen operational aircraft with five squadrons and the training school, while the navy had assigned nine Autogiros into its only operational unit, held two in reserve, and had assigned two to the Autogiro school at Hyères-le-Palyvestre. Of the thirty-one operational Autogiros that began the war on May 10, only seven remained in unoccupied France by the Armistice of June 25, 1940; six of those were captured by Italian forces when Vichy France was overrun in November 1942 and soon became unserviceable. The seventh Autogiro was hidden, was later restored, and is preserved in the Musée de l'Air et de l'Espace at Le Bourget.[16] It was the only French Autogiro to survive the war.

Thus the role of the French Autogiro in war came to an end. Its pilots had not hesitated to go in harm's way. During the brief war Autogiros were used "mainly for short-range liaison duties," and at least two were lost in late May. Captain Guy Briand, previously assigned to the Autogiro Training Unit, was more fortunate while on a reconnaissance mission over German forward lines. He was machine-gunned by a formation of nine Dornier Do 17s flying about 330 feet above his position, but he escaped. Five of the hundred ordered FSNCASE C.301 models[17] had been completed—they were an improved version of the C.30 and featured tabs on the rotor blades that enabled stable high-speed longitudinal flight and pinpoint landings. The latter was significant in that it allowed for instrument flying, and the former proved of value in combat areas, but the coming of war and bombing of the plant curtailed production. They ended up using French rotors when those ordered from Norway could not be shipped to France. It was considered by French authorities to be the best French Autogiro, but it did not survive war, although one was demonstrated at Marignane before a delegation from the German armaments commission.

SOVIET UNION

The Soviet Union had a similar, but much more limited, wartime involvement with the Autogiro. Autogyro development had began in 1929 with the KaSkr-I, an unauthorized copy of the Cierva, named after its designers Nikolai I. Kamov and Nikolai K. Skrzhinsky (Skrzhinskii).[18] It was a modified U-1 trainer, a Soviet copy of the Avro-504K, with an M-2, the Russian copy of the 110-horsepower Le Rhône rotary engine and a wider track landing gear, and it was nicknamed the *Krasnyi inzhener* ("Red Engineer"). It was not flown,[19] as ground tests revealed it to be unstable and prone to overturn sideways, as well as having an inadequate control system and being generally underpowered. It did, however, have a relatively advanced clutch connection to the engine to spin up the four-bladed thirty-nine-foot, four-and-a-half-inch diameter rotor. Modified with a more powerful 230-horsepower French Gnôme-Rhône Titan air-cooled radial engine in a helmeted cowling, it was dubbed the KaSkr-II and was similar to the Cierva C.8L-I. The KaSkr-II first flew in mid-1930 with pilot D. A. Koshits, who made some ninety test-flights, reached an altitude of 1,500 feet, and achieved a maximum speed of sixty-eight mph. The aircraft was subsequently flown with skis during the winter of 1930–31 and was "presented to the state authorities and military commanders at Khodynka airfield in Moscow in May 1931,"[20] who were enthusiastic about its military applications. By October the designers, installed at a special design department known as the OOK (Otdel Oskbykh Konstruktsii) that had been formed within TsAGI at the end of 1926,[21] embarked on the design of an autogyro capable of combat reconnaissance, artillery spotting, and liaison duties. This design/developmental function was made more complicated by the military requirements that the autogyro performance be comparable to that of a light fixed-wing aircraft and be capable of carrying a radio, camera, machine gun, and bombs. This autogyro, called the TsAGI A-7, was from the outset intended as a multi-mission *military* aircraft. But it was not the only Soviet autogyro project.

The second Russian gyroplane was designed by I. P. Bratukhin and Vyacheslav A. Kuznetsov toward the end of 1930, who were then also at the OOK. Dubbed the EA-2 (for "second experimental autogyro") it resembled the Cierva C.19,[22] which had undoubtedly been observed by Soviet agents then in England. The 2-EA was noteworthy in that it was not derived from an existing Soviet aircraft but designed from the beginning as a test platform. Its construction was supervised by A. M. Izakson and featured a welded steel tube fuselage covered by fabric and a four-bladed cable-braced rotor on top of a three strut pylon above the forward open cockpit of the two-seat autogyro.[23] Cierva's earlier influence was apparent in the use of the deflector box-tail to prespin the rotor, and from that standpoint it was not as advanced as the KaSkr I or II models, which incorporated a mechanical drive, but in other respects the 2-EA was a sturdy development platform. Employing the same Gnôme-Rhône Titan engine that had been used in the KsSkr II, it first flew on November 17, 1931, piloted by Sergei A. Korzinshchikov, and its vibrations problems were quickly overcome. After development testing, the single 2-EA was transferred to the Maxim Gorkii Propaganda Squadron *(Makxim Gor'ky propaganda eskadril'ya)* and presented to the Osoaviakim Museum in early 1934, but test program success, however limited, led to further development.

The TsAGI 4-EA (also called the A-4), an autogyro for military pilot training[24] and observation duties, was produced by the TsaGI OOK under the direction of N. K. Skrshinskii, A. M. Cheremukhin, and G. I. Solnitsev. Development commenced early in 1932, and a decision was made in June to go ahead with a limited production run even though the prototype would not fly until November 6 of that year.[25] Under the direction of Pyotr I. Baranov, it had taken only twenty-four days to construct the 4-EA, an amazing feat, but it would take less time for the model to crash, as the production decision proved premature. When Korzinshchikov took to the air on November 6, he immediately encountered vibration problems coupled with low motor rpm, and the second flight, on November 9, resulted in a crash that Korzinshchikov fortunately survived.[26] The 4-EA, employing a locally produced 300-horsepower M-26 engine (a license-built variant of the American Wright Whirlwind nine-cylinder radial engine) enclosed in a Townend ring cowling, upturned wingtips and conventional tail, spun up its rotor by means of a mechanical connection with the engine,[27] similar to the Pitcairn PCA-2. The vibration problem was solved in a pragmatic fashion with testing of several rotor configurations, and the model entered limited military production.[28]

TsAGI OOK continued to develop experimental autogyros with the A-6, under the direction of V. A. Kuznetsov and his team, inspired by the Cierva

C.19MkIV being developed simultaneously as the A-4. The A-6 was a smaller two-seat aircraft employing a three-blade cantilevered rotor that could be folded back for convenient storage.[29] Korzinshchikov served as test pilot for the first flights early in 1933. Even though the A-6 was demonstrated at the Moscow Aviation Festival on August 18, 1934, future development lost in the internal TsAGI power struggle and did not proceed, as it was viewed as being in conflict with the A-4, which was then entering limited military production, but the A-6 made important contributions as an experimental platform to explore issues of stability, control, and ground resonance. Although two additional aircraft were constructed, dubbed the A-8 and A-13/A-14 series,[30] they were limited test platforms during the 1935–36 time frame, with the A-8 first flying on September 17, 1935, and the A-13 on March 13, 1936. Of limited success and even more limited impact, these models represented the final autogyro achievements under the direction of Kuznetsov. Another design team within TsAGI was led by Nikolai Kamov, and his team would develop the most successful Soviet autogyro, the TsAGI A-7.

Begun in 1931, the A-7 was designed from the very beginning as a powerful *military* autogyro fully capable of the expected reconnaissance and liaison duties, but also armed and packing a punch in the form of machine guns and bombs—this was the first rotary wing-aircraft intended as a *combat* aircraft.[31] Originally designated as the EA-7, and later the A-7, it had a fuselage of welded steel tubes with duralumin covering and an integral fin.[32] The best known of the Soviet autogyros, the two-seat A-7 was a "strong and robust machine, made of metal"[33] first flown by Korzinshchikov on September 20, 1934. It was not the first metal-fuselage Autogiro (that had been the French-built Weymann-Lepère C.18[34] featuring a stress-metal skin in 1929[35]) nor was it the most beautiful (that was undoubtedly the Pitcairn jump take-off PA-36 of 1938). But the A-7 was built *tough,* a muscled brute of a machine powered by a radial nine-cylinder 480-horsepower M-22 engine, a proven Gnôme-Rhône Jupiter 9ASB motor that had been built under license since 1930. The motor was streamlined by the addition of a Townend ring cowling and utilized a wooden, two-bladed, fixed-pitch propeller. Comparable in size and weight to the largest Autogiros built by the Cierva licensees, it was the largest autogyro built in the Soviet Union and the most powerful. And to increase speed for this war machine, the tricycle undercarriage landing gear and main rotor supports were encased in streamlined fairings. The three-blade rotor could be folded back for more compact storage, and it was claimed that Kamov included this feature in anticipation of future shipboard deployment.[36]

Given the A-7's multiple combat missions, it is not surprising that it initially weighed in at a hefty 4,530 pounds,[37] eventually increased to 5,070 pounds.[38] This consisted of a 3,416-pound airplane structure, 628 pounds of fuel, 77 pounds of lubricating oil, and a two-person crew weighing a maximum of 396 pounds, with the remaining 553 pounds being allocated to a 13SK-3 radio transmitter,[39] the Potez-1bis camera,[40] and armament.[41] The heavy, durable three-blade cantilever rotor was constructed of stainless steel and prerotated by an engine-driven transmission gear that had become standard since the American PCA-2 in 1931. Kamov's intent was to produce an armed combat aircraft, and he succeeded as no one had before—the machine carried a fuselage-mounted 7.62 mm ShKAS PV-1 fixed machine gun carrying 250 (later increased to 500) rounds, activated by the pilot in the forward cockpit and synchronized with the propeller for effective forward fire.[42] But the A-7's lethal bite did not end there. Kamov had also included a TUR-6 gun post in back of the rear cockpit, on which were initially mounted two, later reduced to one, Degtyarev light machine guns with 10 to 12 magazines. It could carry four 220-pound or two 250-pound bombs suspended beneath its wings, and Kamov later added provision for six RS-82 unguided rockets, and some of the rockets could be reversed under the wing, to be fired to the rear for protection against pursuing fighters. It had a minimum speed of 30 mph and could fly at 130 mph flat out.

As the aircraft employed fixed-wing control surfaces, it was not current with direct control machines that were then flying in England and America, but the A-7 successfully passed through its initial tests after the maiden flight in 1934 by Korzinshchikov and caused a sensation when flown and exhibited at the Soviet Air Display Day on August 18 of the following year. Factory testing was completed on December 9, 1935, and the prototype was turned over to the state aviation authorities for acceptance testing, which continued until April, 1936, with A. A. Ivanovskii as pilot. Additionally, as part of the testing, a C.30A had been imported from England for comparative testing—and the Soviet authorities claimed that the A-7 had better performance and was better suited for a military role.[43] But the testing had revealed deficiencies, including a slight lack of directional stability; rotor, tail, and stick vibrations; and engine overheating, so a second modified prototype was produced in March 1937. Dubbed the A-7bis, the aircraft featured the addition of a vertical fin on each side, a two-strut rotor pylon that resulted in increased side visibility, and greater streamlining that resulted in decreased drag. The A-7bis was tested during the May 1937–July 1938 period and employed under combat conditions during the Soviet-Finn war of 1939–40, when the prototype was tasked with several reconnaissance missions. As a result of the successful testing and experimental military deployment, five military production aircraft were ordered with a slightly lighter airframe and less aerodynamic unfaired landing gear. Production was begun in Smolensk in 1939, with the first delivery of the A-7-3a occurring in early 1940.

A prototype had been used in 1938 in Greenland and employed on board the ice-breaker *Yermak* during an expedition to rescue Papanin's North Pole station[44] from an ice floe. Another was used in 1938–40 by Aeroflot for forestry patrols during the Tien-Shan (Tyan-Shan) expedition in Central Asia, where during April–May 1941, the A-7bis was tested as to its suitability for agricultural spraying, previously demonstrated by pilot George Townson flying a PCA-2 for Giro Associates of Morristown, New Jersey, in 1938.[45] The Soviet tests proved that the rotor disk was highly effective in crop dusting and spraying insecticides and fungicides, as the chemicals were efficiently forced downward, and in 1939 an American author concluded that "in many respects, the autogiro is an ideal machine for dusting. An experienced 'giro pilot can hover his ship at low altitudes and literally push his chemicals into hollows."[46] It was estimated in the Soviet experiments that the A-7bis could achieve an efficiency twice that of the most effective contemporary fixed-wing agricultural aircraft, but the coming of the German war machine to Russia in June 1941 ended these experiments. The Soviet autogyros were about to face the might of Hitler's blitzkrieg war machine. It would prove to be a decidedly one-sided confrontation.

Five A-7bis machines were prepared for combat and deployed to the front lines as a separate squadron within the 163rd Fighter Regiment of the 24th Air Army, under the command of Captain P. Trovimov and with Mikhail L. Mil, who would later achieve fame and honor as a helicopter engineer, as squadron engineer. The designated task at the Smolensk front was reconnaissance and propaganda, chiefly the dropping of leaflets, but the A-7bis did not prove effective in either mission. The aircraft proved vulnerable without fighter protection, and daylight flights ended quickly, as the autogyros were quickly reduced to nighttime close reconnaissance and leaflet dropping.[47] Because the sorties were primarily executed at night, no aircraft were lost to enemy fire, but there were several forced landings, which severely damaged two of the aircraft so that only three were flying by October,[48] and by the end of the month all autogyros were withdrawn from combat for repair. Factory No. 290 had been moved to Bilmby village near Sverslovsk, and it was to that relocated factory that the five A-7bis aircraft were taken, but they would not be returned to combat. In the words of historian John Everett-Heath: "The A-7s were not popular machines to fly, being cast rather in the role of Soviet lambs to the German slaughter."[49]

The withdrawal of the A-7bis marked the end any Russian military role for the autogyro, although Kamov had proposed in 1940 a wingless, direct control autogyro for reconnaissance and liaison duties, but the AK prototype never flew.[50] Additionally Skrzinsky had designed a single-seat autogyro fighter, the A-12.[51] It was intended to meet state-of-the-art military performance standards: a minimum-maximum speed range of 28–186 mph, an altitude ceiling of 23,000 feet, and a ground taxi of no more than 150 feet. The prototype utilized a 670-horsepower Wright Cyclone built under license and designated the M-25, with a NACA cowling, streamlined fighter fuselage and semi-enclosed cockpit. Flight-testing had commenced on May 10, 1936, and actual flight achieved on May 27 by pilot A. P. Chernavsky. That testing program proceeded slowly so that only forty-three flights had been made, for a total of eighteen hours, in the next year. On May 23, 1937, the A-12 crashed after a rotor blade came off in flight,[52] fatally injuring pilot Ivan Kozyrev, ending the project. A speed of 152 mph had been established, but even if this model had succeeded, the Soviet military had become aware of German helicopter development and, as in America, was already beginning to turn away from the autogyro. Mikhail Mil had already designed the two-seat A-15 wingless direct control autogyro, which would have been the largest and most powerful Soviet autogyro, with a 750-horsepower M-25V engine, but it was shelved with the crash of the A-12. It was put into storage and signaled the end of Soviet autogyro development. The abysmal World War II record of the A-7bis did nothing to revive the technology, and an autogyro would not reappear in the Soviet Union until the early 1960s.

The Soviet involvement with the autogyro or gyroplane proceeded down paths already blazed by Cierva and Pitcairn but must clearly be distinguished. Although a C.30A was sold to Russia and flown in comparison with the A-7, the Soviet authorities never became a Cierva licensee (hence never produced an Autogiro), and it is certainly likely that Cierva would not have granted such a license after the death of his only brother, Ricardo, at the hands of the Communists on November 6, 1936, in Paracuellos, near Madrid. But the Soviet development of the rotary-wing aircraft was notable in that, unlike either Cierva or Pitcairn, the Soviets conceived of the autogyro as a weapon of war from the very beginning. But their powerful aircraft were no match for the new kind of air war, and by the end of the conflict the autogyro had all but been forgotten. A similar end awaited the military Autogiros thousands of miles to the east, in Japan.

NOTES

1. Peter W. Brooks, *Cierva Autogiros: The Development of Rotary-Wing Flight* (Washington, D.C.: Smithsonian Institution Press, 1988), pp. 348–51.

2. R.A.C. Brie, *The Autogiro and How to Fly It,* 2nd ed. (London: Sir Isaac Pitman & Sons, 1934); see also Reginald A.C. Brie, "Practical Notes on the Autogiro," *Journal of the Royal Aeronautical Society* 4 (March 1939); reprinted as "Pilot's Notes on Flying the Direct-Control Autogyro in 1939," *Rotorcraft* 34, no. 5 (August 1996): 19–21.

3. Daniel J. March (ed.), *British Warplanes of World War II: Combat Aircraft of the RAF and Fleet Air Arm 1939–1945* (New York: Barnes & Noble Books, 1998), p. 16.

4. Brooks, p. 362.

5. But see March, p. 16, where the editor maintains that by outbreak of war the Rota I had been "struck off charge."

6. Brooks, pp. 191 and 372; but see March, p. 16. And see Flying Officer Norman Hill, "Wingless Combat," *Royal Air Force Flying Review* 18, no. 4 (January 1963): 24–25, 57, where this is related as "Halton" (p. 25). Hill's reference to the 1448 flight is confusing, as the Autogiros were ordered to the 574 Squadron the previous month, June 1943.

7. Norman Hill, p. 25.

8. Hollingsworth Franklin Gregory, *Anything a Horse Can Do: The Story of the Helicopter,* Introduction by Igor Sikorsky (New York: Reynal & Hitchcock, 1944), p. 58.

9. For a photograph of Flying Officer Norman Hill with a 529 Squadron Rota, see Brooks, p. 192.

10. Brooks states that RAF testing of the C.30A had produced a maximum level speed of ninety-four mph (p. 190).

11. For a photograph of the Weymann-Lepère C.18, see Brooks, p. 90 (original form); George Townson and Howard Levy, "The History of the Autogiro: Part 1," *Air Classics Quarterly Review* 4, no. 2 (Summer 1977): 4–18, 13 (later form).

12. For a photograph of the Weymann CTW.200, see Brooks, p. 92.

13. Brooks, p. 92.

14. Bill Gunston, *History of Military Aviation* (London: Hamlyn, an imprint of Octopus Publishing Group Limited, 2000), pp. 52–53.

15. Pierre Riviere and Gerry Beauchamp, "Autogyros at War," *Air Classics Quarterly Review* 3, no. 4 (Winter 1976): 92–97.

16. Bob Ogden, *Great Aircraft Collections of the World* (New York: Gallery Books, 1988), p. 39. Along with the Smithsonian National Air and Space Museum and the United States Air Force Museum at Dayton, the Musée de l'Air et de l'Espace is one of the world's great aviation museums.

17. For a photograph of the SNACSE C.301, see Brooks, p. 202.

18. John Everett-Heath, *Soviet Helicopters: Design, Development, and Tactics* (London: Jane's Publication Company, 1983), pp. 5–6.

19. Brooks, p. 258; but see Everett-Heath, p. 6, where it is asserted that KaSkr-I was "flown for the next two years from Moscow Central Airport by I. V. Mikheyev." And see Lennart Andersson, *Soviet Aircraft and Aviation 1917–1941* (Annapolis, Maryland: Naval Institute Press, 1994), p. 336, where it is claimed, "On 25 September the test pilot Ivan Mikheev managed to fly some 200m at 2–2 1/2 m, but the KASKR was later extensively damaged before getting really airborne. The airframe was repaired at Zavod No 39 and modified into the KASKR-2."

20. "Kamov and Skrzhinsky: Russian Gyroplane Pioneers," *International Autogyro 1/4ly,* no. 10 (October 2001): 18–23, 18.

21. Andersson, p. 335.

22. There is some confusion about which model C.19 the TsAGI 2-EA most resembled. Brooks maintains it was the C.19MkI, while Andersson claims it was the C.19MkII. Compare Brooks, p. 260, with Andersson, p. 335. Everett-Heath stops short and merely claims that the TsAGI model resembled the C.19 (p. 6). Ogden, however, comments that the 2-EA was similar to the Cierva C.19MkIII. Bob Ogden, *Aircraft Museums and Collections of the World,* vol. 9, *Eastern and South Eastern Europe and the C.I.S.* (Woodley, Berkshire, England: Bob Ogden Publications, n.d.), p. 102.

23. For a rendition of the 2-EA, see Andersson, p. 335; for a schematic diagram, see Brooks, p. 259.

24. Brooks also claims that the A-4 was intended from the beginning also for civilian uses (p. 261).

25. For a photograph of the A-4 in flight, see Karl-Heinz Eyermann, *Die Luftfahrt der UdSSR 1917–1977* (Berlin: transpress VEB Verlag für Verkehrswesen, 1977), p. 74.

26. Everett-Heath wryly observes, "It was quite difficult to tell in the early days whether test pilot Korzinshchikov was testing the aircraft or whether the aircraft was testing him" (p. 6).

27. For a photograph of the A-4, see Brooks, p. 261.

28. Brooks claims somewhat ambiguously, "However, the A-4 was never produced in quantity, as had been intended at one stage. Rather more than ten are believed to have been built and the type was briefly evaluated by the military" (pp. 261–62); perhaps Brooks intends to say, "Rather *no* more...." Everett-Heath is not very helpful on the matter, as he states, "*Perhaps* 30 or 40 A-4s eventually found their way into military service" (p. 7, emphasis added).

29. Photographs of the A-6 are found at Brooks, p. 262; Eyermann, p. 75.

30. For a photograph of the wingless A-14, see Eyermann, p. 74.

31. For photos and schematics of the A-7, see Brooks, p. 263 (photo and schematic); "Kamov and Skrzhinsky," pp. 19–23 (photos and highly detailed cutaway engineering drawings).

32. Bill Gunston, *The Osprey Encyclopedia of Russian Aircraft* (Oxford, England: Osprey Publishing, 1995, 2000), p. 67.

33. Everett-Heath, p. 7.

34. For photographs of the Weymann-Lepère C.18, see Brooks, p. 90; Townson and Levy, "History of the Autogiro: Part 1," p. 13.

35. Brooks, pp. 90–91.

36. "Kamov and Skrzhinsky," p. 19.

37. Everett-Heath, p. 8; but see Brooks, p. 264, where the author claims that the initial weight of the A-7 was 4,354 pounds. The difference may be that the former refers to the takeoff weight, while the latter includes only the weight of the aircraft and not the pilot.

38. Brooks, p. 264, which is substantially in agreement with "Kamov and Skrzhinsky," p. 20, where the design takeoff weight for the A-7 is given as 5,069 pounds.

39. Later replaced by a RSI-3 transmitter.

40. Later replaced by the AFA-27A in the production series.

41. "Kamov and Skrzhinsky," p. 20.

42. Everett-Heath, p. 8.

43. See "Kamov and Skrzhinsky," p. 20.

44. Brooks, pp. 264–65; Everett-Heath, p. 8. In Gunston, *The Osprey Encyclopedia of Russian Aircraft*, p. 67, that claim is referenced but resolved with, "Same prototype [A-7] shipped to Greenland 38 to help in rescue of Papanin expedition, *but not needed*" (emphasis added).

45. See George Townson, "Autogiro Crop Dusters," *American Helicopter Museum & Education Center Newsletter* 3, no. 1 (Spring 1996).

46. Hal MacKay, "Bug Fighters," *Popular Aviation* 24 (June 1939): 48–50, 82, 50 (photo caption).

47. For descriptions of the combat role of the A-7bis, see Brooks, p. 265; Andersson, p. 336.

48. "Kamov and Skrzhinsky," p. 21.

49. Everett-Heath, p. 8.

50. Reference to the obscure AK model is found in Everett-Heath, p. 8.

51. For information on the A-12, see Brooks, pp. 266–68; Everett-Heath, p. 9; Andersson, p. 337.

52. Gunston, *Osprey Encyclopedia,* p. 67.

Chapter 10

THE AUTOGIRO GOES TO WAR: THE AXIS

"What is the ultimate, do you think?" Kellett asked.

Together we asserted: "A craft than can go straight up and down. The Autogiro does the job halfway. A successful helicopter would be better."

Kellett smiled: "Yes, perhaps, but we will improve."

The steady improvement of the Autogiro, more than anything else, helped to bring about the first successful helicopter. It was the practical proving ground that gave us the key to controllable vertical flight.

—Hollingsworth Franklin Gregory,
Anything a Horse Can Do

GERMANY

At the start of World War II, a recent engineering graduate, Friedrich von Doblhoff,[1] suggested that a helicopter could be powered with rotor ram jets designed by French engineer Rene Leduk, which would effectively deal with the torque caused by mainframe engine placement. Enlisted in the German war effort as an employee of the Wiener Neustadter Flugzeug-werke (WNF), a Vienna aircraft manufacturer, Doblhoff recruited friends Theodor Laufer and August Stepan in the efforts to design a tip-jet helicopter in a visionary and unauthorized program of research. The test apparatus was constructed of magnesium tubing supporting a rotor with hollow blades, through which compressed air and vaporized gasoline passed to an automobile spark plug positioned at a tip exit nozzle designed for ignition. Although it was destroyed in its maiden test, observing officials were impressed. The machine had managed to lift off with an anvil added to the rig to weigh it down. Destruction had come when it tilted and its rotors struck the floor, but the results led to a half-million-mark authorization for an official project to design a jet tip helicopter.

The world's first tip-jet-powered helicopter, the WNF 342 V1, was flying in the spring of 1943. It was designed to meet a German navy requirement for an observation helicopter to be carried by submarines and small naval vessels. It featured a frame of uncovered metal tubing, with a small twin-finned vertical tail and tricycle landing gear. An Argus As 411 supercharger was adapted as a compressor to provide air to the rotors, an arrangement that would then be employed on all of Doblhoff's prototypes. The V1, slightly damaged in an Allied bombing raid on August 13, 1943, was soon followed by the WNF 342 V2, which added a sail-like rear fuselage fairing with a single fin and an upgraded 90-horsepower Walter Mikron engine. It was constructed in Obergraffendorf, where the WNF development program had been relocated after the bombing. Experience with the first two models convinced its inventor that the high fuel consumption of the tip-jets would make the WNF 342 prohibitively costly to operate, so the decision was made to power the rotors only on takeoff and landing. The rotors would be unloaded in flight, and the craft would then fly as an autogyro.

The resulting WNF 342 V3 was constructed with twin tail booms, each of which supported an oval-shaped vertical fin and rudder with a horizontal stabilizer linking the booms. A BMW-Bramo Sh 14A 140-horsepower engine both provided forward thrust with a pusher propeller and powered the compressor for the jet-tip rotors. During forward flight, power (air and fuel) was cut off from the rotor jet-tips as the engine was declutched from the compressor and power redirected to the propeller—lift was obtained from autorotation. The final model of V3 weighed 1,208 pounds and had flapping and drag rotor hinges—vertical control was achieved by varying the rotor speed. Unfortunately, the innovations incorporated in Doblhoff's third model were not enough to ensure success, and after only a few flights, it was destroyed by ground resonance vibration.

An additional prototype was constructed before the war ended, the WNF 342 V4, the largest of Doblhoff's prototypes. It was in many ways the most significant, although not for any intended reasons. The V4 could carry a crew of two in side-by-side open cockpits, and the fuselage was now faired. It retained the twin-boom layout, but the two verticals were replaced with a single vertical mounted on top of a horizontal tail that connected the booms. Heavier than its predecessors, the V4 weighed 1,411 pounds and had a 32.68-foot diameter rotor, just slightly larger than the V3's. It also innovatively used air pressure to control the collective pitch of the rotor blades—the blades could be pitched for helicopter-powered takeoff and landings and then changed to allow for autogyro flight. Testing of the V4 began in the spring of 1945, with twenty-five hours of flight time having been accumulated by early April, although it was not tested in forward flight over twenty-five to thirty mph. But it was too late—on April 7, 1945, Doblhoff and his colleagues could hear the artillery of the approaching Russian forces as they moved into Vienna, eighteen miles to the east. After some discussion, the decision was made to load the WNF 342 V2 and V4 prototypes on a trailer and flee westward to the Americans and British. For almost twelve days the truck carrying the designers and mechanics, and towing the trailer, moved westward over roads often clogged with refugees and others also fleeing the advancing Russians.

Eventually Doblhoff and his colleagues surrendered to American forces at Zell am See and were quickly interrogated by engineering officers who recognized the importance of the V4 prototype and its designers. The model was crated and shipped to the United States for evaluation—followed quickly by Doblhoff, who eventually went to work for McDonnell Aircraft as chief helicopter engineer and significantly contributed to development of the McDonnell XV-1 compound helicopter convertiplane. Of perhaps greater importance, August Stepan, who had done the structural design and most of the test-flying of the prototypes, joined Fairey Aviation in England as chief tip-jet engineer and contributed to the design of the Fairey Gyrodyne and Rotodyne, which employed the rotor tip-jet technology for takeoffs and landing but flew as an autogyro.

There was an additional irony concerning Doblhoff—he had courted a young Austrian woman in the early 1930s who had spent a summer in Czechoslovakia. The young woman was also courted there by a young man whose family had fled the Russian Revolution—and although neither would win the girl, both Frederick von Doblhoff and Igor Bensen[2] would be instrumental in preserving autorotational technology in the 1950s.

JAPAN

By 1933 five Autogiros were flying in Japan, three Cierva C.19MkIVs (two evaluated by the Japanese navy and one flown by the *Asahi Shimbun* newspaper) and two Kellett K-3 Autogiros being evaluated by the army. It is obvious that the attraction of the Autogiro was, in the majority, its military potential, and neither model proved initially successful. The navy did not find the fixed-spindle Cierva suitable for maritime missions, and one of the naval Autogiros was soon cannibalized to keep the other flying, the end result being that the navy soon lost interest in the Cierva aircraft. The Japanese army had a similar experience with the Kellett K-3 aircraft. One was seriously damaged on June 28, forty-two days later, and army interest soon faded. However, war and the development of direct control models led to renewed interest. The Japanese military may also have been aware of the American, British, and Soviet military interest and evaluation efforts.

In August 1939 Okura and Company imported a Kellett KD-1A for the Japanese military, which was already fighting a land war in China. That war, as many before, was an amalgam of the tactics of previous conflicts and the brutal necessity to innovate brought on by the evolving realities of combat. The Japanese wartime experience had confirmed what others had already realized, namely that the use of observation balloons for artillery spotting and reconnaissance, a tactic seen almost a hundred years earlier, was an increasingly, if not inevitably, fatal assignment. Even fighting the Chinese, in 1939, a relatively low-keyed effort when compared with the later German blitzkrieg, clearly demonstrated the vulnerability of the fixed balloons, which could be downed by an unsophisticated biplane salvaged from the previous war or accurate ground fire or just a lucky hit. The artillery-spotting tests of the imported Kellett were sufficiently successful that a manufacturing license was obtained from the American company, but the Kellett was soon damaged in an accident in February 1940 while being flown at Tachikawa airfield for the army air force.

Before proceeding with manufacture of the Kellett, the Japanese military embarked on a study of the Autogiro in an attempt to improve its military performance. Also, the wrecked Kellett was shipped to Osaka University in August 1940 for repair and a research and development program, which was problematic, because by early 1941 the U.S. State Department made it clear that it would deny export license to any aviation goods, including Kellett spare parts. As there was only one Japanese firm doing research into rotary flight, the Imperial Army Technical Command requested that K. K. Kayaba Seisakusho undertake repair of the Autogiro. It was understood that the Kayaba Company would also develop a Japanese model based on the Kellett for military use, but the informality of this arrangement, and indeed, the selection of a relatively insignificant company, seems to indicate that this effort was not of high priority.

Kayaba completed its repair of the Kellett Autogiro in April of 1941 and commenced flight-testing the following month with pilot Masaaki Iinuma, which would continue until July 1943. The Autogiro achieved outstanding results as various modifications were tried, and a run of only 100 feet was necessary for takeoffs, while near-hovering flight was achieved in a nose-up position with the engine at full power. The results were encouraging, and the military had even drafted the former pilots and mechanics from the *Asahi Shimbun.* The Kellett KD-1A participated in artillery observation, liaison, and rescue work, and its outstanding flight characteristics led to flight-testing on June 4, 1943, from the deck of the *Akitsu Maru,* a light aircraft carrier, of which films still exist. Zero ground-roll landings were regularly achieved, and with the ship underway, the already short takeoff run was reduced to forty feet by pilot Zenji Nishibori.[3] These successful tests led to highly successful experiments in which the aircraft assumed an antisubmarine patrol mode and was fitted to carry a 132-pound depth charge. This antisubmarine role was of particular interest to the Japanese military, which was becoming concerned about protecting home waters from American submarines.

The Kayaba Company also had received an order for two locally built Ka-1 Autogiros[4] in June of 1941, soon after beginning flight-testing of the rebuilt KD-1A. Wartime modifications included the substitution of a Japanese-licensed German air-cooled inverted Argus engine for the original American Jacobs radial engine. These two Japanese Autogiros were completed in November of 1942, but difficulties at adapting the German engine led to a delay in flight-testing until the middle of 1943. However, the military was so impressed with the aircraft that an order was placed in November 1942 with Kayaba for 300 Ka-1A aircraft equipped with the Argus, with the first completed in June of the following year. A total of 35 Ka-1As were manufactured, but 10 deployed Ka-1As were destroyed by the Allies, and the surviving aircraft failed to achieve performance objectives, primarily due to ongoing and unsolved problems with the Argus engine. This caused a return to Japanese versions of the Jacobs engine in subsequent models, called the Kayaba Ka-2, as production accelerated to meet the increasing threat of Allied submarine activity in Japanese waters. The initial Ka-2 models were delivered in the summer of 1944, and final production is estimated at 60 aircraft. The Japanese also experimented with rocket-powered rotor blades in a Ka-1 variant dubbed the Kayaba Ka 1KAI, which utilized the Argus engine but had small solid-fuel rockets fixed to each rotor tip to over-speed the rotor and facilitate jump takeoffs. Although a report exists of a tethered test in April of 1945, during which the rotor achieved 300 rpm in five seconds, it came too late to have any impact, and it is doubtful that jump takeoff capability would have altered the results achieved by the Japanese or any combatant.

Thus Japan had produced a total of 95 military Autogiros, the most of any nation in World War II, with no impact. The Autogiros proved, for all the combatants, inappropriate for artillery spotting, due to extreme vulnerability, and the antisubmarine role created by the Japanese was, at best, inconsequential. It did not lead to any obvious success nor is there any indication that the British or American naval authorities were deterred in submarine deployment or mission profile. In evaluating the use of the Autogiro/autogyro by the combatants, the inescapable conclusion was later stated by Peter Brooks: "the gyroplane had shown itself to be unsuitable as a weapon of war...the gyroplane had shown itself to lack the essential characteristics required."[5] This judgment was due partially to the development of the helicopter that had been accelerated during the war. Although the Autogiro Company of America was publishing a lengthy book on the history of the Autogiro "to clarify certain matters of importance to all concerned, and to offer specific assistance to those who may desire to make use of the store of experience and information we have to offer"[6] in 1944, *Flying Cadet* magazine pronounced the impending demise of the Autogiro in its February issue. Written for aviation-minded youth, its article stated: "Yes, the helicopter is reliable, adaptable, and equal to almost every situation. She seems to have her rival, the autogiro, quite outclassed!"[7] But little-noted wartime developments in Germany and England of the most unlikely of autorotational craft, the rotary kite, would help rescue Cierva's dream.

ROTARY KITES

England

By the mid 1930s, a Pennsylvania company, Captive Flight Devices,[8] had developed a "rotary" kite, borne aloft by the air flowing up through the rotors while being pulled forward, and one had been sent to England in 1937. After brief experimentation by Cierva Autogiro Ltd., the kites had been forgotten as the company turned away from Autogiros to helicopters, but the idea resurfaced in the preparations for war. The admiralty had briefly considered the use of a rotary kite to lift barrage cables above ships to limit attack by low-flying enemy airplanes, but *barrage balloons* were able to accomplish the assigned task. However, the British were developing methods for inserting agents into Europe, and gliders, parachutes, and rotary kites were considered in 1940. The military naturally turned to one

Focke-Achgelis Fa-330 Rotary Kite developed for deployment on German submarines in World War II. Designed to be towed into the wind. (Courtesy of Ron Bartlett.)

of the most knowledgeable rotary-wing pioneers, the Austrian Raoul Hafner, who was a resident in England since 1932 and briefly interned at the beginning of the war as an "enemy alien,"[9] and who then became an English citizen and offered his services to his adopted country. The Hafner Gyroplane Company began development of a rotary kite on October 3, 1940, work that was transferred to the military in December of 1941. Originally taken up by the Central Landing Establishment, the development of the Rotachute 'is most closely associated with the Airborne Forces Experimental Establishment. Almost all experienced Autogiro pilots and engineers were either in uniform or working for the military, and it is not surprising that the first Rotachute[10] was designed by Raoul Hafner, O.L.L. Fitzwilliams, and Dr. J.A.J. Bennett of Cierva Autogiro Ltd.

Hafner had suggested the Rotachute to insert agents into occupied Europe, it being assumed that the Rotachute's controlled descent would allow for greater accuracy. Models were dropped in October 1940 from a Boulton Paul P.75 Overstrand bomber, and a full-size Rotachute was designed in November. The man-carrying model was extremely portable, weighing in at only forty-eight pounds, with lift being provided by a two-blade, fifteen-foot-diameter rotor. It had a sturdy tubular steel frame with a rubber shock-mounted central skid to which wheels were added, and control was achieved by a loop-shaped hanging-stick from the rotor hub,

British Airborne Forces Experimental Establishment (AFEE) Rotachute designed by
Raoul Hafner, O.L.L. Fitzwilliams, and Dr. J.A.J. Bennett of Cierva Autogiro. It was
an individual rotary-kite parachute to insert agents into occupied Europe during World
War II. (Courtesy of Ron Bartlett.)

clearly derived from the C.30 Autogiro. Hafner and his associates intended
this to be launched from an aircraft specifically modified for that purpose,
but this never happened, as the Rotachute was only tested in tow by a moving truck.[11] First tethered manned flights of the Rotachute Mk.I were
accomplished on February 10, 1942,[12] by Flight Lieutenant Ian Little.
Control was enhanced by a rubberized fabric tail, and the pilot, looking
through the loop in the control stick while seated in an open fuselage,
achieved excellent visibility, a Bren gun with 300 rounds of ammunition
by his side. The initial flights revealed a direction instability, which the
designers corrected with the addition of a 50 percent larger semirigid tail.
The Rotachute Mk.II made its maiden flight on May 29, 1942, but was followed immediately on June 2, 1942, by a modified Mk.III.[13] Dr. Bennett
came to America in October 1942 to consult with the ACA regarding a
proposed American Rotachute project then undertaken by the G&A Aircraft Company, the successor to the Pitcairn Autogiro Company, but that
project ended when the prototype crashed during test-flights.

The final model Rotachute, the Mk.IV, first flew in towed flights on
April 29, 1943. It achieved improved stability with the addition of twin

endplate fins on the tail and installation of an instrument panel. Weighing just 85 pounds, the aircraft was designed to carry 285 pounds, but the testing seems to have clearly indicated that its performance would not achieve results substantially better than a parachute, and the project was never fully implemented. However, more than twenty Rotachutes of various models were manufactured by F. Hills and Sons and the Airwork General Trading Company, of which five were eventually brought up to the Mk.IV configuration. The final unmanned flight-tests of the Mk.IV were conducted on October 18, 1943, and the project terminated. It had produced "one of the earliest applications of the seesaw or teetering rotor, in which the two blades, integral with a fixed coning angle, rock on a common hinge at the rotor head,"[14] a mechanical arrangement first tried in 1931 by American Gerald Herrick, further developed by American Arthur D. Young in 1941,[15] and later a seminal feature of Bensen Gyrocopter design.[16] Six Rotachutes were sent to America for evaluation, and at least one ended up at the General Electric facility at the Schnectady (New York) Flight Test Center, where a young engineer named Igor Bensen became interested in its flight characteristics.

There were also two other obscure British experimental derivative rotary-kite programs, distinguished more by daring than success, which represent dead-ends in autorotational development. In April 1942 the AFEE received a proposal to attach Autogiro rotors to a 3,000-pound Jeep (and later in November to a 31,295-pound Vickers Valentine tank), and development commenced in August. It was dubbed the Malcolm Rotaplane, or more popularly the "Rotabuggy"[17] (or "Rotajeep"). The final design centered on an American Willys quarter-ton 4×4 that had, as amazing as it may seem from over a half-century later, been modified for flight. This ungainly craft, certainly one of the strangest to ever leave the ground, was initially tested by the dauntless Ian Little on November 16, 1943, and first flown eleven days later, towed along a runway. On what Brooks later called a "horrific occasion," a Rotabuggy was towed to 1,700 feet by an Armstrong Whitworth Whitley V bomber, and this flying Jeep actually landed successfully.[18] Continued control problems due to excessive vibration doomed the project, however.

Australia

A similar project had been undertaken in Australia, dubbed Project Skywards,[19] occasioned by the military requirement to transport vehicles to the troops fighting the Japanese in New Guinea. Lawrence J. Harnett, head of the Army Inventions Directorate (AID), suggested that rotary-winged gliders be used to deliver vehicles. John L. Watkins, senior aeronautical engineer of the Australian Department of Civil Aviation (DCA) was assigned to study the problem, and he became aware of the rumors of the AFEE Rotabuggy project but was unable to gain any substantive information, as the AFEE project was classified. The Australians then embarked on an independent development program based on the known technology, primarily the Cierva C.30A aircraft that had been imported during the previous decade, which were then not in flying condition. Experimentation demonstrated that the C.30A rotor became unstable at approximately 115 mph, far less than that required by the military to insure personnel safety, so an ambitious program of rotor development was undertaken, based on a recent series of articles in *Aircraft Engineering* on rotor design authored by Dr. Bennett. The project, however, was cancelled after six months, as the New Guinea campaign reached a successful conclusion and the perceived need for a flying Jeep, or "Fleep," no longer existed. At the time of project cancellation, the prototype was almost completed, and its developer had no doubt that "it would have worked, given just a little more time." But it never flew, and it is only known from a few references and surviving photographs.[20] The fully-loaded 3,300-pound Fleep never flew, but the more well-known 375-pound Focke-Achgelis Fa-330 rotary kite[21] did and was to have a significant impact on the survival of Cierva's technology.

Germany

In Germany the Focke-Achgelis company developed a large gyroplane glider called the Fa-225 in 1942 by combining a rotor pylon on top of a glider fuselage. The prototype[22] successfully flew in 1943, but changing operational requirements and Allied threats insured that it never went into production. Of much greater importance, however, the German submarine campaign utilizing its sizable U-boat fleet against Allied shipping was in full force in the fall of 1941, leading the British (and later American) naval authorities to begin development of effective convoy techniques. The U-boat commanders requested a means by which visibility could be increased, to more effectively identify targets. Focke-Achgelis GmbH, then a subsidiary of Weser Flugzeugwerke, which had previously developed the Fa-61 helicopter, began construction of a rotary kite (glider) that could be easily transported on a submarine, quickly launched while the boat cruised on the surface, and recovered prior to an attack. The rotary kite was constructed of steel tubing with a rudder extending from the rear of the airframe, lifted aloft by a three-blade rotor mounted on a pylon behind the open pilot's seat. Control was achieved by means of a ground-adjustable rotor pitch, a pedal-control rudder, and a floor-mounted control stick that tilted the rotor head for longitudinal and lateral control.[23] The rotor was started with a rope, or, if conditions permitted, the submarine would cruise into the wind and the kite would lift off, by the air flowing through its rotor blades.[24] The small rotary kite was nicknamed the "Bach-stelze" (Water Wagtail) or "Ubootsauge" (U-boat's eye).[25]

The Fa-330 was attached by a cable, and information was relayed by the pilot using a telephone link. The boat carried 492 feet of cable to raise the Fa-330 to a height of 392 feet,[26] but its operational height was about 400 feet, which enabled an observer with naval binoculars to view twenty-five miles. Prior to commencing an attack, the kite would be recovered by a manual winch, but if an emergency arose, the pilot could release the entire rotor assembly by means of a control lever placed near the rotor head, disconnecting the rotor and ejecting a parachute stored on the pylon behind the pilot. The pilot then released his seat belt, and the fuselage followed the rotor assembly into the sea while the pilot floated to the sea, hopefully to be picked up by the submarine before it submerged. The Fa-330 could easily be dismantled and assembled by a few men—when not in use it was stored in a watertight container attached to the deck. Two or three U-boat seamen per boat, who had generally never flown before, were trained in a wind tunnel at Chalais-Meudon, France.[27]

Initial test-flights were made in the spring of 1942, with experimental flights from ships later that summer. By early fall 1942 production began, and an estimated two hundred Fa-330s were built before the end of World War II. Placed aboard an unknown number of Type IX–U-boats[28] from mid-1942 on, little is known about the scope of Fa-330 operations, but it is thought that it was probably deployed primarily, if not exclusively, in the South Atlantic and Indian Oceans,[29] and even then only rarely saw action. The Allies had developed radar to the extent that the Fa-330 would have created a *signature* and rendered the boat vulnerable. It is perhaps understandable, then, that a U-boat skipper exchanged a rotary kite for a Japanese floatplane at a naval base at Surabaya, Java, for naval patrol.[30]

The Allies had become aware of the Fa-330 when they were found aboard captured submarines in 1943,[31] and more of the Fa-330 rotary kites survive in museums than any other autorotational aircraft, as British forces had seized completed and crated models at the aeronautical factory of Weser Flugzeugwerke, Hoykenkamp. American JIOA (Joint Intelligence

Objectives Agency) personnel subsequently entered the factory and discovered several completed Fa-330s. The few remaining factory workers demonstrated how quickly a crated model could be assembled, and it was so impressive that several were sent to the CIOS (Intelligence) Secretariat in London. At least one was sent to America for examination by the United States Army Air Corps, where it was flown in 1949 under contract by George Townson,[32] and it was transferred by the Department of the Air Force on July 7, 1950, to the National Air and Space Museum of the Smithsonian Institution.[33] A motorized version of the rotary kite utilizing a 60-horsepower engine, designated the Fa-336, was proposed but never built.[34]

In 1945 the future of the Autogiro was bleak: Cierva and Weir were out of the Autogiro business and the Autogiro Company of America existed only as a licensing company, with no one clamoring for a license. Pitcairn's manufacturing company had, after several reorganizations, become the G&A Aircraft Company and been acquired by Firestone in 1943. By 1946 it would develop two models of helicopters, but these aircraft failed to gain military support. An abortive attempt to introduce the larger model into the civilian market doomed the company, which ceased business in 1948.[35] Its rival changed its name to the Kellett Aircraft Company to reflect a wider interest in aviation. By 1945 Kellett had also shifted focus to the helicopter and produced the XR-8, described as a "flea topped canted side-by-side rotors with intermeshing blades."[36] It survives in the collection of the National Air and Space Museum and was the first American synchropter rotary-wing aircraft based on Anton Flettner's eggbeater configuration. Its success led to a follow-on military contract for a larger model. The XR-10[37] was delivered to the army air force in 1947, but the program came to an end with a crash in 1948, in which the pilot was killed when his parachute became entangled in the rotors. It was the last straw for the company—having declared bankruptcy in 1946 and operated by trustees, it lacked the resources to continue and was forced to sell off its assets, including a large prototype XH-17 flying crane. The purchaser was Howard Hughes, but the turbojet-powered crane did not prove successful and was abandoned in 1952. But both Kellett and the Autogiro Company of America made unsuccessful attempts to bring back the Autogiro.

The story of the unsuccessful attempt in 1961 of Skyway Engineering Company, Inc., to produce and market a modernized version of the roadable Pitcairn AC-35 Autogiro under an ACA license has been told, and the Kellett attempts met a similar fate. In 1949 Kellett was awarded a developmental contract by the U.S. Navy Bureau of Aeronautics for a research convertiplane, and Kellett subsequently proposed to modify the KD-1B that had previously been flown by John Miller of Eastern Air Lines on the 1939–40 Philadelphia, Pennsylvania–Camden, New Jersey, experimental airmail route. The Autogiro had been placed in storage in a hangar at New York's LaGuardia Field when the experimental route service had ended in July 1940 and was subsequently sold in December 1940 for $23,000 to Miller's friend Tex Bohannon. The KD-1B, having passed through several hands,[38] had been damaged in a minor ground accident when acquired by George Townson in 1953, who subsequently sold it back to Kellett while simultaneously accepting employment as engineer and test pilot. The intent was to equip the wingless direct control Autogiro with short, stub wings and 150-horsepower Lycoming 0-320 engines with propellers that rotated in opposite directions to counter rotor torque, much as the Bratukhin 11EA had done almost three decades before, but the project never advanced beyond refurbishing the original Autogiro and a few test-flights. It was designated the KH-17A.[39] The original tilting rotor control system had been replaced by a feathering rotor control mechanism and, as its original rotor blades were no longer functioning, with a three-blade rotor taken from a Piasecki HUP navy helicopter. This hodge-podge arrangement proved unstable, and the severe control-stick vibration encountered made the aircraft unsafe—the project was quickly abandoned.

Kellett made one final attempt to revitalize the Autogiro,[40] this time involving a reconstructed[41] direct control KD-1A, during the 1958–60 time period. The airframe came from the model originally sold to the army, called the YG-1B and then reconfigured as the XR-3. It had been sold to General Electric[42] after the war as surplus for use in the development of helicopter rotor blades at the company's flight-testing facility[43] in Schenectady, New York, and subsequently passed to a private owner in Harrisburg, Pennsylvania, who leased it back to Kellett. The XR-3 had not been maintained, and Kellett was forced to refurbish the airframe and construct new rotor blades. The decision was also made to recreate a direct control tilting rotor head from the original specifications, which allowed certification under the original ATC No. 712. The recreated KD-1A was called the Cropmaster,[44] intended for agricultural uses as well as "pipeline inspection, geological and mineral exploration duties that were being performed by helicopters that cost almost twice as much and had high maintenance costs."[45] Bob Kenworthy served as project engineer, with Townson as assistant project engineer and test pilot along with assistant test pilot Roland "Blackie" Maier.[46] Townson had flown a PCA-2 as a crop duster in Pennsylvania, New Jersey, and Florida for Giro Associates of Morristown, New Jersey, in 1938.[47] He was undoubtedly of monumental importance to the project, as it became readily apparent that vital Autogiro construction and flying skills had been lost by 1958 when the project commenced. Fortunately, shop foremen were located who had worked on the original Kellett, and Townson was able to teach Autogiro flying to Kellett's helicopter pilots and three Federal Aviation Administration (FAA) flight-test engineers—to no avail. The craft failed to find a market, even priced at $25,000, far below the price of a comparable helicopter. Light planes specifically designed for crop dusting, and the helicopter with its hovering capacity, were the preferred agricultural vehicles, and Kellett, forced to abandon its attempt to reintroduce the Autogiro, terminated the lease and returned the KD-1A to its owner. Thus all the attempts to reintroduce the traditional Autogiro ended in failure, and were it not for the XR-3 and the Rotachute sent to General Electric, historical obscurity logically would have been the fate of Cierva's "flying windmill." But the XR-3 *had* been sold as surplus to General Electric, where Igor Bensen first encountered the Autogiro. It was a fateful encounter.

Igor Bensen, born in 1917, was the son of Alexandra P. Bensen and a Russian agricultural scientist, Basil Mitrophan, whose ancestors had originally migrated to Russia from Sweden.[48] As Bensen's father had been educated at the University of Minnesota in the first decade of the twentieth century[49] and his mother was a graduate of the University of Kiev, the home environment created a rich intellectual atmosphere, influencing him to pursue engineering (his older brother Vladimir studied medicine). It was also a deeply religious home, and that value would significantly influence the inventor's later course of life. His father, having returned to Russia to help develop drought-resistant crops, was posted to Czechoslovakia in 1917, at the beginning of the Russian Revolution, while the rest of the family remained behind. The battling Red and White forces in the ensuing Russian civil war lead to harsh times, and the Bensen family was soon reunited in Prague, far from the Russian turmoil. At the age of seventeen Bensen was sent to the University of Louvain in Belgium, from which he received a B.S. degree.

Always seeking educational opportunities, he accepted a scholarship from the Stevens Institute in New Jersey in 1937 to study mechanical engineering—a truly daring move, as he only spoke a rudimentary English. Even though Bensen later claimed to understand only one out of every three words when he started engineering school,[50] he graduated with honors in 1940.[51] Because he was not a citizen, Bensen had been forced to turn down a job offer to work for Igor Sikorsky, who was at that time engaged in historic helicopter development under government contract, and his first job was as an engineer with General Electric at the age of twenty-three. Initially he worked on nonaviation projects, but by the time he became an American citizen in 1944, he was seriously interested in helicopter development. General Electric executives took notice and assigned the young engineer to the company's helicopter development efforts. It was a fateful assignment.

General Electric, influenced by the wartime frenzy of helicopter development occasioned by the Dorsey Bill, was developing a jet-tipped rotor helicopter. While Bensen was working on the helicopter project, he encountered the Kellett XR-3, acquired as salvage by General Electric for rotor and control system development, and he eventually gained almost exclusive use of the aircraft as other engineers lost interest in the surplus Autogiro. Beginning in 1942, Bensen became a highly skilled pilot in the XR-3 and, in the process of improving the aircraft's control system, gained a deep understanding of the dynamics and theory of autorotational flight. The United States Army Air Force had received some of the recovered Fa-330 rotary kites and were experimenting with George Townson as pilot. They also had obtained a Rotachute from England at the close of the war, and General Electric vice president David C. Prince first saw the British aircraft at an air force open house at Dayton, Ohio. When he returned to Schnectady and expressed enthusiasm for the small rotary-wing aircraft, Bensen asked his boss to acquire the Rotachute for evaluation.[52] The military agreed to loan the Rotachute, *provided* that General Electric agreed not to fly it.

Bensen, placed in charge of evaluating the Rotachute, eventually ignored the military's requirements and actually flew the rotary glider. It was first flown in a brisk upstate New York wind augmented by the prop wash of the XR-3, a challenging "kiting" experience given the facts that no one in America had ever flown the Rotachute and that landings were accomplished, in the absence of any installed gear, with the pilot's feet! But the flights went well and soon progressed to a towed format behind an automobile. In 1967 Bensen wrote of that first flight that the Rotachute "weighed barely 100 lbs. In full flight gear I weighed then 220 lbs, so the pilot could be described not as sitting in this machine, but more properly as wearing it."[53] The Rotachute, with its overhead stick and welded steel tube fuselage, as observed later by Paul Bergen Abbott, "undoubtedly was a strong influence" on the Bensen Gyroglider and later the Gyrocopter.[54] Bensen even launched the unmanned Rotachute from the bomb rack of the XR-3 in what proved to be successful test-flights.

Perhaps because Bensen's early work was clearly based on the English and German rotary kites, he wrote very little about the creation of the

Gyroglider in his autobiography but concentrated on the Gyrocopter, the achievement for which he was most famous. But in 1993 he wrote about the Bensen B-1, his "first full-scale" rotorcraft creation,[55] which had been created while he was still at General Electric after experiments with the Rotachute. It was an amateur-built, 120-pound glider based on the Rotachute capable of carrying a load of 300 pounds while being towed behind a vehicle, but it differed, with the addition of nose and tail wheels, a semirigid rotor in place of the Rotachute's individual flapping rotor blades, and a control-stick "reverser" to allow more effective direct control of the rotor. Little came of the model, but Bensen was gaining a great deal of knowledge about rotary-glider performance and design that would contribute directly to later designs. The B-1 was destroyed in an accident that fortunately did not injure Bensen, and the crash led directly to the B-2, which was of an all-metal construction. Under the direction of Prince, the B-2 led to the G-E Gyro-Glider[56] in November 1946, but little came of the G-E model. And subsequently in Schenectady, the Helicraft Equipment Company developed a 60-pound variant of the Rotachute called the Heliglider[57] in 1949. An extremely simple design that flew with a fourteen-foot rotor that achieved 550 rpm, the lack of weight made it difficult to fly with an overhead stick control, and the project was soon abandoned. And while the Schenectady efforts seemed to lead nowhere, the world was just beginning to hear from Igor Bensen.

NOTES

1. See J.R. Smith and Antony L. Kay, *German Aircraft of the Second World War* (London: Putnam, 1972), p. 589, for a discussion of Doblhoff; Mal Halcomb, "Vertical Lift, German Helicopter Development through the End of World War II," *Airpower Magazine,* March 1990.

2. For a description of the courting of the Austrian girl by both Doblhoff and Bensen, see Igor B. Bensen, *A Dream of Flight* (Indianapolis, Indiana: The Abbott Company, 1992), pp. 7–9.

3. For a fuller description of the flight testing of the Kellett KD-1A by the Japanese military authorities and a picture of it on the deck of the *Akitsu Maru,* see Peter W. Brooks, *Cierva Autogiros: The Development of Rotary-Wing Flight* (Washington, D.C.: Smithsonian Institution Press, 1988), pp. 276–78.

4. As the Ka-1 was presumably built under the Kellett license, it and all other Kayaba autorotational aircraft are correctly referred to as *Autogiros.*

5. Brooks, pp. 281–82.

6. *Some Facts of Interest about Rotating-Wing Aircraft and the Autogiro Company of America* (Philadelphia: The Autogiro Company of America, 1944).

7. "Helicopter or Autogiro?" *Flying Cadet* 2, no. 2 (February 1944): 46.

8. Brooks, p. 283.

9. Wing Commander Kenneth Wallis, "From Wing Commander Wallis" *Rotorcraft* 38, no. 8 (November 2000): 12.

10. See "The Hafner Rotachute" *Fly Gyro!* no. 3 (February 2001): 18–19; "Design Classroom," in *Collected Works of Design Classroom* (Anaheim, California: Popular Rotorcraft Association, 1974), p. 11; Brooks, pp. 286–87; Ron Bartlett, "Gyroplanes: The Early Years," *Autogyro Quarterly,* no. 5 (n.d.):13 (author's photograph of the Rotachute exhibited at the Army Air Corps Museum at Middle Wallop, England).

11. For a rare photograph of the Rotachute being pulled as a gyroglider, see "Hafner Rotachute," p. 18.

12. Brooks, p. 326; but see p. 287, where Brooks claims that the first flight was on February 2.

13. For a photograph of the Rotachute Mk.III, see Brooks, p. 287.

14. Ibid., p. 288.

15. Ibid., p. 301.

16. Bensen, *Dream of Flight,* p. 10; Brooks, p. 288.

17. Brooks, p. 289.

18. Ibid., pp. 287–89; for a rare photograph of the Whitley test-flight, see Brian Johnson, *Classic Aircraft: A Century of Powered Flight* (London and Basingstoke, England: Channel 4 Books, 1998), p. 103; rare film footage can be found in The History Channel: *Heavy Rigs of Combat: Jeep* (video).

19. For a description of the Australian Project Skywards, see Group Captain Keith Issacs, "Project Skywards," *Rotorcraft* 32, no. 4 (June–July 1994): 6–9; Brooks, p. 287.

20. Issacs, pp. 8–9.

21. For a description of the Fa-330, see Roger Ford, *Germany's Secret Weapons in World War II* (Osceloa, Wisconsin: MBI Publishing Company, 2000), p. 58; "Flying a Kite: The Focke-Achgelis Fa-330 Rotary Wing Kite," *Rotor Gazette International* 1, no. 2 (July–August 1992): 1–2, 4; "Focke-Achgelis FA-330 Bachstelze (Wagtail)" *Fly Gyro!* no. 3 (February 2001): 8–9; *Aircraft of the National Air and Space Museum,* 4th ed. (Washington, D.C., and London: Smithsonian Institution Press, 1991), see "Focke-Achgelis Fa-330"; Smith and Kay, pp. 606–8; Johnson, pp. 101–2; Martin Hollmann, "The Focke-Achgelis Fa 330 Gyroplane Kite," *Gyroplane World,* no. 14 (November 1977): p. 1; Bill Gunston, *History of Military Aviation* (London: Hamlyn (Octopus Publishing Group Limited), 2000), p. 126; Bartlett, "Gyroplanes," p. 13 (author's photograph of the FA-330 exhibited at the Musee de l'Air at Chalet Meudon); Bryan Philpott, *The Encyclopedia of German Military Aircraft* (New York: Park South Books, 1981), 95 (rare photo of British test at sea); and see Brian Ford, *German Secret Weapons: Blueprint for Mars* (New York: Ballentine Books, 1969), which has an incorrect illustration of rotary kite on pp. 92–93 marred by misidentification as the FA-230; compare with actual photographs of Fa-330 cited previously. It may be that the Ford illustration is of an earlier developmental version of the Fa-330, but it is not

identified as such. For an excellent rendering of the Fa-330, see Bill Gunston, *Helicopters at War* (London, England: Hamlyn, 1977), p. 29.

22. See Brooks, pp. 284–85; Smith and Kay, p. 603.

23. For a detailed description of the Fa-330 control systems, see Smith and Kay, pp. 606–8.

24. For photographs of the Fa-330 lifting off into the wind from a German submarine, see Brooks, p. 284; "Focke Achgelis," p. 8; Gunston, *Helicopters at War,* p. 29.

25. Brooks, p. 383; Smith and Kay, p. 606.

26. See Smith and Kay, p. 606, and Gunston, *History,* p. 126; but see Brooks, p. 284 ("The single-seat FA 330 was said to be capable of lifting its pilot-observer to a maximum height of between 150 and 300 m (500 ft to 1,000 ft), but the normal operating height was about 120 m (400 ft)."); *Aircraft of the National Air and Space Museum* ("The U-boat carried enough cable to raise the Fa-330 to a height of 700 feet; most often it flew at between 200 and 500 feet."); "Focke Achgelis," p. 8 ("The unpowered rotor glider would be towed behind the submarine, rising to an altitude of 300 to 500 feet.").

27. For photographs of the wind tunnel training of Fa-330 pilots, see Gunston, *History,* p. 124; Martin Hollmann, *Helicopters* (Monterey, California: Aircraft Designs, Inc., 2000), p. 68.

28. Brooks, p. 284; Smith and Kay, p. 608 (authors also maintain that the Fa-330 was possibly used on the Type IX–D/2 supply U-boat).

29. It is recorded that the captain of U-861 employed the Fa-330 while on patrol off Madagascar. Smith and Kay, p. 608.

30. Ibid.

31. See Combined Intelligence Objectives Sub-Committee "German Submarine Rotary Wing Kite." (British intelligence report, London, England, 1945). A model captured on a German submarine was analyzed in this twenty-six-page British intelligence report, complete with photographs and technical drawings.

32. Paul Hengel, "Portrait of a Pioneer Rotary-Wing Pilot," *American Helicopter Museum & Education Center Newsletter* 2, no. 3 (Spring 1996): 3.

33. *Aircraft of the National Air and Space Museum,* "Focke-Achgelis Fa-330" entry.

34. Brooks, p. 284; Smith and Kay, p. 608; for reference to the proposed powered version of the rotary kite, see "German Submarine Rotary Wing Kite."

35. Frank Kingston Smith, *Legacy of Wings: The Story of Harold F. Pitcairn* (New York: Jason Aronson, 1981), pp. 303–20; Jay P. Spenser, *Whirlybirds: A History of the U.S. Helicopter Pioneers* (Seattle and London: University of Washington Press, 1998), p. 381.

36. Spenser, pp. 380–81.

37. Ibid., p. 382 (photograph of the Kellett XR-10).

38. Brooks claims that the KD-1A that was converted to the KH-17 was initially sold to the Charles H. Babb Co. in December 1941 (p. 353).

39. For a description of the unsuccessful Kellett convertiplane development and picture of the Kellett KH-17, see Howard Levy, "Kellett Gyrations," *Aeroplane* 24, no. 1, issue 273 (January 1996): 32–34, 34; George Townson, *Autogiro: The Story of "the Windmill Plane"* (Fallbrook, California: Aero Publishers, 1985), pp. 122–23.

40. James G. Ray, "Is the Autogiro Making a Comeback?" *Flying* 66, no. 1 (January 1960): 34–35, 91–92.

41. Townson maintains that the Kellett aircraft also incorporated some parts from the KD-1B that had transported the mail for Eastern Airlines and later been converted into the KH-17. Townson, *Autogiro,* p. 138.

42. Brooks, p. 353.

43. For a photograph of the XR-3 at the General Electric Schenectady facility, see Townson, *Autogiro,* p. 139.

44. For a photograph of the Cropmaster, see Levy, "Kellett Gyrations," p. 34; "Rotary-wing Aircraft," *Flying* 67, no. 4 (October 1960): 24–26, 100–101.

45. Levy, "Kellett Gyrations," p. 34.

46. Townson, *Autogiro,* pp. 118 (photograph of Kenworthy) and 138 (photograph of Maier).

47. George Townson, "Autogiro Crop Dusters," *American Helicopter Museum & Education Center Newsletter* 3, no. 1 (Spring 1996); Paul Hengel, "Portrait of a Pioneer Rotary-Wing Pilot," *American Helicopter Museum & Education Center Newsletter* 2, no. 3 (Summer 1995); Hal MacKay, "Bug Fighters," *Popular Aviation* 24 (June 1939): 48–50, 82.

48. Bensen, *Dream of Flight,* p. 6.

49. See Paul Bergen Abbott's introduction to Bensen, *Dream of Flight.* Abbott, a long-time associate of Bensen and former editor of the PRA *Popular Rotorcraft Flying/Rotorcraft* magazine, fails to name Bensen's father. It is likely that, although Abbott was undoubtedly familiar with Bensen's family story, his failure to name the father is the result of Bensen's request. It should also be noted that Bensen himself fails to mention his wife, Mary, in his book, even though she reportedly played a significant role in the Bensen Aircraft Company.

50. Bensen, *Dream of Flight,* p. 8.

51. "The Reverend Igor B. Bensen," *Popular Rotorcraft Flying* 7, no. 4 (July–August 1969): 18.

52. Igor B. Bensen, "Design Classroom: Bensen Model B-1," *Rotorcraft* 31, no. 4 (June–July 1993): 22.

53. "Design Classroom," pp. 14–15 (see p. 14, fig. 2, for Bensen flying the maiden flight of the Rotachute).

54. See Bensen, *Dream of Flight,* pp. 9–12, for a complete description of Bensen's involvement with the Kellet XR-3, the Focke-Achgelis Fa-330, and the British Rotachute; see also Paul Bergen Abbott, *The Gyrocopter Flight Manual.* Introduction by Dr. Igor Bensen (Indianapolis, Indiana: The Abbott Company, 1983, 1986); "Design Classroom," pp. 14–15; Igor Bensen, "Rotachute, Rotary

Wing Glider-Kite," report no. 33200 (Schenectady, New York: General Electric Co., 1946).

55. Bensen, "Design Classroom," p. 22.

56. "New G-E Gyro-Glider," *Rotorcraft* 39, no. 5 (August 2001): 22; reprinted from *Wings* 1, no. 12 (November 1946).

57. For extremely rare photographs of the Heli-glider, see "Design Classroom," p. 14; "Glimpses of History," *Popular Rotorcraft Flying* 2, no. 4 (Fall 1964): 13.

Chapter 11

IGOR BENSEN AND THE
DEVELOPMENT OF
THE GYROCOPTER

Conversation overheard by a PRA-er's wife after a gyrocopter demonstration: "Hell, this looks like more fun than chasing women!" said one spectator. "Cheaper, too, I'll bet"...reflected the other.

Popular Rotorcraft Flying

Igor Benson, now firmly committed to rotary-flight development, was quick to accept an offer in 1951 to join Kaman Aircraft, the fourth-largest helicopter manufacturer in the world, and Bensen was undoubtedly thrilled to be part of the cutting edge of rotary-wing research and achievement. But after almost two years of work at Kaman, during which he had organized and directed the research department and flown air force and navy helicopters, Bensen, borrowing money from his brother, left in 1953 to found his own company in Raleigh, North Carolina.

In 1953 Bensen Aircraft Corporation introduced the B-5 Gyro-Glider, a single-seat rotary-kite towed behind a vehicle and deriving its lift from the autorotation of an unpowered rotor. It featured a light tubular aluminum frame resembling a cross with two pieces, a longer keel crossed by a shorter perpendicular section. A lightweight aluminum-frame web set was attached to both the keel and a reinforced metal mast extending upward from the keel. Control was initially achieved with a hanging-stick control attached directly to the rotor hub, which was positioned on top of the mast with a two-blade rotor. A nose wheel was attached directly to the front of

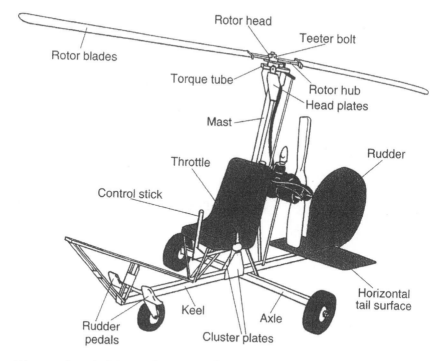

Diagram of standard Bensen Gyrocopter. (Courtesy of Paul Bergen Abbott.)

the keel, while landing wheels were affixed to each end of the perpendicular crosspiece.[1] The keel, behind the seat and mast, carried a plywood fin and rudder much as had the Rotachute. It flew well when towed by even a small automobile, did not require any license, and was relatively safe. It was also distinguished by ease of construction, and the builder could either purchase a kit or build from plans. The materials were readily obtained, and fabrication could be completed by the moderately skilled in three to four weeks. It would become the homebuilt B-6, and the first model would be accepted into the National Air and Space Museum of the Smithsonian Institution, the national aviation collection, on July 22, 1965.[2]

Bensen subsequently developed a Reynolds aluminum prototype, the B-7 Gyroglider, which flew on June 17, 1955, at Raleigh-Durham Airport, the sight for future Popular Rotorcraft Association (PRA) fly-ins. From B-7 came the B-7M (for "motorized") (N75355), which first flew on December 6, 1955, with Bensen as pilot and Charles "Charlie" Elrod and Tim Johnson as ground crew. It weighed 188 pounds, as the airframe was made of rounded aluminum tubing and had a wooden propeller attached to a modest forty-two-horsepower Nelson two-stroke engine, with the wooden rotor attached to a spindle-type tilting-head cyclic pitch rotor with a hanging control-stick.[3] Bensen called his creation, clearly derived from the Rotachute, a *Gyrocopter,* a term he subsequently trademarked much as Cierva had previously done with *Autogiro.* It derived its lift from the unpowered rotor but received forward thrust from its engine attached to the frame in back of the flimsy seat. After three days of successful flight-testing with Bensen as pilot, the B-7M crashed as its pressurized fuel tank failed. Bensen, a highly experienced Autogiro pilot, set the aircraft down in woods adjacent to his North Carolina factory. He later ascribed the safe landing to "much luck and the good Lord's will."[4] The B-7M, rebuilt in three days, was flying by December 17, 1955, a particularly moving experience for Benson, as that was the fifty-second anniversary of the Wright brothers' first powered flight. Ever the aeronautical engineer and pragmatic scientist, Bensen relentlessly analyzed the flight performance of the B-7M, particularly those factors that had led to the accident, and the result was an improved control linkage to the rotor head.

The subsequent B-8M model,[5] incorporating the improvements developed and tested in the B-7M and powered by a more powerful, seventy-two-horsepower McCulloch two-stroke piston engine that had been used on drones for the military, was placed into production in 1957 and became the most produced and copied aircraft design in history. It provided, in kit form and plan-built, the most popular way to fly. The *Spirit of Kitty Hawk,* a B-8M Gyrocopter in which, on December 17, 1966, Bensen had personally duplicated the Wright brothers' historic first flight at Kitty Hawk and with which he had set twelve world and national Gyrocopter speed, distance, and altitude records between May 1967 and June 1968, was accepted into the Smithsonian Institution aviation collection on May 14, 1969. The Bensen, and its variants and local adaptations, was to dominate the American Gyrocopter movement for almost twenty-five years. But although Bensen became the most well-known and successful of the early American developers, he was not the only one.

The work of Harris Woods has been virtually forgotten in the honors and iconic status rendered to Bensen, for, flowing from his fascination with the Kellett KD-1B used in 1939–40 in the experimental airmail route, Woods invented a gyroglider[6] in 1945, a development apparently unknown to Bensen. Although Woods later referenced the Kellett, he may also have been inspired by an article entitled "Gyro Cars for Fun" that appeared in the November 1945 issue of *Mechanix Illustrated,*[7] which not only featured an artists' rendition of the German Fa-330 rotary kite high above a submarine but also depicted a gyroglider that bore a resemblance to Woods's first design. It had a welded steel tube frame covered by fabric, and utilized an original rotor design, a two-bladed, eight-inch chord with a nineteen-foot-diameter employing, like the Kellett, vertical and flapping hinges, and an overhead stick connected directly to the rotor head. It was, in Woods's evaluation, a "fair flying machine." His second aircraft was constructed in 1951 of a welded chrome-moly tube construction, and employed the early Bensen symmetrical airfoil rotor blades, but it eventually crashed in a gusty wind. Woods's third machine was constructed in 1954 of welded tubing and aluminum covered with aircraft fabric. This aircraft now utilized a Bensen rotor head and clearly showed Bensen's influence. Woods's fourth glider crashed in its first flight, and by 1956 he built a Bensen Model B-7 Gyroglider. Woods went on to become chief engineer for the Bensen Company and became PRA member #298 in 1962, the first year of the organization's existence.

While Woods's early development of the gyroglider had been independent of Bensen, it had become completely absorbed by Bensen's impact on the market within a decade. And Woods's gyroglider is even more impressive because, unlike that of Bensen, it did not proceed from the German and British World War II predecessors but rather was an evolution from the Kellett tractor Autogiros, which may have even served as the basis for Woods's 1963 tractor autogyro.[8] Nevertheless, it was Bensen who brought the Gyrocopter to fruition and thrust it aggressively into the public imagination, a superb achievement, even though he may have been aware of the World War II German plans to mount an engine on a Fa-336, a subsequent model never built.

Bensen seriously advertised in such magazines as *Popular Mechanics,* with the invitation to "Build this machine, teach yourself how to fly, fly in your own backyard."[9] Thousands responded and ordered the Bensen plans and began construction of their own aircraft. One builder, writing of the experience many years later, described receiving the requested Bensen plans in Chile as feeling like "holding the originals of the Dead Sea Scrolls."[10] Most were attracted by the modularity of Bensen's concept—they could begin by constructing a Gyroglider, capable of flight while being towed, and then, after flight proficiency had been acquired, add a motor. He also aggressively sought out publications geared toward men, to whom he had targeted his aircraft.[11]

Bensen's approach was to prove the beginning for others who would significantly advance this technology and make it the most popular aviation pursuit in the world. Lacking either Bensen's vision, simplicity, zeal, success, or luck, most have been either forgotten or relegated to footnote status. But in unique ways, each contributed to the survival of Cierva's vision and to application of the innovative autorotational technology. Arliss Riggs, who would achieve the accolade of the "grand old man of gyroplanes"[12] and "a legend in his own time,"[13] was designing and attempting to fly Cierva-type tractor gyroplanes during the mid-1950s–mid-1970s, with much attention and little success. But Riggs's efforts at constructing a wingless Autogiro (which he called a gyroplane, avoiding any legal difficulties from the use of Cierva's term) kept that legacy alive,[14] and others such as Galen Bengston,[15] Joe Kirk with his *Gizmo,*[16] Jim Eich with his JE-2[17] and later XNJ 790,[18] Ron Herron with his series of Little Wing autogyros,[19] and John VanVoorhees with his *Pitbull* have continued the tradition.[20] And the tractor configuration would be carried on by Afro-American artist-designer David Gittens, who would design and build the *Ikenga* 530 Z, a strikingly original, award-winning gyroplane[21] currently in the collection of the National Air and Space Museum at the Smithsonian Institution. It is the only Afro-American designed aircraft in the national aviation collection. And Groen Brothers Aviation has recently introduced its Hawk 6G, a modified Cesna Model 337 Skymaster with its wings removed and the Groen Brothers Hawk 4 rotor system added. The result is a "heavy hauler" Cierva-type tractor gyroplane specifically designed for patrol and repair of oil pipelines.[22] But the mainstream of Gyrocopter success was to be found in those who began with Bensen kits and then pursued their own directions.

Bensen had been described as "a burning-eyed Russian immigrant who speaks with the fervor of a revival preacher."[23] Having earned a Doctor of Divinity degree from the University of Indiana, he was ordained a priest in the Russian Orthodox Greek Catholic Church of America and Canada on June 1, 1969, in New York City by His Eminence Archbishop Ireney.[24] Bensen's missionary zeal for the Gyrocopter led to the creation of the PRA in 1962, operating out of the North Carolina factory and consisting of Bensen's employees and associated dealers.

Ed Trent, who was one of the earliest employees of Bensen Aircraft, had the idea to form an organization so builders of Bensen Gyrocopters would have a way of connecting with other builders. He set up a publication called *Popular Rotorcraft Flying* as a quarterly magazine for the members. This was in

English registered G-BIGU Bensen Gyrocopter utilizing an Air Command–type nose pod and long-range fuel tanks—motor test without rotor blades. (Courtesy of Ron Bartlett.)

1962, and the first issue of the magazine was in winter, 1963. . . . the magazine was more like a newsletter. It was a communication vehicle.[25]

The PRA popularized the Bensen Gyrocopter and, becoming the premier gyroplane association in the world, it would be instrumental in the preservation of Cierva's legacy, which was at grave risk in the early 1960s. The British government ended the Fairey Rotodyne project in February of 1962 and ordered the only flying model destroyed. It was a genuine tragedy for autorotation aviation, as the Rotodyne represented the most technologically advanced embodiment of autorotation ever achieved and, as such, will be discussed in the following chapter. But that period also saw the death of Harold F. Pitcairn, and circumstances surrounding the tragic events of April 23, 1960, reveal much about the growing obscurity of the Autogiro.

Pitcairn's offer to forego wartime patent royalties, accepted by the Wartime Royalty Adjustment Board for the duration of the hostilities plus six months, had expired on March 2, 1946. The ACA then became legally entitled to collect 5 percent of the cost of each helicopter manufactured by

Cricket Bensen-Type Gyroplane (G-BTVB) lifting off from Henstridge Airfield, England, flown by Jim Pumford. (Courtesy of Ron Bartlett.)

those making use of its numerous patents, but these fees did not materialize. While United Aircraft Corporation, corporate parent of Vought-Sikorsky, renewed its ACA license, Kaman Aircraft Corporation, Bell Aircraft, Piasecki Helicopter Corporation, and Hiller Helicopters ignored the Pitcairn and ACA patents, challenging Harold F. Pitcairn to take legal action. Internal memoranda circulated at the new rotary aircraft companies clearly demonstrated that they were well aware of the seminal nature of the patented technologies.[26] Pitcairn undertook an evaluation of his standing vis-à-vis patent law and a detailed investigation of a Bell-47 helicopter, concluding that there was indeed serious patent infringement and that his patents, drawn by Synnestvedt & Lechner since 1924, gave him a strong position from which to seek legal relief. His original strategy was to sue Bell Helicopters and then to use a court victory to force the other infringing companies to agree to royalties, so a suit was filed in United States District Court of Northern New York in 1948, but that suit would soon be withdrawn.

Military contracts had, since 1910, contained language by which the supplier assumed all responsibility for patent infringement and indemnified the United States; that is, they insured the government against claims of patent infringement. Thus anyone selling to the military could be sued if patent infringement were determined, and this was precisely the goal of Pitcairn's lawsuit, but it rapidly became apparent that it would not nor could not achieve its goals of industry compliance. The other helicopter manufacturers, taking a public wait-and-see attitude, embarked on an intense behind-the-scenes lobbying of the military to change the defense procurement policy. Having made his point with commencement of the lawsuit, Pitcairn quietly withdrew the suit and sought to negotiate an industrywide settlement—what he got shocked him and led to a legal entanglement that would result in the longest patent suit in history. And although it would end in victory, Harold Pitcairn would not live to see its conclusion.

The navy announced that it was reformulating the standard procurement contract to now hold the suppliers harmless. It meant that any suit for patent infringement for military goods would have to proceed against the government itself, and anyone embarking on such a course of action would have to face the might and power of the United States of America. It was a daunting task for the ACA and Pitcairn. The object of any ACA lawsuit was to enforce basic rotary-flight patents, the intellectual product and property of Harold Pitcairn's life, all that remained of his foray into Autogiro development; to validate what remained of the Autogiro, he would have to sue his country. Pitcairn had had little difficulty surrendering his airfield to the navy for military training or forgoing patent royalties during World War II as patriotic gestures; the decision to sue the United States was both difficult and distasteful. Pitcairn later wrote, "If I had been aware that the government would become so hostile to the payment of reasonable royalties, our Company certainly would not have embarked on the development of rotary-wing aircraft."[27]

After much preparation, the lawsuit was filed against the United States of America on September 21, 1951. *Autogiro Company of America vs. The United States of America* found Synnestvedt & Lechner opposing a seemingly inexhaustible supply of government lawyers, ready, willing and apparently encouraged to object to every point, and able to bury the claimants in monumental flurries of legal briefs. The *discovery* part of the trial, during which each side seeks to gain information from the other to fully understand the issues and strengths of the other and which may lead to settlement, took over four years of constant contest. It was an exhausting process, wearing on all parties, but finally the case actually began on May 18, 1958. The position set forth by the Autogiro Company of America was strong, methodically presented, each point reinforced by expert, convincing testimony. Jim Ray, H. Franklin Gregory (now a brigadier general), Pitcairn, Franklin Institute Professor Ralph H. McClarren, and Frank N. Piasecki testified—to a successful conclusion, for on March 5, 1966, the trial commissioner issued a 232-page opinion containing a three-part finding: the ACA/Pitcairn patents were valid, they had been infringed, and the United States government was liable to the Autogiro Company of America for an unspecified amount of damages. The government appealed the finding, and the Supreme Court eventually affirmed the verdict. Only then could the damages determination phase begin, which commenced on July 9, 1973. On July 12, 1977 the court awarded the plaintiff $14.4 million in unpaid royalties and $17 million in "delay compensation," and the Supreme Court refused to hear an appeal on July 23, a fitting end to the longest patent suit in United States history. Harold F. Pitcairn, however, was dead.

On Friday, July 22, 1960, almost six years before the first successful ruling in the lawsuit, Harold Pitcairn was preparing for a gala occasion, the celebration the following night of his brother Raymond's seventy-fifth birthday in the large assembly hall of the nearby Bryn Athyn Cathedral. He was undoubtedly thinking of the festive evening for more than 450 guests, the remarks he would make as master of ceremonies, and the honors and celebratory telegrams that had arrived from dignitaries, including former president Eisenhower and industry and civic leaders. But it's not unlikely that Pitcairn also thought of that day twenty-nine years earlier when he stood next to another president and received the Collier Trophy on the White House lawn. By all accounts Raymond Pitcairn's seventy-fifth birthday party was a glittering occasion on Saturday evening and Harold was in a particularly jovial mood. After returning home Pitcairn went into his study, which was located on the ground floor. Ever since the kidnapping of the son of Charles Lindbergh in February, 1932, in Hopewell, New Jersey, it had been Pitcairn's nightly practice to check all the accessible windows and doors to insure security for his family, and he was accustomed to do so with a loaded Savage .32-caliber semi-automatic pistol. His wife, Clara, assumed he would come to bed after making those rounds, but soon she was drawn back to the study by the sound of a gunshot. Having rushed down the stairs, she found her husband dead at his desk from a gunshot to the head.

The circumstances of Harold Pitcairn's death remain unclear even after four decades. The death was immediately reported as a suicide by the media,[28] a rushed verdict seconded by at least one later author.[29] As asserted by Pitcairn admirers Frank Kingston Smith[30] and Frank Anders Jr.,[31] "police investigation disclosed that two shots had been fired; one had penetrated the ceiling directly over the desk in the first floor study, another had struck Pitcairn in the eye."[32] The widowed Clara requested that the inquiry be ended, and she claimed that she never wished to speak of it again. Harold Pitcairn became, in the words of Frank Anders, the "Forgotten Rotorcraft Pioneer"—the only remaining part of his legacy being the check from the United States Treasury, and that was for contributions made to helicopter development. Although Steve Pitcairn, flying "the only remaining flying PCA-2 Autogiro, *Miss Champion*"[33] at air shows, never fails to elicit the crowd's attention and admiration, Cierva's vision was preserved in the humble Bensen Gyrocopter and advanced by the Popular Rotorcraft Association. And others around the world were taking note of Bensen, with profound implications for the proliferation of the Gyrocopter.

In Canada, Bernard Haseloh followed his youthful fascination with windmills into an involvement with Gyrocopter aviation, and with PRA #202, he was an early first-year member of the fledging organization. The family legend is that he inquired of the Canadian Ministry of Transport in the 1950s about building a helicopter. When he was informed that private individuals could not build helicopters with powered rotors, "without engineering courses, never having flown, using whatever materials he could find, he invented a free-wheeling rotor."[34] He later stated that "I figured it was pretty safe…But I took a chance, too, and sure I broke my nose and all those things but that's part of the game." Rotary Air Force (RAF) was founded in 1987 to manufacture and sell a two-place, enclosed-cabin Gyrocopter when Bernard's nephews Dan and Peter Haseloh, his niece and her husband, Linda (Haseloh), and Don LaFleur acquired the initial gyroplane plans. Establishing a modern manufacturing plant and securing government certification of their craft in Canada, England, the United States, Germany, Russia, New Zealand, Australia, and parts of South America and the Middle East, RAF has emerged as a world-class gyroplane company although its lack of a horizontal stabilizer has generated controversy and concern about flight stability.[35]

In 1957 Chuck Vanek began his odyssey of gyroplane development and design. His company, Vancraft, whose name changed to Sport Copters, "single-handedly invented, produced, developed and marketed his own aircraft."[36] Vanek's significant contributions to gyroplane technology include the "tall-tail gyroplane," the first experimental two-phase gyroplane, and later the first enclosed ultralight gyrocopter.

While on a tour with the American Air Force in 1958, Royal Air Force Wing Commander Ken Wallis purchased plans for a Bensen B-7

Wing Commander Ken Wallis flying one of his autogyros.
(Courtesy of Ron Bartlett.)

Gyroglider and began construction after he returned to England. Wallis experienced the usual difficulties of fabricating a Bensen from plans but also immediately began to devise ways of improving the glider. Realizing that he would need government permission, he sought aid at the Ministry of Aviation and was fortunate to encounter an old friend, Michael Vivian, who was deputy director of flight safety. It was an unexpected meeting, as Wallis did not know that his friend worked at the ministry, and it was of great significance for future autogyro development in the UK. Vivian was instrumental in securing permission from the Air Registration Board

(ARB) for Wallis to develop what would be the first postwar powered Gyrocopter in 1959 in the UK, but of greater importance, the ARB gave Wallis the power "to experiment within reason, function as [his] own design organization and building under their formal authority."[37] It was extraordinary giving Wallis "developer" status, which he has maintained throughout his career, made possible by Vivian's vouching for Wallis as a skilled pilot and self-taught engineer, and it would allow him to develop a seminal series of designs that influenced the European gyrocopter movement into directions away from the American adherence to the basic Bensen configuration. This was to result in a series of innovative models and patents on mechanical improvements that continue to influence both European and American designers, and Wallis emerged as the first international autogyro/gyroplane celebrity since Cierva.

Kenneth Horatio Wallis[38] was born in 1916 at Ely, Cambridge, and received his education at the King's School, Ely. Upon graduation, he entered into his father's motor and cycle business and raced high-speed boats (which employed air and underwater propellers). His family had an aviation orientation from the early part of the century, his father and uncle having built in 1910 the *Wallbro* monoplane in Cambridge,[39] thought to "have the had the first steel-tubed airframe in the world."[40] And the family legend is that when Wallis's father and uncle were moving the Wallbro monoplane through town to a field, it caused a commotion at the local school, as children dashed to windows to see the strange craft coming up the street. The teacher later met the pilot of the aircraft, and the product of that union between Miss Emily May Barker and Horace Wallis was Ken![41] Ken's first boat-racing success came in 1934—the last coming in 1957, his fifty-sixth. In 1937 he obtained his pilot's "A" license and joined the Civil Air Guard. During World War II he served as operational pilot on *Lysanders* in the No. 268th Squadron and flew Wellington bombers over Germany once the war had commenced. Remaining in the RAF after the war, he would retire in 1964 with the rank of Wing Commander (Lieutenant Colonel). Very inventive, he created subminiature cameras (1944) and firearms, a bomb-loading trolley during World War II, and the first electric slot cars and racing circuit (1942), an idea commercially developed by others.

Wallis approached the Bensen design with an experienced aviation innovator/pilot's experience and, as a foreigner, almost totally lacking the devotion that would later characterize Bensen's self-nurtured iconic status among Americans. Wallis may also have felt free to modify the design because he was clearly aware that "Bensen had based his design closely on the wartime Rotachute, a towed autogyro glider."[42] He later stated that "I found a number of things with the Bensen design that I didn't like. Some things I modified because I simply would not fly it that way, but I was careful to stick largely to the original design in case things had been made thus for a particular reason."[43] Wallis built a modified version with improved controls, including replacing Bensen's overhead control with a conventional floor-mounted stick that was immediately more effective than any Bensen development, a more effective rotor head, more extensive instrumentation, and fuel tanks on either side of the keel to balance the load, unlike the Bensen with its single side-mounted tank. It flew well and Wallis responded to many invitations for public demonstrations, including the Battle of Britain Days in 1959. That first Wallis autogyro would be placed on exhibit in the Manchester Air and Space Museum, but it proved to be only the beginning, for even before its completion, Wallis was designing his own aircraft.

In designing his own flying machine, Wallis turned from Bensen, although not Bensen's scale, and began with a blank sheet of paper. As the army had become interested, perhaps based on its earlier experience with the Rota Autogiros but more likely flowing from Wallis's well-received public appearances, the new machine was designed with potential military applications. Wallis had experienced frustration with the Bensen's lack of a prerotator, as takeoff often required waiting until a nearby fixed-wing aircraft was taking off so that the small Gyrocopter could catch the prop wash and spin up its rotors, so an effective prerotator was a necessity. And it was ingeniously developed utilizing a flexible shaft and belt arrangement that spun up the rotor with power from the engine, made out of old surplus Jumo engine starter units from his military armament testing days. (He is rumored to never throw anything away.) Additionally Wallis, who had experienced various stability problems with his first machine, set out to design an improved rotor head, the result being the "offset gymbal, a fundamental mechanical fix for which Ken was granted various patents,"[44] which has been found on almost every gyroplane since, leading to a dramatic increase in gyro stability. This design and testing in 1960 of a new model attracted the interest of Frederick and George Miles and, thorough them, Beagle Aircraft Ltd. This British company dedicated to the development of light aircraft agreed to construct the aircraft with the Wallis-built model as prototype and to undertake an expensive certification process, a necessary step for future military duties. The resulting aircraft, the sixteenth single-engine Beagle aircraft, was called the Beagle-Wallis 116, and while it revolutionized European autogyro design, it failed, as so many gyroplanes before, to achieve a viable military role.

Beagle constructed four WA-116 aircraft,[45] some of which were tested by the British military during the winter of 1962–63. With a Bensen-style open fuselage, the little autogyro was not yet fitted with a cabin or even a front nacelle, and the unfortunate British pilots found it uncomfortable to fly on winter days. The engine also experienced carburetor-icing problems, leading to a severe impact on performance. Additionally, the military required the small aircraft to carry a wet battery and a heavy, unreliable tank radio for communications, further degrading flight capabilities. It was not unsurprising, then, that the military opted for enclosed-cabin Bell 47 helicopters over the WA-116. With the waning of military interest, Beagle's interest waned as well, and their commitment to the Wallis WA-118 lapsed. Although it was the only certified autogyro in England, Beagle did not see a civilian future, and Wallis had made "an early decision not to aim at the recreation and sport market, as he felt that pilots might not take them seriously enough—a recipe that could easily generate a rash of mishaps and the inevitable bad press."[46] In 1964 Wallis and Beagle parted upon his retirement from the RAF, and it looked as if the autogyro would remain his personal pursuit—but the WA-116 was about to become an international film star.

Wallis first encountered the world of motion pictures when engaged to provide the sound effects of airplane engines for the 1964 film *Those Magnificent Men in Their Flying Machines* by the film's consultant Air Commodore Wheeler. The sound technicians recorded his McCulloch engine and then played it back at speeds designed to simulate earlier aircraft motors. In 1966 Wallis was engaged to fly the WA-116 in an Italian film, *Dick Smart Agent 2.007,* which was filmed in Brazil and Italy in 1966.[47] This film, apart from the four Wallis autogyro sequences, is eminently forgettable; had his film career rested on its success, no one would have ever heard of him, as the film has achieved a thoroughly deserved obscurity.[48] But fortunately for Wallis, he was interviewed on radio just prior to departing for Brazil, and the interviewer asked the retired RAF wing commander if he would like to go up against a helicopter in his autogyro. Perhaps still smarting from the army decision to buy Bell 47 helicopters over his WA-116, Wallis answered enthusiastically, "Give me half a chance!"[49] This resonated with Ken Adam—the art director for the latest James Bond film, *You Only Live Twice,* which was completing its preproduction and preparing to begin filming in Japan in six weeks—who contacted Wallis. Ken's WA-116 had been shipped to Brazil, and the only autogyro available had been "pranged"[50] by his cousin, so he was at first resistant to a flying audition and sent a film instead. This served to whet the film crew's interest, but they demanded a demonstration, not unreasonable for a multimillion-dollar international film production.

Wallis took an available WA-116 to Pinewood Studios for a flying audition just three days before he was to leave for Brazil, only to discover that the film's aviation consultant, Group Commander Hamish Mahaddie, thought it was a miniature helicopter. But the screen test went well, and he was engaged to fly in the Albert R. "Cubby" Brocolli and Harry Saltzman production in Japan six weeks later. They informed Wallis that his autogyro had to be painted in a dramatic scheme and fitted with prop weapons, including machine guns, rocket packs, air-to-air guided missiles, parachute-launched grenades, and simulated flamethrowers. And all of this had to be done, and the aircraft had to be in Japan in six weeks! Wallis agreed and went off to Brazil—the WA-116 that was to emerge as *Little Nellie* six weeks later went to the special effects shop. They would be reunited when filming began.

Wallis doubled for Sean Connery as James Bond in *You Only Live Twice,* the first major autogyro film role since 1934. As Wallis later explained, there had been a famous music-hall performer from the 1930s named Nellie Wallace, so "you were apt to be nicknamed 'Nellie.' "[51] This tradition was undoubtedly known to the film's English producers, so the WA-116 became the most famous autogyro in history, *Little Nellie.*[52] While filming in Japan and Spain,[53] Ken Wallis brought autogyro derring-do to a global audience, becoming in the process the only internationally recognized autogyro pilot in history. The thrilling six minutes of screen time actually required forty-six hours of flight time spread over eighty-five flights and two countries. The shots of *Little Nellie* flying over the sea of Japan to the Sakurajima volcano on the volcanic island of Kyushu remain among the most beautiful autogyro scenes ever filmed, but what are most remembered are the aerial combat sequences, which were filmed in southern Spain over the Sierra de Mijas mountains. This was necessary because Japanese law prohibited the firing of guns, even theatrical props, in the air over Japan,[54] and these Spanish scenes presented unique problems. The rockets were apt to turn around after launch, and the parachute grenades tended to tangle in the autogyro's landing wheels. The flamethrowers presented a particular hazard, and Wallis insisted upon dropping them after their use—and later wondered if any filmgoers noticed that they were missing for the rest of the combat sequences.[55]

Wallis flew *Little Nellie* extensively in promotion of the film, and as in the movie, the little autogyro traveled in the cargo hold of large airliners. Promotional flights were made at Morristown, New Jersey, Wallis and *Little Nellie* appeared on the *Today* and *Tonight* television shows, toured Western Europe and, during Christmas 1967, appeared in Australia. Wallis continued to fly *Little Nellie* around the world, eventually appearing at over 750 air shows, thrilling audiences and introducing additional thousands to the autogyro. His first two-seater, a WA-116-T, with *Zeus III* on its tail, was featured in two separate episodes of "The Martian Chronicles,"[56] filmed in 1978 and 1979 on Malta and Island of Lanzarote in the Canary Islands, and also served as a camera platform to film the volcanic landscape that served as Mars for the production. Additionally, one of his autogyros was used for airborne camera work in the British World War II series on Bomber Command entitled *The Path Finders.*

Wallis withdrew from public view after a disastrous fatal crash of a WA-117 on September 11, 1970, at Farnborough, Hampshire, England, in front of a large crowd and television and film cameras. In 1970 the Airmark firm, seeking to enter the British civilian market with an autogyro, purchased a WA-117/R powered by a 100-horsepower Rolls-Royce Continental engine[57] from Ken's cousin to use as a prototype for Certificate of Airworthiness testing. It was issued a Permit to Fly for an air show and was flown by experienced test pilot John W. C. "Pee Wee" Judge for three of the four days in which it flew prior to the fatal accident. Judge was forty-eight years old, a licensed British commercial pilot with ultralight single-seat gyroplanes endorsements and a private pilot's license (gyroplanes), and had been granted a Certificate of Test for the Wallis-117R on July 15, 1970. Although he had over 9,300 flying hours, Judge had only 5 hours, 30 minutes (22 flights) in the WA-117 during the months before his death, and 14 hours, 25 minutes (137 flights) in the WA-116 between November 1961 and December 1963. Prior to the certification flights of the WA-117, Judge had not flown in an autogyro in six and a half years, and his inexperience was partially blamed in the official report for the accident. He exceeded the maximum recommended speed, and the aircraft began to climb in a nose-up position and experienced difficulty, as the autogyro could be adjusted ("trimmed") for low-speed flight but not for high-speed flight.[58] As Judge leaned forward to reach the trim controls while flying past the reviewing stand, he relaxed his hold on the control-stick, and the autogyro began to climb. At that point his fixed-wing instincts took over[59]—Judge shoved the stick forward, and the aircraft dove into the ground, rolled to the right, bounced, and settled on the engine. The official accident report concluded that "all damage, except that sustained by the rotor blades, propeller, fin and rudder, was caused by ground impact."[60] Only 6.25 seconds had elapsed from the first climb to the fatal impact!

Wallis cooperated fully in the subsequent British Board of Trade (equivalent to the FAA) Aircraft Accident Investigation Bureau investigation with the remaining WA-117, and although the conclusions vindicated the design and workmanship, those conclusions reinforced his belief that his autogyros were not for the amateur builder.[61] But his aircraft were employed in professional endeavors that were varied and sometimes exotic. He presented the WA series as platforms for remote sensing, and the WA-117R was employed with experimental silencers in the 1970 search for the Loch Ness monster. His aircraft have also been used for police surveillance, detection of graves and murder victims, coastal pollution assessment, pipeline observation and maintenance, archaeological survey, and military reconnaissance and liaison/communication.

It was in the reconnaissance role in 1978–84 that Wallis came closest to military acceptance, the first such acceptance of this role in England since the World War II Rota. The WA-117R was fitted with an innovative panoramic aerial reconnaissance camera, and Wallis took 180 dramatic photos of Central London, some of which were printed in the Farnborough air show brochure. This international exposure renewed interest in a military reconnaissance autogyro by the camera manufacturer, but this was not realized. Of greater possibility was the interest shown by the German military as a result of Wallis's participation in 1983–84 UK-NATO joint military exercises in England and Germany. The German military was so impressed with the autogyro role in day and night all-weather reconnaissance that a contract was actually signed for 100 aircraft to be manufactured by a licensee. The manufacturer subsequently so modified the design that the prototype would not fly and the contract was cancelled. This not only ended the most realistic possibility for the autogyro to garner a military role but also reinforced Wallis's conviction that others could not, and perhaps should not, construct his autogyros, which were then and which continue to be regarded by some as the finest in the world.[62]

Held in renown, Wallis's is the most honored name in British rotary-wing aircraft.[63] And perhaps mindful of Igor Sikorsky's admonition that "[e]very designer should have to fly what they design. That will get rid of bad designers," he set every autogyro world record between 1968 and 1998. And it was apparent that the advent of the new century and achievement of his ninth decade had not slowed Wallis down in his dedication to gyro achievement, as he set a new record for speed over three kilometers on November 16, 2002![64] He has influenced autogyro/gyrocopter/gyroplane design because, although he refused to release plans or sell kits, his designs have been analyzed and imitated worldwide and his professional efforts in applications of reconnaissance technologies have pioneered new roles and reinvigorated old ones for the aircraft. The British Rotor Association has honored Wallis by naming its annual fly-in "Wallis Days" (imitating the Popular Rotorcraft Association's naming of the American annual fly-in as "Bensen Days"), a fitting tribute to this autogyro giant.

In a similar manner, his fellow designer in Scotland, Jim Montgomerie,[65] also began with a Bensen kit and was so committed to the Gyrocopter that he became a Bensen dealer, eventually selling forty-three kits and becoming a major parts distributor for the UK, but he soon began innovating his own designs. Performing some of the earliest market research in the gyroplane industry, Montgomerie realized that many of his kits remained unfinished due to the basic nature of the Bensen materials—basically just a box of metal tubing with no finishing, shaping, welding, or even drilled holes! His innovation, a finished kit, became the B-8MR Merlin, an adaptation of the Bensen design but with the eventual addition of a fully fitting nacelle, an instrument panel, and internal fuel tank, which could be assembled in eighty hours. Montgomerie's involvement with Gyrocopter design, manufacturer, and flight would span four decades and extend into the twenty-first century. But in many ways, the greatest European influence of Bensen would be seen in the person of Jukka Tervamäki, the greatest gyroplane designer in Finland, whose influence continues to be felt.

NOTES

1. For photographs of the Bensen GyroGlider, see *Aircraft of the National Air and Space Museum,* 4th ed. (Washington, D.C.: Smithsonian Institution Press, 1991); Paul Bergen Abbott, *The Gyroplane Flight Manual* (Indianapolis, Indiana: The Abbott Company, 1996), pp. 54–55.

2. *Aircraft of the National Air and Space Museum,* "Bensen Gyro-Glider" and "Gyro-Copter" entries; "Gyroglider in Smithsonian," *Popular Rotorcraft Flying* 3, no. 3 (Summer 1965): p. 19.

3. For photographs and rotor head diagram of the B-7M, see Igor B. Benson, *A Dream of Flight* (Indianapolis, Indiana: The Abbott Company, 1992), pp. 60–67.

4. Benson, *A Dream of Flight*, p. 63.

5. For photographs of the B-8M, see Ibid., p. 29; "One-Man Rotary Wing Craft," *Air Progress* 16, no. 3 (June/July 1964): cover and p. 78; *Popular Rotorcraft Flying* 4, no. 1 (Winter 1966), cover.

6. Harris Woods, "My 7 Gyros," *Popular Rotorcraft Flying* 2, no. 2 (Spring 1964): 6–7, 10–11.

7. Ed Yulke, "Gyro Cars for Fun," *Popular Mechanix,* November 1945, 74–76, 148.

8. *Popular Rotorcraft Flying,* fall 1963, p. 7 (photo).

9. Paul Bergen Abbott, "Meet Ken Brock," *Rotorcraft* 31, no. 6 (September 1993): 12–17, 14.

10. Bolek Brunak, P.E., "Visions," *Popular Rotorcraft Flying* 14, no. 3 (June 1976): 25–28, 26.

11. See, for example, "Icarus II," *The Dude* 3, no. 2 (November 1958); Richard Ashby, "Come Fly with Me," *Knight* 5, no. 7 (July 1966): 63–65; "Ride with Micky [Dolenz] in his Gyrocopter," *FAVE!* 1, no. 7 (March 1968): 25.

12. Jim Eich, "P.S. to 'Arliss Riggs and Twenty Years of Gyroplanes,'" *Popular Rotorcraft Flying* 16, no. 2 (April 1978): 25.

13. Jim Eich, "Arliss Riggs and Twenty Years of Gyroplanes," *Popular Rotorcraft Flying* 15, no. 6 (December 1977): 14–16, 16 (photo caption).

14. Ed Alderfer, "A Study of the Tractor Gyroplane," *Rotorcraft* 26, no. 6 (December 1988–January 1989): 17–18; Martin Hollmann, "Pusher Gyroplanes, Increasing Interest," *Gyroplane World,* no. 26 (November 1978): 1.

15. "Tractor Gyros," *Popular Rotorcraft Flying* 3, no. 4 (Fall 1965): 32 (photographs of tractor models by Galen Bengston, PRA member #60, and Arliss Riggs).

16. Joe Kirk, "Gizmo," *Gyroplane World,* no. 10 (July 1977): 1–3.

17. Jim Eich, "The Eich JE-2 Two Seat Gyroplane," *Popular Rotorcraft Flying* 15, no. 2 (April 1977): 18–21; *Homebuilt Rotorcraft* 12, no. 11 (November 1999): cover photograph.

18. Jim Eich, "The XNJ 790 Autogyro," *Rotorcraft* 29, no. 3 (May 1991): 10–13.

19. For a description of Herron's Little Wing tractor autogyro, see Ron Herron, "First Flight of a New Tractor Autogyro," *Rotorcraft* 34, no. 8 (November 1996): 6–7; Ron Herron, "Flying Backwards in a Tractor Autogyro" *Rotorcraft* 33, no. 5 (August 1995): 32–34; Ron Herron, "History of Little Wing Autogyros," *Fly Gyro!* no. 4 (March–April 2001): 12–14; Ron Herron, "Little Wing 'Roto-Pup,'" *Homebuilt Rotorcraft* 8, no. 11 (November 1995): 12–13; Ron Herron, "Bringing Back the Autogiro," *Rotorcraft* 33, no. 1 (February–March 1995): 12–13 (citing inspiration by the 1930's [David] Kay Gyroplane).

20. John VanVoorhees, "*The Pitbull Autogyro,*" *Rotorcraft* 34, no. 3 (May 1966): 16; *Rotorcraft* 38, no. 8 (November 2000): 12 (photograph of the inventor with his tractor gyroplane).

21. David Gittens and Kia Woods, "Ikenga, An Artist's Approach to Gyroplane Design," *Rotorcraft* 27, no. 1 (February–March 1989): 44.

22. Don Parham, "Bensen Days and Sun 'N Fun 2001," *Homebuilt Rotorcraft* 14, no. 5 (May 2001): 9–13; *Homebuilt Rotorcraft* 14, no. 5 (May 2001): 14.

23. "The Reverend Igor B. Bensen," *Popular Aircraft Flying* 7, no. 4 (July–August 1969): 18.

24. "A Fun Vehicle That Flies," *Mechanix Illustrated* 66, no. 502 (March 1970): 60–61, 145–46, 60; "The Reverend Igor B. Bensen," *Popular Rotorcraft Flying* 7, no. 4 (July–August 1969): 18.

25. "A Conversation with Ken Brock: From the First PRA Convention to the Latest," *Rotorcraft* 34, no. 6 (September 1996): 31–32, 31; Paul Bergen Abbott, "Ken Brock: A Full and Wonderful Life," *Rotorcraft* 39, no. 9 (December 2001–January 2002): 5.

26. Frank Kingston Smith, *Legacy of Wings: The Story of Harold F. Pitcairn* (New York: Jason Aronson, 1981), p. 323 (citing a memorandum from Frank Piasecki's patent counsel stating, "We find that we apparently infringe ten patents held by the Autogiro Company of America...The validity of these patents cannot easily be questioned as they appear to be very well drawn up by counsel not only skilled in patent practice but very familiar with the rotary-wing art.").

27. Quoted in Richard Aellen, "The Autogiro and Its Legacy," *Air & Space Smithsonian,* December 1989/January 1990, 52–59.

28. Smith, *Legacy of Wings,* p. 334.

29. See, for example, Kathryn E. O'Brien, *The Great and the Gracious on Millionaires' Row* (Utica, New York: North Country Books, Inc., 1978), p. 89. It should be read with great caution, however, as the author asserts that Pitcairn "had told his wife he was going upstairs to do some work," which clearly incorrectly locates Pitcairn's study; additionally, O'Brien had incorrectly asserted that Harold Pitcairn landed the Autogiro on the White House lawn, a flight clearly flown by Jim Ray.

30. Smith, *Legacy of Wings,* p. 334.

31. Frank Anders, "The Forgotten Rotorcraft Pioneer: Harold F. Pitcairn," *Rotor & Wing International* (May 1990): 34–37.

32. Frank Kingston Smith adds, *neatly* accounting for all the circumstances, "The next morning it was discovered that the semi-automatic pistol was defective; when cocked, it had a supersensitive 'hair trigger,' and it had a faulty disconnector so that it would fire more than one shot at a time, a condition known as 'doubling.' None of this information ever came out. Mrs. Pitcairn declared that she never wants to hear another word about the tragedy." Smith, *Legacy of Wings: The*

Story of Harold F. Pitcairn, p. 334; but see Frank Kingston Smith, "Mr. Pitcairn's Autogiros," *Airpower* 12, no. 2 (March 1983): 49, where the author asserts, "At age 75 Harold Pitcairn died," obviously confusing Harold's age with that of his older brother Raymond.

33. For a stunning photograph of Pitcairn flying *Miss Champion,* see Aellen, pp. 58–59.

34. David Brownridge, "A Sycamore Seed for Grownups," *Western People,* December 5, 1996, 5–6.

35. Jim Sottile, "RAF Makes Dreams Come True," *Rotorcraft* 32, no. 6 (September 1994): 11–13.

36. Jim Vanek, *Sport Copter* (Scappoose, Oregon: Sport Copter, Inc., 2000), p. 6.

37. "Profile: Wing Commander K H Wallis," *Popular Flying,* April–May 1996, 13–21, 15.

38. For a comprehensive view of the many careers of Ken Wallis, see Ian Hancock, *The Lives of Ken Wallis: Engineer and Aviator Extraordinaire,* 2nd ed. (Suffolk, England: Norfolk & Suffolk Aviation Museum, 2001; 2nd ed., 2002).

39. "Wing Commander Wallis to be Honored," *Popular Rotorcraft Flying* 26, no. 1 (February 1988): 26.

40. "Profile," p. 13.

41. Hancock, p. 139.

42. Phillip Whiteman, "Pilot Profile: Ken Wallis," *Pilot,* January 1998, 22–25, 22.

43. "Profile," p. 16.

44. Whiteman, p. 23.

45. For a discussion of the WA-116, see "Beagle Wallis WA.116 Autogyro: Origin and Development," *Rotor Gazette International,* no. 13 (May–June 1994): 7.

46. Peter Lawton, "Flying a Wallis Autogyro," *Pilot,* January 1998, 26–28, 26; Whiteman, p. 25 ("The Wallis autogyro not in Ken's ownership is retained by Nigel de Ferranti. Ken refuses to supply amateur builders with plans or drawings of his autogyros for fear of someone coming to grief through modifying his designs.").

47. For a description of the filming in Brazil, see Wing Commander Kenneth H. Wallis, "Movie Flying in Brazil," *Rotorcraft* 33, no. 9 (December1995–January 1996): 10.

48. See Wing Commander Kenneth H. Wallis, "I Was 50 before I Was '007,'" *Rotorcraft* 28, no. 2 (April 1990): 8–14.

49. Recounted in "Profile," p. 17.

50. He described this model as then suffering from "a bout of 'Macitis'" [engine difficulty with the McCulloch engine]. For a description of the filming of the Bond film, see Wallis, "I Was 50," pp. 8–14.

51. Ibid., p. 9.

52. For a detailed technical discussion of *Little Nellie,* see Alastair Dougall, *James Bond: The Secret World of 007* (London: Dorling Kindersley, 2000), 46–47; additionally, for a description of Wallis's role in the filming of *You Only Live Twice,* including the flying accident that lead to the loss of a leg by cameraman Johnny Jordan (who would recover from the *YOLT* accident but perish when he fell out of a camera plane during the filming of *Catch 22*), see Steven Jay Rubin, *The Complete James Bond Movie Encyclopedia* (Chicago, Illinois: Contemporary Books, 1995), pp. 241–42.

53. Wallis described the Bond-movie filming in Wing Commander Kenneth H. Wallis, "The Longest Spin," *Rotorcraft* 33, no. 3 (May 1995): 18–24, and in "I Was 50," pp. 8–14.

54. Rubin, p. 242.

55. Wallis, "I Was 50," p. 11.

56. "Martian Chronicles," *Cinefantastique* 10, no. 1 (Summer 1980): 19–23; "Ken Wallis Keeps Busy," *Rotorcraft* 27, no. 4 (June–July 1989): 4.

57. For a listing of the Wallis's autogyros, see Wallis, "I Was 50," p. 13.

58. "Negative Gravity: An Accident Analyzed," *Gyroplane News & Small Helicopter,* no. 1 (Spring 1990); Wallis, "I Was 50," p. 19.

59. See Mel Morris Jones, "Talkshop: A Conversation with Jukka Tervamäki," *Fly Gyro!* no. 1 (September–October 2000): 4–10, 18, 9 (commenting on the Judge accident in the WA-117).

60. "Negative Gravity."

61. "Profile," p. 21 ("My autogyros are not intended for homebuilding. Some of the techniques are a bit special, such as electron beam welding.")

62. "The series of small autogyros built by Wing Commander K. Wallis represent not only Britain's top designs but probably the finest of their type in the world." Michael J. H. Taylor and John W. R. Taylor, *Encyclopedia of Aircraft* (New York: G. P. Putnam's Sons, 1978), p. 224.

63. His honors include: 1963, the Alan Marsh Medal, awarded by the Royal Aeronautical Society and the Helicopter Association of Great Britain; 1969, the Segrave Trophy, awarded by the Royal Automobile Club and the Royal Aeronautical Society to the individual who has most contributed to British aviation; 1975, the Silver Medal, awarded by the Royal Aero Club; 1975, the Rose Trophy, awarded by the Helicopter Club of Great Britain; 1980, Honorary Fellowship, awarded by Manchester Polytechnic; 1982, the Reginald Mitchell Medal, awarded by Stoke-on-Trent Association of Engineers; 1984, the Rose Trophy, awarded by the Helicopter Club of Great Britain (second award); 1985, the Segrave Trophy, awarded by the Royal Automobile Club and the Royal Aeronautical Society (second award); 1989, the Salomon Trophy, awarded by the Royal Aero Club; 1995, Rotorcraft Gold Medal, awarded by the Federation Aeronautique Internationale; 1996, Member of the Order of the British Empire, awarded by Her Majesty Queen

Elizabeth II; 2003, Ph.D. *(honoris causa)* and designation as a Gyroplane Pioneer by Hofstra University.

64. See Wing Commander Kenneth H. Wallis, "Autogyro World Records: Past, Present and Future..." *Fly Gyro!* no. 6 (July–August 2001): 14–18; Ian Hancock, email to author, November 17, 2002.

65. Jim Montgomerie, "Autogyro Basics and World Record Flights," *Rotorcraft* 27, no. 2 (April 1989): 12–13; "The Jim Montgomerie Story: A Man and His Gyros," *Rotor Gazette International,* no. 17 (January–February 1995): 3–5, 10–11; "Latest Merlin GTS," *Fly Gyro!* no. 4 (March–April 2001): 16–17.

Chapter 12

BENSEN, TERVAMÄKI, GROWTH OF THE PRA, AND THE FAIREY ROTODYNE

[I]t's not the pilot, it's the machine that is a good flier.

Igor Bensen, *Popular Rotorcraft Flying*

I suspect there are some bureaucrats in Washington who would like to swat your mosquito-like whirlybirds out of the skies, or spray them with DDT, and hope they would die. But we can't let that happen. We must permit these new designs to develop and to become more useful. I think it's absurd that these beautiful machines that I saw flying all day today can't be sold here in the U.S. for commercial and many other uses. I can think of many uses for it in agriculture. I am sure that there are others. So that's something I am certain that FAA will have to look at.

Congressman Harold Cooley from North Carolina, Chairman of the House Agriculture Committee, at 1965 PRA Fly-In, Raleigh-Durham, North Carolina

The article entitled "Brave New Aircraft" appeared in the June 14, 1954, issue of *Life* magazine presented the Bensen Gyroglider (N3785C). Bensen, wearing goggles and white shirt complete with tie, was shown piloting the $295, eighty-six-pound glider, which was being towed by a truck. The article was to inspire many but immediately gained the rapt attention of young Jukka Tervamäki in Finland. Tervamäki had become interested in aviation from the earliest period of his life—at the age of four he drew a postcard of himself flying an airplane and sent it to his father, who was fighting against the Russian invaders during World War II. He was struck by the fact that although aviation was generally beyond the reach of the average person, the Bensen aircraft was "such a simple machine that even a schoolboy could build it."[1] Responding to a Gyrocopter advertisement in *Popular Mechanics,* Tervamäki began a correspondence with Bensen, who was delighted to nurture his enthusiasm. Tervamäki eventually ordered a set of Gyroglider plans, purchased the Bensen factory-manufactured rotor head, and in 1957 started to design his own powered Gyrocopter with a Triumph engine. Even though his approach was innovative, it was unsuccessful due to design inexperience. But Igor Bensen was so impressed that Tervamäki came to America and began to work at Bensen's North Carolina factory in 1959.

Although Tervamäki's time with Bensen was primarily spent producing standard B-8M machines, Bensen asked him to test the Triumph engine installation. This second (and final) attempt did not fare better than the first, as even with the consultation and aid of the Bensen factory engineers, connecting rods were not sufficiently tightened and the engine exploded. But upon his return to Finland, Tervamäki, by now familiar with the origins of the Autogiro and the work of Juan de la Cierva, decided to specialize in rotary-wing aircraft in his studies at the Helsinki University of Technology, from which he graduated in 1963. He took a job at the Finnish air force headquarters as a maintenance engineer for helicopters, learned to fly, and met a skilled helicopter pilot/engineer with whom he would go on to design autogyros, Aulis Eerola. Tervamäki's engineering background and experience were to lead to significant innovations in gyroplane design.[2]

The two young enthusiasts initially began their design efforts with fiberglass rotor blades, an innovation unknown even in helicopter technology of that period. Tervamäki tested the blades on their unpowered gyroglider, the ATE-2 (Autogyro-Tervamäki-Eerola), which was a modified version of Tervamäki's 1958 JT-1, serving as a towed testing platform for blade development. He and Eerola began in 1966 to develop a new autogyro utilizing the fiberglass blades, the ATE-3. After much development and some truly hair-raising experiences, including blade flutter during the first testflight, the design was proved and successfully flown in 1968. In addition to its airframe of welded steel, which differed from the more commonly used square aluminum tubing, the ATE-3 was innovative in the use of glass-reinforced plastic rotor blades, propeller, cockpit, and tail surfaces. The fiberglass cockpit had been appropriated (Tervamäki would some-

(Courtesy of Jukka Tervamäki.)

times use the word "stolen" or claim it had been scrapped)[3] from a Finnish UTU fiberglass sailplane under development at the same time. A side benefit of using a sailplane cockpit was that it afforded the pilot great protection, as it had been stressed to 6 Gs, while the ATE-3 could only achieve 2.5 Gs. And while Ken Wallis had saved the world in *Little Nellie* in 1966, the ATE-3 (OH-XYV) had a flying role in the 1969 Spede Pasanen and Ere Kokkonen Finnish film *Leikkikalugangsteri* (*Toy Gangster*). That film began with perhaps a knowing tribute to Ken Wallis in that the film's hero, a toy company executive/playboy, is first shown engaged in slot-car racing. Although both inventors flew for the movie, Eerola did most of the flying. The film shows off the Gyrocopter's abilities but clearly lacks the production values, stunning photography, scenery, thrill, and world-saving rush of James Bond.

On January 7, 1973, Tervamäki first flew the JT-5, a single-seat autogyro featuring an enclosed fiberglass cockpit with a sideward opening Plexiglas canopy. The design made extensive use of internal structures of glass fiber reinforced epoxy resin. For maintenance and preflight checks, the instrument panel cover and pilot seat back, which formed a firewall to the rear engine compartment, opened together with the canopy. This innovative use of fiberglass came from Tervamäki's experience in reinforced fixed-wing design and plastics technology gleaned as the project manager for the development of an all-fiberglass glider-towing aircraft for his alma mater, Helsinki University of Technology, in 1972. As his interest shifted to motorgliding, Tervamäki sold the JT-5 manufacturing rights to Italy's Vittorio Magni[4] in 1973, but Tervamäki continued to sell JT-5 plans and to pioneer in the creation of computer programs to aid in gyroplane development.[5] In 1980 Magni, now a close friend, asked Tervamäki to design a two-seat cabin autogyro. The MT-7, for Magni-Tervamäki, was a stunning two-passenger, side-by-side composite aircraft with an aerodynamic shape, sweeping Plexiglass canopy, and twin tails similar to those on the first JT-5. The MT-7, called the Griffon, first flew in 1985. It was reported to have crashed some years later in France, but its influence lives on in subsequent Magni models.

Even as Europeans were engaging and then adapting Bensen's designs and taking autogyro/gyroplane development in new and exciting directions, Bensen's zeal and enthusiasm and continued development were leading to constant growth. The eventual growing pains would radically change his position within the movement he had created, and 1972 was to prove a watershed year for the American gyroplane community. Igor Bensen (he was not yet "Dr. Bensen"), at the urging of his employee Ed Trent, had founded the Popular Rotorcraft Association in 1962 with his wife Mary, family members, and close associates. The first members included Igor Bensen, #1; Mary T. Bensen, #2; Charles W. Elrod, #3; Edgar B. Trent, #4; Donald Dean, #5; Morton Roberts, #6; J. Blake Self, #7; David I. Bensen (son), #8; Ricky I. Bensen (son), #9; Mark V. Bensen (nephew?), #10; and Dr. V.B. Bensen (brother?), #11[6] and a group of associates, employees, and Bensen dealers.

The association's stated goal, undoubtedly written by Bensen, was to be "a voluntary, non-profit, non-partisan organization, whose members are dedicated to the advancement of knowledge, public education and safety of privately owned non-commercial rotorcraft.... The principal goal of PRA is to serve as an instrument of unification of men with the common interest of advancing progress of rotorcraft for personal flying."[7] The in-house rationale for support of this organization was "so that builders of Bensen Gyrocopters would have a way of connecting with other builders."[8] As Bensen was already developing helicopters and lifting platforms, the association was not limited to Gyrocopters but included all rotorcraft. It clearly was an in-house effort—headquartered in the Bensen factory in Raleigh, North Carolina—and featuring the Bensen machine, it was pretty much the only game in town. Bensen's own aircraft (N2588B), in whole or in part, decked out with pontoons, rolling down city streets, standing next to what appears to be a large stuffed black bear, and so on, graced almost every issue of the association's publication, *Popular Rotorcraft Flying*.

The PRA became the voice of the American gyroplane community, speaking with a growing confidence and authority. As the aviation scene was changing in America with the growth of private aircraft and the emergence of homebuilt planes, the government was pressed to produce new Federal Aviation Regulations (FARs) to promote individual involvement. The Civil Aeronautics Act of 1938, the product of a congressional committee chaired by Senator Harry S. Truman and later the revised Federal Aviation Act of 1958, did not recognize the category of amateur-built experimental aircraft and totally ignored rotary aircraft—understandable since there were no homebuilt helicopters and Bensen had been selling his kits and plans for only a few years. The regulations of 1938 and 1958 required that only aircraft with "air-worthiness certificates" could be flown, and as observed by Bensen in an editorial in 1963, those who built and flew his Gyrocopters were technically violating the law.[9] Gyrocopters could not be legally sold in the United States, but Bensen effectively skirted the law by selling legal Gyroglider kits. If the enthusiasts desired to add an engine, Bensen would provide plans, an engine mount, controls, and either a new ninety-horsepower McCulloch engine or a used engine that had been factory rebuilt and modified. Additionally, by 1966, at least fifteen sales and service organizations, chiefly Bensen dealers, advertised parts and assembly services. There was even an insurance agency offering specialized insurance for pilots and machines.

The first PRA fly-in took place on June 15, 1963, only a year after its founding, and had a registered attendance of 275 members from fourteen states and Canada. With fifteen rotary aircraft present, it was "the world's largest gathering of private rotorcraft at any one time and place."[10] Beginning with that very first gathering, the PRA established a favorable and effective relationship with the FAA, eventually lobbying for the creation of an official "experimental" category (more than 51% homebuilt required federal registration [N] number, an FAA airworthiness inspection prior to flight, and no commercial uses) and later the additional "ultralight" category (254 pounds or less), both of which included gyroplanes (which was the inclusive term adopted by the FAA). The experimental aircraft was to prove of tremendous benefit to the amateur market, as it effectively insulated the designer, kit manufacturer, and marketer from the liability accompanying certified aircraft. Legally, the amateur builder *is* the manufacturer and is solely responsible for the ship's airworthiness. In fact, each builder signs an affidavit that is submitted to the FAA in the registration process naming [himself] as the manufacturer.

It was significant and extremely fortuitous that the FAA official who would become most responsible for encouraging the PRA and a knowledgeable and enthusiastic advocate of amateur rotorcraft,[11] fellow North Carolina native Juan K. "Jay" Croft, was also a "homebuilder" who encouraged amateur-built aircraft. Representing FAA administrator Najeeb Halaby (who was made an honorary member of the PRA), Croft was a rated Autogiro and helicopter pilot and ideally suited to understand this new amateur aviation movement. Coming with not only a sympathetic view, he gained an immediate hands-on involvement such that Igor Bensen presented him with a bottle of champagne at the concluding banquet for being the first FAA official to check out in a dual-seat Gyroglider. He soloed that same day.

On September 5, 1963, in Washington, D.C., Croft presided over the first meeting of PRA and FAA officials. President Igor Bensen, Vice President Charles W. Elrod, Vice President J. B. Self, and Director Edgar B. Trent represented the PRA. Juan K. Croft, W. B. Masden, M. W. Leaphart, F. M. Kelly, R. J. Scholtz, and T. D. Sheehan represented the FAA. Accompanying the PRA delegation were consultants M. J. Joyce and C. W. Williamson. The PRA-certified minutes of that meeting reveal a wide-ranging discussion of issues of concern to the PRA membership. The FAA officials "stressed the need of maintaining communications between PRA membership and FAA's Washington offices and praised the tone and quality of PRA's publication, the 'Popular Rotorcraft Flying.' "[12] Although these minutes may be viewed by a jaundiced eye as self-serving and -aggrandizing, subsequent events clearly demonstrate that a viable and productive relationship had been forged and that the PRA had become the voice of the gyroplane community in America, a role it continues to play four decades later. Croft would continue as an advocate for the PRA until his death at his desk in Washington, D.C., in 1978—his voice had been heard and his influence felt for the first sixteen years of the PRA. It would be missed.

The 1957–74 period saw the appearance of several unique aircraft and serious, but ultimately unsuccessful, attempts by well-intended groups to

revive the Autogiro in advanced forms. Each has been largely all but forgotten, ironically so since the British Fairey Rotodyne, the Russian Kamov
Ka-22, the America Umbaugh (later Air and Space) 18A and McCulloch
J-2 and Canadian Avian 2/180 represent significant technological achievements. There are currently only a few remaining flying 18-As and a few
museum exhibits of the Avian and J-2, and little remains of the single Rotodyne prototype. They deserve better—much better. Of these, the Rotodyne
and its Russian counterpart, the Ka-22, are the most unique. Of the American and Canadian aircraft, three of the four certified gyroplanes were adaptations of helicopter rotor and control systems, without the accompanying
complexity, to rotary-wing flight by means of the gyroplane mode. But they
represented an attempt to create an autorotational aircraft that Cierva
would have recognized and to bring it to a mass market—and are worthy
of attention. The McCulloch J-2 is particularly so, as it had the greatest
nongovernmental financial backing—the McCulloch company eventually
lost $30 million dollars,[13] more than six times what Harold Pitcairn had
spent thirty years before. But the most amazing aircraft was clearly the
Fairey Rotodyne, an aircraft with roots in the past and so far ahead of its
time that the rare films of its flight bring amazement even in the twenty-
first century!

FAIREY ROTODYNE

The Fairey Rotodyne was first flown on November 6, 1957, at the Fairey
facility at White Waltham with Chief Helicopter Test Pilot Squadron
Leader W. Ron Gellatly at the helm, along with Assistant Chief Helicopter
Test Pilot Lieutenant Commander John G. P. Morton as second pilot. The
Rotodyne[14] Y form (XE521) carried a crew of two and forty passengers,
and receiving lift from fixed wings of forty-six feet, six inches, was pro-
pelled forward by two wing-mounted 3,000-shaft-horsepower Napier
Eland N.E.1.3 turboprop engines. But this was no ordinary aircraft—its
fifty-eight-foot, eight-inch fuselage could lift off and cruise as a helicopter
with four tip-mounted pressure jets powering rotors that provided a disk
ninety feet in diameter. The tip-jets of the stainless steel rotors were powered by the same Eland engines, which were coupled with compressors to
force air into the tip-rotor pressure jets. Fuel was mixed with the compressed air and then ignited to create thrust capable of turning the rotor.
But once aloft, the Fairy Rotodyne would disengage its rotor, which would
then unload and autorotate to provide approximately 65 percent of the aircraft's lift. The first transition from vertical to horizontal flight was on

Fairey Rotodyne. (Courtesy of Agusta Westland.)

April 10, 1958, the realization of a quest to merge the benefits of the auto-gyro, helicopter, and airplane.[15] But films and a few surviving components in the British Rotorcraft Museum in Weston-super-Mare—after Westland Aircraft Ltd. (having acquired Fairey Aviation on May 2, 1960) cancelled the Rotodyne project with the ending of official funding on February 26, 1962—are all that remain of this magnificent achievement. In the words of Derek Wood, author of *Project Cancelled*, "so died the world's first verti-cal take-off military/civil transport."[16]

The Rotodyne, the most impressive application of the Autogiro princi-ples and autorotation technologies developed by Cierva and Pitcairn, was created by Fairey Aviation Ltd. Its roots are found in the most unlikely of places, namely the turning away of the Cierva Autogiro Company Ltd. from development of the Autogiro to the helicopter. And the beginning of the untimely end of the Rotodyne is to be found in an equally unlikely and remote event, the successful 1946 negotiation and January 1947 acquisi-tion by Westland Aircraft Ltd. of a license to build a modified version of the four-seat Sikorsky S-51 (reengineered and produced as the Dragonfly in 1948) helicopter by Mr. (later Sir Eric) Menforth and Mr. E. C. Wheeldon. The eventual government decision that British helicopter development would be based on Sikorsky engine and rotor technology in the late 1950s would doom the Rotodyne in favor of Westland.

Although Cierva's death in 1936 did not stop Autogiro development in England and America, the former continuing for three years and the latter hanging on for seven years, the focus of the Cierva Autogiro Company under Dr. J.A.J. Bennett, who had become technical director after Cierva's death, shifted primarily to the design and testing of jump takeoff rotor heads and the helicopter, a direction that would inevitably lead to the Rotodyne.

Helicopter development in post–World War II England resumed after a general hiatus, and the Rotodyne began with the concept for a compound helicopter developed by Dr. Bennett and Captain A. Graham Forsyth of Fairey Aviation, based on 1947 studies. During World War II Dr. Bennett, along with former Cierva pilot Wing Commander Reggie Brie, had served as principle technical officer to the British Air Commission in Washington, D.C. As such he was well aware of Allied and German rotary-wing developments, including the work of Flettner and Doblhoff, and Bennett had advanced the idea for an aircraft that could take off and land vertically in helicopter mode with a power-driven rotor using a controllable-pitch propeller for yaw control. In horizontal flight, however, power would be transferred to the propeller for forward movement while lift was generated partially by the autorotating rotor and partially by small wings. The rotor's collective pitch controlled vertical lift in the helicopter mode and would change as the throttle was opened or closed. Roll and pitch were controlled by tilting the rotor head, while a single tractor propeller in the starboard wingtip was used to control yaw as well as to provide forward thrust in the autogyro mode. In this manner the Gyrodyne,[17] as it was named, was also reminiscent of the control approach taken by Bratukhin in the 11-EA, with its small wing propellers and helicopter and autogyro modes of flight.

The first of Fairey Aviation's compound helicopter prototypes, the Gyrodyne, first flew on December 7, 1947, based in part on the 1938 design for the Cierva S-22/38, in response to a Royal Navy specification for a ship-based helicopter. It was powered by a relatively powerful 520-horsepower Alvis Leonides radial engine and established a new world's helicopter speed record of 124.3 mph on June 28, 1948. The technology, although derived from known sources, embodied a leap into the unknown in terms of metal fatigue, for which the previous experience of other developers did not provide—both the pilot and the observer died when the rotor head disintegrated and the Gyrodyne crashed ten months later. This tragedy made Fairey Aviation realize that while it was attempting to incorporate known technological achievements into an innovative aircraft form, such a combination would require a great deal of new research and development, and they embarked on four years of effort. The result was a second prototype, which featured a completely redesigned transmission system and strengthened rotor designed to withstand the stress of helicopter takeoff and landing, autogyro flight, and the in-flight conversion between the two. And although the company attempted to convey to the public the new and innovative nature of its second compound helicopter by naming it the Jet Gyrodyne, it was actually powered by the same type of Alvis radial engine that powered the original. That is where the similarity ended, however—the original gearbox transmission that shifted power from the rotor to the engine was replaced by a pair of engine-driven modified Super-marine Spitfire superchargers that served as compressors to force air into miniature jet nozzles located at the tip of each rotor. Fuel was forced into the nozzle by the centrifugal force of blade rotation and then ignited—effectively a Doblhoff tip-jet powered helicopter. Much development and testing had gone into the technology of this new model, which first flew in January 1954. The first in-flight transition involving autogyro mode was on March 24, 1955, by test pilot John N. Dennis. It proved to be both underpowered and, reminiscent of the fuel-consumption problem previously confronted by Doblhoff that led to his decision to power the rotor only in takeoff and landings, could carry enough fuel for only fifteen minutes. However, the Jet Gyrodyne was not designed for production but as a proof-of-concept testing platform for the technologies that would power a much grander vision that was simultaneously being developed—the Fairey Rotodyne.

Dr. Bennett and Captain Forsyth had begun articulating the Rotodyne concept in 1947, before the potential of a larger transport helicopter was recognized. A turbine-powered design was submitted to the British government on January 26, 1949, for a compound craft capable of carrying twenty passengers, with a four-blade rotor powered by two Armstrong Siddeley Mamba engines, but the research on the Jet Gyrodyne would result in an eventual design that incorporated jet-tip rotors. However, even though the Jet Rotodyne was still years away, Fairey began almost immediately to modify the Rotodyne design and by March of that year had formally submitted three alternative new designs: a model now powered by Mamba or Rolls-Royce Dart engines for forward flight and providing pressurized air for jet-tipped rotors in helicopter flight mode; a model with three Mamba engines, two of which would be for forward flight and one to power the jet-tipped rotors with compressed air; and a third design with two wing-mounted Mamba or Dart engines for forward flight and auxiliary air compressors for the jet-tipped rotors. An initial development contract was awarded in October 1950 for a model based on the Dart engine, but that was modified when Lord Ernest Hives, who had originally been head of the Rolls-Royce experimental shop and chief test driver, complained that the Rolls-Royce engine design team was overcommitted. The government then decided in late 1950 that the Rotodyne project would go forward with Arm-strong Siddeley Mamba engines with auxiliary compressors, a power plant then dubbed the "Cobra." However, by July 1951 Fairey itself had com-pleted a redesign of the Rotodyne to meet, in part, the requirements of the British European Airways (BEA) articulated in 1951 for a ten- to twelve-passenger helicopter to provide service between British cities. Two new designs were submitted, one with two Mamba engines and a four-blade rotor and a second with three Mamba engines and a five-blade rotor. The former had an all-up weight of 20,000 pounds and the latter, due to the increased lift of its rotor, was projected at 30,000 pounds. However, neither of these designs was to be—for, like Rolls-Royce previously, Armstrong Siddeley complained to the government that its production facilities were also then overloaded.

There can be little doubt that the design difficulties experienced by Fairey with regard to engine procurement were an accurate reflection of the frenetic research and development effort in postwar British aviation. But there is also another possibility that is far better reflective of the poli-tics that would eventually doom the Rotodyne project and the abandon-ment of jet-tip power in the British rotor industry. In 1949, just after Fairey's first design submission to the government, the director of engine research had publicly strongly objected to support being given to the Roto-dyne project[18]—eventually total support would be withdrawn and the proj-ect killed in favor of Westland and its application of its licensed Sikorsky technology. But while the opposition was known in the early 1950s, the eventual end was not, and Fairey pressed on with an alternative design.

By June 1952 the Rotodyne design now featured a de Havilland H.7 tur-bine engine combined with auxiliary compressors, the former for forward flight and the latter for rotor power. But again this was not to be, as agree-ment could not be reached with de Havilland, and Fairey complained to the Ministry of Supply that it was being neglected. Subsequently, after consultation with the ministry, Fairey settled on the just-introduced Napier Eland engine in April 1953. Napier, under the developmental direction of

A. J. Penn and Bertie Bayne, had entered into the development of gas turbine engines, and its Eland engine would be produced from 1952 until 1961.[19] The Rotodyne Y prototype then went forward, with two Eland N.E.1.3 engines with auxiliary compressors and a now-enlarged section four-blade rotor with an all-up weight projected to be 33,000 pounds. Projected, but never realized, was a cargo version, with the larger Eland N.E.1.7 engine and a flying weight of 39,000 pounds.

The British government had continued to fund Rotodyne development, but funding for the proposed Eland prototype was not approved until April 1953 and itself was not free from controversy. Fairey had suggested that £710,000 would be sufficient for development of the airframe, but its estimate was met with great skepticism as a result of its inability to effectively project (or control) developmental costs of the Gyrodyne. The remaining Jet Gyrodyne (XD 759) had been converted to jet-tip rotor propulsion using auxiliary compressors powered by the Leonides engine to test the Rotodyne concepts, but the costs of this conversion had escalated from a projected £75,000 to a spectacular £192,000, a cost overrun of 156 percent! Nevertheless, even though the government had reportedly been staggered by the cost overrun of the converted Jet Gyrodyne, the Ministry of Supply entered into a contract with Fairey Aviation to construct the Rotodyne Y prototype in July 1953. The contract specified a forth-to-fifty-passenger model with a 150-mph cruising speed and a range of 250 nautical miles, and it was understood that a larger machine would follow.[20] Construction was under the direction of Captain Forsyth, because in April 1952, as a result of a disagreement with Fairey, Dr. Bennett had left to join Hiller Aircraft in the United States. The man who had spanned the golden age of rotary aviation, from Cierva to the Rotodyne left the project, but there was every reason to be optimistic in 1953 and to believe that the Rotodyne would indeed be the most successful application of Cierva's autorotational principles. It was not to be.

Officially called an experimental compound helicopter, the Rotodyne Y (XE521—a military registration number) featured a single four-blade main all-metal, primarily stainless steel rotor carried above its fifty-eight-foot, eight-inch-long fuselage on a large, fully faired dorsal pylon structure. The ninety-foot-diameter rotor was driven by pressure-jet units at the rotor tips. In horizontal flight the rotor was unloaded as an autogyro and allowed to autorotate, providing approximately 65 percent of the aircraft's lift, the remaining lift coming from a forty-six-foot, four-inch cantilever high-wing on either side of the fuselage. The wings were all-metal two-spar construction. The boxlike rectangular cross-section fuselage, featuring double clamshell doors at the rear to allow for efficient cargo and vehicle loading, was an all-metal semimonocoque structure. The all-metal tail, boxlike, was a braced monoplane type mounted on top of the fuselage and originally featured two endplate fins and two rudders, but a third central fin was added in early 1960 to improve control and increase stability. The Napier Eland N.E.1.3 turboprop engines were seated below each wing in underslung nacelles—each driving a de Havilland four-blade propeller. Fuel was carried in tanks located within each wing. The landing gear was a tricycle type, with the main wheels retracting into the bottom of the engine nacelles. Taxiing and landing were cushioned by the use of Oleo-pneumatic shock absorbers. The cockpit was set forward in the nose of the fuselage and featured dual controls and an unrestricted view. With an overall height of twenty-two feet and a design gross weight of 33,000 pounds, but sometimes flown at 38,000 pounds, it was the largest helicopter of its day, with the tail assembly being built at Fairey's Stockport factory and the fuselage, wings, and rotor assembly being constructed at Hayes.[21] Assembly of the component units took place at the airfield at White Waltham, with a full-scale static test rig having been constructed at the Aeroplane and Armament Establishment at RAF Boscombe Down, Wiltshire. There the rotor and power plant were assembled and tested, including a twenty-five-hour test of the tip-jets, in conformance with a required ministry approval test. The aircraft's weight was proving to be an issue, for by 1956, the rotor system weighed in at 5,503 pounds, which was 68 percent above the originally estimated 3,270 pounds. But despite the weight and also excessive noise produced by the tip-jets, the testing went well and the civilian aviation community began to take interest. Fairey, anticipating the civilian market and the Rotodyne's ability to land in helicopter (or autogyro) mode in the midst of cities, took the noise issue very seriously and by late 1955 had tested forty different types of noise suppressors. Based on the measurements taken with the Jet Gyrodyne fitted with tip-jets, Fairey anticipated that the Rotodyne noise level at an altitude of two hundred feet from both engines would be 106 decibels (dB). Fairey hoped to reduce this to a more acceptable 96 PN dB, but although the estimated sound level was correct, the reduction was never achieved and the Rotodyne remained a loud machine. The claim, however, that the Rotodyne failed because of its noise level is overly simplistic.

BEA was consistently updated as to the Rotodyne testing program and the general expectation was that, as London-to-Paris center-city service had been discussed in December 1954, the Rotodyne would enter civilian service with an eighty-eight-minute flying time. Additionally, Fairey had proposed that the second anticipated, larger prototype would be capable of carrying vehicles for the military. Things were definitely looking up, but in early 1956 the government imposed serious budget limitations and there was a marked reduction in enthusiasm for programs such as the Rotodyne, which had been funded as part of the defense budget. It had been a logical way to further the developmental program, given the RAF and army interest, but it also left the Rotodyne vulnerable to government cutbacks—which now came. The Defense Ministry withdrew further financial support by stating that there was no further military interest in the Rotodyne, effectively throwing the entire project into the civilian sector, which now had to bear not only the whole developmental costs but also the funding to develop the Eland N.E.1.3 engines for Rotodyne application. After much discussion during 1956, the government agreed to fund the Rotodyne and Eland projects until the end of September of 1957, a deadline subsequently extended through the end of the year, subject to three conditions: (1) the Rotodyne had to be a technical success; (2) Fairey Aviation must secure a firm order from BEA and intents-to-purchase from other air carriers; and (3) Fairey Aviation and English Electric, the corporate parent of Napier, must fund a proportion of the developmental costs. Fairey agreed because, although the first flight had slipped from its projected 1956 date, the company was confident that the Rotodyne would finally take to the air in 1957 and that commercial success would surely follow.[22]

The Rotodyne first flew on November 6, 1957, and its first successful transition from vertical/horizontal/vertical flight was on April 10, 1958. There was much publicity of the Rotodyne achievement, and it was the center of industry attention and public acclaim at the SBAC Show at Farnborough in September, where it performed vertical and horizontal flight, a successful demonstration resulting in an order for a Rotodyne and options for two more from Okanagan Helicopters Ltd. of Vancouver, Canada. This was not surprising given that Fairey Aviation had previously expanded in Canada[23] in 1948 with a plant at Eastern Passage, Nova Scotia. At the time it was viewed as a savvy business move, as the air component of the Royal Canadian Navy (RCN) was being expanded, with the acquisition of Canada's first fleet carrier, the *Warrior,* and its Fairey *Firefly* aircraft. It was correctly perceived that significant opportunities for repair and conversion work would result. The company then went on to purchase a hangar at Patricia Bay, in response to the 1954 RCN commissioning of VU33 and VC922 in late 1954. In January the local newspaper, the *Victoria Daily Times,* quoting unnamed Fairey officials, speculated that "construction of a jet-powered helicopter of revolutionary design, with rotor, fixed wings and speeds equal to that of a DC-3, likely will be one of the first long term projects of Fairey Aviation Co. at its Patricia Bay plant."[24]

Okanagan officials had followed the public announcements of Rotodyne developments with great interest and avidly reported on its application to intercity transport with enthusiasm in its 1958 annual report, announcing that the company would proceed with applications for permission to operate the Rotodyne in "triangle-service" between Vancouver, Victoria, and Seattle. In September 1959 Mr. Alf Stringer, vice president of engineering, and Carl Agar, vice president of operations of Okanagan, had spent two weeks at Fairey examining the Rotodyne. While impressed with its flight characteristics, Stringer commented on the noise created by the tip-jets, observing that they could be heard ten miles away and that this would be a significant obstacle to intercity service, as few, if any, municipalities would be prepared to allow such noise in town. Indeed, the Rotodyne's noise was to prove fatal to the project, but that was not evident at the time. In fact, developing commercial interest pointed to a bright future for the Rotodyne.

Fairey Aviation had been engaged in negotiations with New York Airways (NYA) in 1958 for potential purchase of the Rotodyne for intercity service, while Japan Airlines had arranged to visit Britain to evaluate the prototype. Indeed, on January 5, 1959, the Rotodyne established a world speed record for convertiplane-type aircraft of 190.89 mph, bettering the old record by 30 mph. It also delivered an outstanding performance in June at the 1959 Paris Air Show, complete with safe autogyro landings. Kaman Aircraft Corporation had negotiated a sales and service contract and a Rotodyne manufacturing license for military[25] and civilian production of the Rotodyne in the United States, and NYA was so impressed by the Rotodyne's performance that it had signed a letter of intent to purchase five Rotodynes at $2 million each, with an option for fifteen additional aircraft. The latter order was subject to successful testing of the initial five, and the cost was then anticipated to be $1.5 million per aircraft. This was of great significance, as NYA was recognized as the world's first scheduled airline to exclusively use rotary-wing craft, flying commuter flights from the top of the Pan American Building in the heart of Manhattan to various outlying airports. This successful company, well versed in such commuter service and its hard-nosed economic realities, had calculated that the Rotodyne, with an enhanced sixty-five-passenger capacity, could reduce the airline's operating cost per seat per mile by at least 50 percent over the helicopter.[26] Interest was then also expressed by Chicago Helicopter Airways and Japan Airlines, the latter who now stated that it was considering the Rotodyne specifically for travel between Tokyo and Osaka.

It was obvious, then, that the commercial desirability of the Rotodyne depended on its economic viability, and development of the larger model would require an additional £8–10 million expenditure. Fairey, with orders, options, and economic analyses in hand, approached the British government of Prime Minister Harold Macmillan for help. The prime minister had written on June 6, 1959, to the Honourable Aubrey Jones, minister of supply, in support of the Rotodyne project, stating that "this project must not be allowed to die."[27] Fairey justifiably felt that it had cause to believe that its appeal for additional funding would be favorably received. Although such appeared to be superficially the case, the conditions attached to an offer for half of the additional funding were to prove onerous and ultimately fatal.

The British government made its offer of continued support contingent on a firm order from BEA, but both BEA and NYA made their orders dependent on a larger prototype flying by fall 1961. Such a larger model was also perceived as desirable because the military was pressing for a model capable of carrying seventy-five troops. It was to prove a daunting task, as Fairey was encountering ongoing difficulties at reducing the noise produced by the Rotodyne's tip-jets. And BEA, having announced in January 1959 intent to purchase six Rotodynes, insisted that *all* its requirements, including noise reduction, be met.[28]

Fairey, having encountered difficulties working with the Napier Company, had also become dissatisfied with the Eland engines, and it was readily apparent that a sixty-five- to seventy-five-passenger model would require significantly more powerful engines. Although the prototype Eland N.E.1.7 engine was to have started at 3,000 horsepower and to have improved to produce 4,200 horsepower, it never achieved more than 2,550 horsepower! This caused Rotodyne pilots to enrich the fuel mixture to achieve necessary power, which, in turn, resulted in unacceptably high fuel consumption and was responsible, in part, for the increased noise level that concerned all potential purchasers.[29] The company, giving up on Napier, turned back to Rolls-Royce, with its Tyne engine. This second-generation turboprop had been designed by Lionel Haworth in 1954–55 to take over where the Dart ended, at 2,500 horsepower, but the engine had proved far more powerful and, by the time that Fairey considered it, was comfortably rated at 4,220 horsepower. Fairey, confidant that the Tyne could be pushed to 5,000 horsepower, felt it would be sufficient for the projected Rotodyne Z. The Ministry of Supply promised to finance 50 percent of the Tyne-Rotodyne development costs (to a defined maximum), but this was also on the condition of a BEA order.

Fairey, then, confidently approached the design of the Rotodyne Z. What emerged from the first design efforts was a craft with a 56-foot, 6-inch span and with a rotor diameter now increased to 104 feet. Fairey, in an attempt to spur both civilian interest and military support, circulated the new design, and for a time it was rumored that the United States Army was interested in acquiring two hundred of the new, larger Rotodynes. But this never materialized, as such an arms procurement required evaluation, and even if Fairey had entered into production of the larger model, military import into the United States was prohibited. Although it was then suggested that Eastern Airlines (having had Autogiro experience with the 1939–40 Philadelphia 30th Street Post Office–Camden, New Jersey, Autogiro mail flights) purchase a civilian model and lease it to the army for trials, nothing came of this. Fairey also sought Mutual Aid Money, which was conditioned on an RAF order for twenty-five Rotodyne Zs, but the RAF publicly stated that it would not commit to more than twelve. Other events were even then coming to a head, however, that would shortly doom the Rotodyne and insure that the larger model would never be built. While the Tyne engine would continue to be made for over thirty years, the Rotodyne had less than three years left.

NOTES

1. Mel Morris Jones, "Talkshop: A Conversation with Jukka Tervamäki," *Fly Gyro!* no. 1 (September–October 2000): 4.

2. See, for example, Jukka Tervamäki, "Some Thoughts of Autogyro Design: Part 3," *Sport Aviation* 15, no. 4 (April 1966): 36–37; Jukka Tervamäki, "Some Thoughts of Autogyro Design: Part 2," *Sport Aviation* 15, no. 2 (February 1966): 11–13; Jukka Tervamäki, "Some Thoughts of Autogyro Design: Part 1," *Sport Aviation* 14, no. 11 (November 1965): 6–8.

3. Jukka Tervamäki and A[ulis] Eerola, "The ATE-3 Project," *Popular Rotorcraft Flying* 7, no. 4 (July–August 1969): 21.

4. Jukka Tervamäki, "The Sleek New JT-5 from Finland," *Sport Aviation* 23, no. 2 (February 1974): 61; Howard Levy, "Italian Import: The Magni Gyroplane Makes Its U.S. Debut," *Kitplanes* 18, no. 2 (February 2001): 41–44.

5. Jukka Tervamäki, "Losing Faith in Autogyros and Gaining It Back Again," *Sport Aviation* 20, no. 5 (May 1971): 40–41.

6. *Directory of PRA Members: 1966* (Raleigh, North Carolina: Popular Rotorcraft Association, Inc., 1966).

7. As stated in every early issue of *Popular Rotorcraft Flying*. See, for example, *Popular Rotorcraft Flying,* spring 1963, 2.

8. Paul Bergen Abbott, "From the First PRA Convention to the Latest: A Conversation with Ken Brock," *Rotorcraft* 34, no. 6 (September 1996): 31–32; see

also Paul Bergen Abbott, "Meet Ken Brock," *Rotorcraft* 31, no. 6 (September 1996): 12–17.

9. Igor B. Bensen, "Wanted: Teamwork," *Popular Rotorcraft Flying,* summer 1963, 2.

10. Edgar B. Trent, "PRA Fly-In: Let's Do It Again!" *Popular Rotorcraft Flying,* summer 1963, 3–4.

11. "The First Fly-In–PRA International Fly-In: Raleigh-Durham Airport, June 15–16, 1963. (We Did It!)" *Popular Rotorcraft Flying,* summer 1963, 8–11.

12. Edgar B. Trent, "FAA meets PRA," *Popular Rotorcraft Flying,* fall 1963, 5–7.

13. The $30 million is quoted by Bill Hines, McCulloch sales manager (1970–74) and air show demonstration pilot, who delivered his first J-2 to Ben Parker of Carson City, Nevada, on July 2. "Return of the Flight of the Phoenix," *Popular Flying,* January–February 1971, 10. Martin Hollmann, however, maintains that the actual McCulloch loss was only $8 million dollars, but this is otherwise undocumented. Hollmann, *Flying the Gyroplane* (Monterey, California: Aircraft Designs, Inc., 1986), p. 50.

14. For "rotary aerodyne." See Jean-Pierre Harrison, "Fairey Rotodyne," *Air Classics* 22, no. 44 (April 1996): 44–47, 60–62, 64–66, 79–80, 47.

15. The amazing abilities of the Rotodyne were routinely reported in the world's aviation press. See, for example, "Rotodyne Demonstrates VTOL Features," *Aviation Week* 69, no. 14 (October 6, 1958).

16. For a description of the Fairey aircraft, see H. A. Taylor, *Fairey Aircraft Since 1915* (London, England: Putnam Aeronautical Books, 1974; Annapolis, Maryland: Naval Institute Press, 1988).

17. For "Gyratory aerodyne." See Harrison, p. 46.

18. See Derek Wood, *Project Cancelled: British Aircraft That Never Flew* (Indianapolis and New York: Bobbs-Merrill Company Inc., 1975), 109–29.

19. The Eland would be almost the last hurrah for D. Napier & Son—in 1960 the company was divided, and Napier Aero Engines Ltd. became a subsidiary of Rolls-Royce in 1962, a year after it had ceased production of the Eland power plant.

20. The Rotodyne Y was much smaller than the projected commercial version, the Rotodyne Z, which was to be powered by two Rolls-Royce Tyne turboprop engines and to carry fifty-four to seventy passengers. It was never built.

21. For a history of the evolutionary steps in Rotodyne design, see George S. Hislop, "The Fairey Rotodyne" (paper presented before the Helicopter Association of Great Britain and the Royal Aeronautical Society, London, England, November 7, 1958).

22. Indeed, in November 1956, Fairey Aviation's advertisements for the Rotodyne announced that it "will be the first large transport aircraft to offer high cruising speed with the ability to operate from small landing sites" and that "[t]he Rotodyne, being independent of conventional runways, will bring the advantages

of air transport to almost every locality." See the *Aeroplane,* November 23, 1956, 26; see also the *Illustrated London News,* August 30, 1958, in which the Fairey Aviation advertisement, touting "Rotodyne travel" and showing the Rotodyne being boarded by passengers in Paris with the Eiffel Tower in the background, announced that "[f]orty-eight people will settle themselves in the wide, comfortable cabin of the Fairey Rotodyne, as some small open space in the middle of a town. The Rotodyne will lift them vertically far above chimney-smoke and church spires—and then, gradually transferring the power of its two turbine engines from the big rotor to the forward propellers, it will whisk them across land and water at nearly 200 m.p.h. Over the destination—the center of a city, not some airport far outside—the rotor will lower them, straight down, to a safe arrival. A new conception in aircraft design has brought this kind of travel into plain sight— the Rotodyne which is neither aeroplane nor helicopter, but something of both, and the world's first Vertical Takeoff Airliner."

23. For a description of the Fairey Aviation Canadian involvement, see "The Fairey Rotodyne: Nearly the Answer," *West Coast Aviator,* September/October 1995, 35–37.

24. As cited in David Parker, "The Fairey Rotodyne—Nearly the Answer," *West Coast Aviator* 5, no. 1 (September/October, 1995): 36.

25. Harrison, p. 60 (photo caption); Charles H. Kaman, *Kaman Helicopters and the Evolution of Vertical Flight* (General Harold R. Harris "Sight" Lecture, presented before the Thirty-third Wings Club, New York, May 15, 1996), 23.

26. NYA calculated that the break-even load for the Rotodyne, with enlarged capacity, would be 45–50 percent whereas that of the largest passenger helicopters, the Sikorsky S-61L and Vertol 107 II (seating twenty-five and twenty-eight, respectively) rose to 80 percent. The projected Rotodyne operating cost was estimated to be four cents per seat per mile, while the Vertol would cost at least twelve cents and the Sikorsky even more. The Rotodyne's economic efficiency was derived, in part, from its ability to land in the inner city and avoid the necessity of extensive ground travel to and from an outlying airport. See Harrison, p. 62.

27. Wood, *Project Cancelled,* p. 120.

28. Wood, p. 121.

29. In fact, noise had already proven an insurmountable problem in the mid-1950s for an American predecessor convertiplane, the McDonnell XV-1 compound helicopter, which was similar in concept but much smaller (twenty-six feet long; 4,277 pounds weight empty; 5,505 pounds gross weight) than the Rotodyne. In 1949, while the Rotodyne vision was first being articulated, the Convertible Aircraft Congress in America sought to stimulate development of an observation and reconnaissance convertiplane. Of the designs that were submitted and selected for development, the McDonnell XV-1, a compound helicopter (more properly, convertiplane) was one of the most impressive. Chief engineer was jet-powered-rotor pioneer Friedrich von Doblhoff, and the XV-1 was powered by pressure jets and tip-burning rotors. It carried two passengers and derived its lift

in takeoff, landing and hovering from a three-blade pressure-jet-driven rotor. In forward flight lift came from an unloaded rotor, small wings with thrust coming from a pusher propeller located between a twin-boom configuration. The XV-1 made successful conversions between helicopter and autogyro modes in 1955 and achieved speeds of 200 mph (initially a record for helicopters, but later rescinded as the craft was redesignated a convertiplane); "configuration-induced aerodynamic problems and *the excessive noise of the tip jet burning* prompted the phase-out of the XV-1." See John J. Schneider, "Rotary-Wing V/STOL," in Walter J. Boyne and Donald S. Lopez, *VERTICAL FLIGHT: The Age of the Helicopter* (Washington, D.C.: Smithsonian Institution Press, 1984), pp. 178–79 (emphasis added). The XV-1 was discontinued in 1956, and today the McDonnell XV-1 Convertiplane (SN 53-4016) resides in the United States Army Aviation Museum, and National Air and Space Museum of the Smithsonian Institution, Washington, D.C.

Chapter 13

FAILURE TO REVIVE THE AUTOGIRO: VARIOUS COMPANIES TAKE THE RISK

"What is the ultimate, do you think?" Kellett asked.

Together we asserted: "A craft that can go straight up and down. The Autogiro does the job halfway. A successful helicopter would be better."

Kellett smiled: "Yes, perhaps, but we will improve." The steady improvement of the Autogiro, more than anything else, helped to bring about the first successful helicopter. It was the practical proving ground that gave us the key to controllable vertical flight.

Hollingsworth Franklin Gregory,
Anything a Horse Can Do

While Fairey's attention was justifiably focused on the Ministry of Supply, events in the Ministry of Aviation were even then darkening the skies for rotary aircraft. The Honourable Duncan Sandys, minister of aviation, wanted to consolidate, or "rationalize," the British helicopter industry, and he felt that Westland was the logical choice, as it was a major supplier of Sikorsky-licensed and -derived helicopters for the military and civilian markets. Wielding government subsidies as a scalpel to dissect and reorder the British helicopter industry, Sandys, son-in-law of Winston Churchill, selectively withdrew government support and forced company mergers.[1] In 1959 Westland took over Saunders-Roe Ltd. Then, on March 23, 1960, Westland acquired the Helicopter Division of Bristol Aircraft Ltd., based at

Weston-Super-Mare, Somerset, renaming it the Bristol Helicopter Division of Westland Aircraft Ltd. Less than seven weeks later it was Fairey's turn.

Sandys had withheld further contracts and support for Fairey Aviation, and the company found itself without further funding for fixed-wing aircraft or guided weapons. It was made abundantly clear that the price for further development funding for the Rotodyne would be the sale of Fairey's aviation interests, apart from specialized manufacturing such as hydraulic components, to Westland. With a monthly bill of £70,000 just to keep the smaller Rotodyne prototype flying and mounting developmental costs for the larger, Z version, Fairey had no choice and agreed to the sale. On May 2, 1960, it sold the Rotodyne and its aviation interests to Westland. At the time of the sale the Rotodyne had flown a total of 120 hours and had made 350 flights and 230 transitions between helicopter and autogyro—without any accident. This safe aircraft continued to amaze all who saw it fly, and there was no reason, now that Sandys had successfully engineered the consolidation of the helicopter industry, not to proceed with its development. Accordingly, Westland received a government development contract in the amount of £4 million and a promise of an additional £1.5 million to facilitate the larger Rotodyne entering BEA service.

As the development of the Rotodyne Z continued, the design became even larger. The final version of the aircraft was to weigh 58,500 pounds, with an increased rotor diameter of 109 feet and an equally impressive 75-foot wingspan. The military version of the revised design would have been capable of carrying seventy-five troops with operational equipment, armored cars and trucks (via the double clamshell doors in the rear of the fuselage), missiles or the fuselage of a small aircraft and would have been able to function as a flying crane, capable of lifting a 100-foot bridge span, vehicles, and disabled aircraft. As admirable as these projected capabilities were, the RAF was not then interested in a compound helicopter—its focus and budget were on nuclear deterrence. So even though there was some mild military interest, it never reached a critical mass. It was readily apparent that the civilian market would have to carry the entire project—and it was not to be. All that was left was to write the final chapter in late 1961.

The linchpin was always the BEA order, and its precondition that all its terms be met now came back to haunt Fairey in what proved to be the Rotodyne's death throes. Although concerns were voiced about the increase in weight and rising costs, the continuing noise issue became the focus of growing criticism. But before this resulted in a final cry to end the project, the issue of engine design arose one last time, this time not to be resolved successfully. It was obvious that the enlarged design would require more power than even that which would be produced by the Rolls-Royce Tyne power plant. It was suggested that power be increased by fitting a Rolls-Royce RB. 176 auxiliary booster engine in the rear of each Tyne nacelle. Government policy continued to be that development costs be shared by industry, a policy that would have required Rolls-Royce to invest an estimated £9 million—an expense the company was not then willing to assume.

The government then pounded the final nail in the Rotodyne coffin, when it rejected a requested Westland quote for delivery of twelve Rotodyne Zs for the RAF and an additional six for BEA, stating that the military was no longer interested. Official funding was withdrawn on February 26, 1962, and the Rotodyne was dead, as the British government and Westland were fully committed to those aircraft derived from its Sikorsky license. Thus the lead in compound helicopters and the most successful application of Juan de la Cierva's autorotational flight ended. The government, which owned the prototype, had the Rotodyne Y dismantled and almost completely destroyed, with only a few components surviving at the British Rotorcraft Museum, Weston-super-Mare, Avon, accessible to serious researchers with prior appointment. Even the tooling used to create the Rotodyne was destroyed, but a film record remains, a rare visual record that even today, four decades later, never fails to awe the few viewers who happen upon it.[2]

Analysis of the noise issue clearly indicates that it is incorrect to ascribe the downfall of this incredible aircraft to decibels.[3] By February 1962 the noise cancellation project had resulted in ninety-six dB at 600 feet, but those intent upon making this a seminal issue either failed to note or deliberately ignored the fact that the rotor would be powered only for approximately one minute at takeoff/climb out and one minute during landing. And to additionally minimize the Rotodyne noise, it was estimated that a vertical climb upon takeoff of 250 feet before acceleration to 600 feet and a standard approach angle of fifteen degrees for landing would further reduce its noise. And to make a point about the reality of Rotodyne flight in the inner city, Chief Pilot Gellatly twice flew over downtown London and made multiple landings and takeoffs at the Battersea Heliport on a calm morning, with no complaints raised. Only two comments were received as a result of the Battersea flights: A lady inquired, on behalf of her son, if that indeed had been the Rotodyne; and a second woman commented that her "light sleeping baby" had not in the least been disturbed by its flight.[4] In fact, the Eland engines on the Rotodyne prototype produced less noise than the DC-8, and at the time of project cancellation, "the continuing development of the silencers had further reduced the noise level by another 16 dB."[5]

Politics doomed the Rotodyne even though almost a thousand passengers had participated in demonstration flights, including a significant number of the world's airline leaders and military and government officials, with no accidents reported. This compound helicopter flew in every Farnborough and Paris air show from 1958 through 1962 to constant acclaim—truly a machine far ahead of its time. It is now unfairly cited as one of "the world's strangest aircraft,"[6] but twenty-five years earlier, the Rotodyne had been included in *Milestones of the Air: JANE'S 100 Significant Aircraft.*[7]

The Fairey Rotodyne was the most developed application of autorotational technology but was not the only one. Its death signaled a precarious future for the rest, however, and the Kamov Ka-22, which most closely resembled the Rotodyne and came the closest to success, suffered a similar fate. The other attempts were viewed either as developmental craft not destined for production or as the product of limited vision and even more limited capitalization. The creations of Bruno Nagler, the Nagler Heli-Giro Aeronca conversion and VG-Vertigyro have all but been forgotten, and the late 1960s VFW (Vereingte Flugtechnische Werke) H-3 three-seat heli-gyro did not prove a viable proposition. Additionally, Anton Flettner again reappeared, now the founder of his own company in New York after the war. The Fl 201 Heligyro, evolving out of his earlier work on the Fl 185, was a thirty- to forty-passenger twin-rotor helicopter designed to take off and land as a helicopter but fly as an autogyro. Under United States Navy sponsorship the Fl 201 Heligyro was tested at New York Naval Air Station, Floyd Bennett Field, but the model never advanced beyond the testing phase.[8]

KAMOV KA-22 (THE "RUSSIAN ROTODYNE")

Known in the Soviet Union as the Vintokrulya (Vintokryl) ("Screw Wing") and dubbed "Hoop" by NATO, this impressive aircraft also was called the "Russian Rotodyne,"[9] and it suffered a similar same fate to that of Fairey Rotodyne. It was a transport convertiplane equipped with a rear ramp for cargo loading. The Ka-22 was larger than even the proposed Rotodyne Z and could easily carry 80–100 passengers or 36,500 pounds. It had an all-metal fuselage with a flight deck raised high above the glazed nose to allow excellent vision for landing in small areas. It was powered at the end of each wing by two Ivchenko AI-20V propeller-turbines, one mounted at each end of a ninety-foot tapered wing, alternatively reported as a 6,500-horsepower Soloviev D-25VK engine, a nine-stage single-spool turboshaft modified from the D-25V engine previously used on the Mil Mi-6, Mi-10, and V-12 helicopters.

Each engine drove both a conventional four-blade propeller for forward flight and a four-blade rotor for takeoff, hovering, and landing, like the Rotodyne. In forward flight each rotor, which was at the end of each wing rising from the turbine engine, was unloaded, and the plane derived lift from its wings and the unloaded rotor functioning in autogyro mode. Flight testing began on April 20, 1960, with a crew consisting of D. K. Yefremov, V. M. Evdokimov, V. B. Alperovich, E. I. Filatov, and Yu. I. Emelianov. The Ka-22 made its only public appearance at the Soviet National Aviation Day display at the Tushino Airshow on July 9, 1961. The Ka-22's Class E.II speed record of 221.4 mph over a fifteen- to twenty-five-kilometer course, set on October 7, 1961, and its load record of 36,343 pounds to a height of 6,562 feet, set on November 24, 1961, still stand for convertiplanes.[10]

Four aircraft were built in 1959–63, one at the Lubertsy experimental plant and three additional models at Tashkent. Despite their impressive performance and appearance, the program was apparently cancelled in 1964 after a crash in 1964, although its designer, Nikolai Kamov, maintained that the configuration was "still active" in 1966.[11] There is one ironic note to the Ka-22, the echo of a voice heard before in the field of helicopter/autogyro/fixed-wing compound aircraft—that of Professor Ivan Pavel Bratukhin. Prior to the disbanding of his engineering bureau in 1950, he proposed a twin-rotor, ten-seat convertiplane, dubbed the B-11, in a configuration that strongly resembled the later Kamov Ka-22. Although Bratukhin's proposal was officially ignored, the logical speculation is that his work "may have been passed on to Kamov's staff."[12]

NAGLER HELI-GIRO AERONCA AND VG-VERTIGYRO

The work of Austrian Bruno Nagler remains generally unknown.[13] After moving to America and settling in White Plains, New York, Nagler embarked on several areas of aviation research and development. One of these was a pressure jet rotorcraft similar to the Fairey Rotodyne. Although Nagler is usually cited for his helicopter designs incorporating this technology, there are photographic records of at least two different attempts to evolve a convertiplane that could take off as a helicopter and fly as an autogyro. The obscure Nagler Heli-Giro Aeronca conversion was a pressure jet–powered rotor affixed to what appears to be a converted Aeronca K Scout with modified control surfaces, the wings removed, and a reconfigured tail.[14]

Less obscure, the Nagler Vertigyro VG-1 was a converted Piper Colt adapted to prove Nagler's Vertigyro,[15] which resembled the earlier Pitcairn and Kellett direct control Autogiros in that it was an airplane fuselage topped by a direct control pressure-tip rotor and a tractor propeller mounted in front. It could fly as a helicopter and as an autogyro and could convert between the two. And using cold-pressure compressed air, the craft avoided the noise issue that recently brought down the Rotodyne. Although it was stated in 1965 that the VG-1 prototype (N5395Z)[16] would be followed by a definitive VG-2 model, there is no record of that happening, as Nagler was invited by investor Darrow Thompson to set up a facility near Phoenix, Arizona, to develop a single-seat homebuilt helicopter. Nagler worked on this project until his death in 1979, but his conversions and use of pressure-tip technology for a low-cost, off-the-shelf convertiplane remains a product of the same creative impulses and fascination with the flight possibilities of autorotation that motivated the Ka-22 and Rotodyne, and the fate of Nagler's creations was equally fatal.

VFW H-3

Designed by German engineer Christian Fischer, and manufactured by VFW GmbH, the H-3 three-seat heli-gyro of the late 1960s and early 1970s represented an additional attempt to combine autogyro and helicopter technology to gain the benefits of autorotational flight. Fischer, an admirer of the "ingeniously simple design of the Bensen gyrocopter,"[17] improved, like Nagler, the cold jet–powered rotors. The prototype, developed in 1967, was a sleek aerodynamic design with an enclosed single-seat cabin and a variable control three-blade cold jet-tipped rotor to achieve helicopter flight. In autogyro mode the power from its Allison 250 gas-turbine engine was shifted to side-mounted ducted fans on each side of the fuselage forward of its V-tail, while the unloaded rotor provided lift. By 1971 VFW was flying the H-3 in three-seat configuration in tie-down and hovering flight-tests, with VFW-Fokker test pilot Heinz Hoffman. After testing, VFW decided not to continue the program, and this adaptation of compound technologies, like the Rotodyne, Kamov Ka-22, and others, disappeared.

UMBAUGH (AIR & SPACE) 18A

The largest market for autorotational aircraft had been the United States, however, and since the founding of the Popular Rotorcraft Association, that market had continued to develop, albeit with much smaller Gyrocopter aircraft. It should not be surprising that attempts were also made in America and Canada to create a larger, jump takeoff gyroplane that was commercially viable. One was the Umbaugh 18A,[18] and it ultimately fared no better than its predecessors, contemporaries, or successors. Its fate, moreover, was the product of strikingly similar events to those that had doomed Pitcairn and others—bad business decisions and a bit of government interference—and this stunning gyroplane, an outstanding performer, has all but disappeared. Its fate was shared with two other certified gyroplanes, the American McCulloch J-2 and the Canadian Avian 2/180, but it represented the boldest attempt to realize Cierva's dream.

Raymond E. "Ray" Umbaugh, founder of Umbaugh Chemical Fertilizer Co., was successful in the specialized agricultural custom-blend fertilizer market. These commodities were sold by salesmen who traveled from farm to farm, a costly form of so-called missionary marketing. Looking for an easier, more cost-effective manner for his personnel to visit farms, Umbaugh became interested in the Bensen Gyrocopter, after trying fixed-wing wing aircraft and helicopters. He even unsuccessfully experimented in the late 1950s with an enclosed-cabin configuration of a Bensen,[19] and then decided to manufacture an enclosed two-passenger jump takeoff model. After studying the history of Autogiro design, Umbaugh hired noted aeronautical designer/engineer Gilbert Devore to design a low-cost, easy-to-fly, safe autogyro that could be produced in volume for an anticipated mass market by the Umbaugh Aircraft Corporation, which had been created by Ray and his financial backers. As this was, from the beginning, designed to be manufactured, Umbaugh knew that the aircraft would have to be certified. Devore was to oversee the FAA certification process, eventually assisted by former Pitcairn pilot Slim Soule and chief test pilot Ken Hayden. Manufacture of the prototypes for the certification process was subcontracted to Fairchild Engine and Aircraft Corporation of Hagerstown, Maryland, but the seeds of its failure had already been sown.

In many ways, the failure of the Umbaugh project was, with hindsight, readily evident in the grandiose vision of its founder—Umbaugh stated that he intended to become the "Henry Ford of aviation"[20] and announced the expansive goal on December 1, 1958, of delivering 1,000 aircraft between January 9, 1959, and October 16, 1959.[21] All this came from a man who had not previously manufactured anything. It sounded too good to be true, and that is exactly what the reality proved to be, but nothing stopped Ray from gaining publicity and arousing great public interest, including an interview in *Business Week*.[22]

The Model 18 was announced in press releases in April 1958 during its design, and early sales brochures and advertising copy listed the price as $7,995 to $9,995, depending upon options, with a base price to dealers of $6,500, and an anticipated production of 1,000 per month. It was a stunning tandem two-place jump takeoff cabin gyroplane powered by a tested Lycoming 0-260 A1D 180-horsepower engine mounted behind the cabin in a pusher-mode topped with a streamlined pylon and a three-blade rotor. The design also featured a long boom extending to the rear of the cabin and three vertical tail surfaces on top of a horizontal for increased stability. The public responded enthusiastically, and 132 dealerships were quickly sold for what was represented as exclusive sales territories. The initial FAA certification application was filed with the Forth Worth, Texas, office on January 9, 1959 (the same day that the 1958 press release had designated for delivery of the first aircraft!), but the announcement of the Fairchild subcontract was not announced until eight months later, on August 14. Type certification was received in September 1961 based on the five Fairchild prototypes, but Fairchild was no longer associated with the project, having severed its contract with Umbaugh and his associates via arbitration in May of that year. It had become apparent that, as finally produced, however technologically sophisticated and advanced, the Umbaugh 18A was no longer economical at the announced price. This did not stop Umbaugh, and by January 1962 he announced that the company had received "8,000 orders based on the $9,995 figure and this is not subject to change."[23] It also received a favorable review in the January 1971 *Flying,* when pilot A.C. Bass, after describing the gyroplane's performance, stated that as "a business tool it has a vast potential and should attract a great deal of interest in this still relatively new area of short haul business transportation."[24]

At this point Umbaugh had relocated the project to Muncie, Indiana, but the production line was still far into the future. And Umbaugh's house of cards was about to come tumbling down, a process begun in 1963 when he informed his dealers that the $6,500 price was actually subject to change and would in fact have to be raised, as the actual production costs for the 18A were $6,657. As Umbaugh sold dealer franchises, he colored in a Rand-McNally U.S. map—and by now that map looked like a checkerboard.[25] Dealers, frustrated by the seemingly endless delays in production, called for accountings, went into court to seek legal remedy in the form of receiverships, and charged that overlapping franchises and territories had been misrepresented as being exclusive. After a year of litigation, during which the company had not made any progress in production, the federal district court of Tampa, Florida, ordered the assets sold to the Air & Space Manufacturing, Inc. (A & S), a group of about one hundred former Umbaugh dealers and distributors who either still believed in the 18A, or who just wanted to get their money out of the deal and saw A & S as they only way to do so.

The model was renamed the Air & Space 18-A, and it finally looked as if the dream would become a reality for the first certified gyroplane since Pitcairn and Kellett. A marketing executive with the new company, Mort Linder, mindful of its corporate predecessor's reputation, stated in an interview in the April 1965 issue of *Flying*.[26]

> The general image now is a mixed one of fast talk and a lot or promises and a lot of dates set that were never made. The new image is predicated solely on the idea that we are here to correct whatever has to be corrected to make the aircraft in volume quantity. We hope to keep the price below $14,000.

Production began in June 1965, and dealers eagerly took delivery of newly improved and recertified aircraft with a completion of one aircraft per day. The first model was delivered to Florida dealer John T. Potter in June, and he was soon booked solid with appearances and was delighted with its improved performance. The improvements had made recertification necessary, and the delivery price had risen to $18,540, but sales slowed considerably after the first aircraft[27] had been produced and delivery actually commenced. This was the product of an economic decline in the private aviation market and a series of fatal accidents that plagued the 18-A. The FAA temporarily withdrew the aircraft's type certification, effectively grounding all models while conducting an official investigation. The certification was returned several weeks later when the FAA determined that the accidents had been alcohol-related[28], but the damage had been done, and many prospective purchasers now considered the 18-A unsafe. Additionally, A & S management, given the confirmed orders, had decided to expand to a production rate of three per day and to that end, went forward with a successful $2.5 million dollar stock offering to raise capital for expansion of the production facility.

However, the company's aeronautic expertise did not extend into the world of finance, and they soon fell afoul of the Securities and Exchange Commission (SEC), which charged management with an illegal sale of securities across state lines without the necessary national registration. Additional federal charges of misrepresentations and fraud quickly followed, and the company found itself embroiled in costly litigation when its naive offer to repurchase the stock and start over with the required registration was rejected. It would take two years for the lawsuit to be dismissed, but by then the company was effectively out of business, as purchasers withdrew their deposits and the company slid into bankruptcy. It seemed that the Air & Space dream of a certified, technologically advanced jump takeoff gyroplane was dead, but there were a few more chapters to the 18-A saga, and they would prove worthy of soap opera.

Normally, in a bankruptcy the owners become "debtors-in-possession" and, under the scrutiny of the federal bankruptcy court, seek to reorganize the company, but given the charges against the A & S managers, it is not surprising that the company was placed under the direction of a federal trustee. The trustee, understandably knowing little of the gyroplane market, closed the factory for a year, rejected proposed reorganization plans, and finally disposed of the assets in 1967 when the government dropped the securities fraud lawsuit. The assets were sold to the Weldon Stump Company of Toledo, Ohio, who sold A & S to Pearl Equipment, Inc. of Nashville, Tennessee.[29] The Crown Tool Corporation of Dayton, Ohio, owned by Les Smithhart, purchased at auction the manufacturing rights, certification, tooling, and a few completed machines, with the hope of restarting production. But this never materialized, for Crown soon experienced its own hard times and had little experience with this kind of production. Smithhart eventually sold the production rights and tooling to an investor group that wanted to get the 18-A flying again. Smithhart, eager to be free from the expenses of storage and insurance of the tool-and-die and parts, agreed to send them to a warehouse leased by the purchasing group while they arranged for financing, but delays ensued and payment was not forthcoming. Smithhart, an experienced businessman, finally became suspicious and demanded that the delays end and the deal go forward, only to learn that one of the group had embezzled the purchase money and that the remaining members had been stalling for time, hoping to get alternate funding. But it was worse than merely a deal gone bad— the rent had not been paid on the warehouse, and its contents were to be sold at public auction! It seemed that the 18-A had come to the end of the road, but another white knight appeared on the horizon, and for a while it appeared that the gyroplane would soar again. That man was to become one of the most famous and genuinely liked members of the American gyroplane community—Don Farrington.

Ira Donald "Don" Farrington Jr. was born in 1931 and graduated from Purdue University with a degree in aeronautical engineering in 1952. He came to aviation when he joined the air force after graduation, and he would remain in the reserves after leaving active duty four years later, eventually retiring with the rank of lieutenant colonel. Although he spent some time in the family automobile business, the lure of aviation continued to call and Farrington embarked on a flying career with Pan American Airlines in 1965. He would continue to fly for the airlines until reaching the required-retirement age of sixty, and that paycheck allowed him to pursue other aviation interests. He had been one of the original investors in Umbaugh's dream, and even though he had grimly watched his money disappear, he had not lost faith in the 18-A. He had purchased various 18-A aircraft when they became available and now bid and won stored 18-A parts and tooling at the warehouse auction. Smithhart, who now had little use for the remaining parts and tooling, sold them to Farrington Aircraft Corporation, located at the airport Don established southeast of Paducah, Kentucky, where he flew and maintained the 18-As.

Farrington teamed with Umbaugh alumnus chief test pilot/designer Gil Devore and made several significant improvements to the 18-A—including greater noise reduction by means of redesign of the engine exhaust system and engine cowling, substitution of a lighter propeller, improved mechanics and landing gear, refined instrumentation, and greater fuel capacity—in anticipation of industrial uses. Each of these improvements were signed off by the FAA, a continual, sometimes tedious, always expensive, process of gaining supplemental type certificates, but the end result was an improved flying machine, and Farrington aggressively attempted to market his refurbished models. Farrington Aircraft Corporation sent two aircraft to England, at least one of which was for experimental flights with the London Metropolitan Police Department, which did not result in a sale. By August 1973 these models were for sale,[30] as Farrington sought to avoid the cost of returning them to America, but at least one (G-BALB) remained unsold in a hangar in Biggin Hill until 1984, where it was discovered by Ireland's Pat Joyce and Jon Todd.[31] Farrington had removed the instruments, and the model, lacking rotor blades, was definitely a fixer-upper, but its lines excited Joyce, who entered into protracted negotiations with its owner. Farrington sold Joyce the 18-A in 1984, and Joyce finally restored it to flying condition in 1997, partially by using parts taken from the other 18-A that had remained in England. An Air & Space 18A flown by John Potter was briefly featured in the obscure[32] 1972 New Television Workshop production of *Between Time and Timbuktu*, loosely based on a collection of Kurt Vonnegut Jr. short stories. Ron Menzie flies a second Gyrocopter in the film, a B-8M (N3891) Gyrocopter. And while it must have seemed ironic that the 18-A was supposedly flown by the hated police of an authoritarian state in the film, given that the federal authorities of the SEC had previously put the company out of business, the image of a police gyroplane was exactly what Farrington was then marketing in England.

The only remaining part of the picture was the most important—the manufacturing rights, which remained with Smithhart, as the sale had never been completed—eventually, even this was sold to Farrington. It was 1991, and Farrington had finally retired from Pan Am; Farrington Aircraft Corporation had by then grown since 1971 to become the West Kentucky Airpark, employing twenty people and offering flight training[33] in his two-seat certified gyroplane and other rotorcraft, aircraft maintenance services, fuel and hangar rentals and continued refurbishing of the 18-A models[34] as they came on the market. Farrington had gained a reputation as a solid citizen in the American rotorcraft movement, and his airpark flight training facility was considered the best available for the aspiring gyroplane pilot. By 1994 the company announced the production of an improved 18-A, now dubbed the Heliplane, and an intent to engage in worldwide sales.[35] Additionally Farrington Aircraft manufactured and marketed a two-place open-cockpit trainer called the Twinstar[36] in kit form. Both the Twinstar and the 18-A were well received, and Farrington had reason to be optimistic. Their success was, however, not to be.

By 1997 Farrington's holdings were worth an estimated $10 million; he was finally ready to ramp up to serious production of the Heliplane and Twinstarr (Farrington added the additional *r* when the French Aérospatiale company complained that it already had a model called the Twinstar), and he was prepared to raise capital to fund this expansion. Perhaps mindful of the previous financial debacle that had sunk the operation, a separate investor group was incorporated as Air & Space Holdings, and this group was to acquire Farrington Aircraft Corporation and its various divisions by means of a purchase option and was to raise the necessary millions of dollars in the stock market. That public offering was only partially completed when Don Farrington died (as the result of a heart attack while flying a demonstration in the 18-A at the Experimental Aircraft Association Sun 'n Fun Fly-In at Lakeland, Florida) on April 13, 2000, and its purchase option subsequently expired. Farrington's estate, lacking his familiarity with the highly personal business and the confidence to carry his dream to realization, and needing funds for settlement of outstanding obligations, auctioned off the company's assets on April 21, 2001.[37] The sale, it was rumored, had barely covered the outstanding debts, for although noted European instructor and gyroplane enthusiast Bart "Woody" de Saar acquired the tooling, parts, and manufacturing rights to the Twinstarr, the 18-A certification and manufacturing rights went unsold. Whether Don Farrington's dream of reviving the 18-A remains ultimately unfulfilled or merely deferred remains to be seen, but what is certain is that it represented perhaps the most sophisticated embodiment in America of the vision of Pitcairn and Cierva, and as a certified two-place aircraft, it was used for many years to train new pilots.

MCCULLOCH J-2 GYROPLANE

The second American certified gyroplane was also created during this same period but had even less success and, by all accounts, little to recommend it. Talented and considered a genius by some,[38] its designer was Drago K. Jovanovich of El Segundo, California, a highly skilled engineer,[39] who had previously designed the rotor hub and blades for the Hughes 300 series helicopters and the small JOV-3 tandem-rotor helicopter. Its development and eventual commercial certification were achieved by the McCulloch Aircraft Corporation's helicopter division, and in 1949 Jovanovich became its chief designer. His own company, Helicopter Engineering and Research Corporation, was renamed Jovair in the middle/late 1950s and turned its attention to the design of advanced tandem-rotor helicopters and an enclosed-cabin two-passenger tandem autogyro, the J-2. It first flew in June, 1962, the same year that Igor Bensen and his associates founded the PRA. Jovanovich then began a search for an aviation manufacturing company for his enclosed-cabin, three-blade rotor gyroplane and at various times announced that it would be produced under the names of Jovair and Lear, but in 1969 he finally settled on McCulloch, with which he was familiar and which was willing to commit to a major investment. After laying out the manufacturing facility in Lake Havasu, Nevada, and successfully completing the FAA certification process, the McCulloch J-2 Gyroplane,[40] as it was now called, entered production at the beginning of 1971, having received FAA certification, with a basic price tag of $20,000.[41] Production continued between 1971 and 1974, with an estimated eighty-three to ninety aircraft produced.[42] It met with a decidedly mixed response from the aviation public.

In February 1971 Peter Garrison, writing in *Flying,* which had nine years earlier praised the Umbaugh 18A, published an article entitled "Everybody Loves an Autogyro" but quickly informed its readers that that was not true. Claiming that the J-2 was too expensive and underpowered, the author concluded:

Mind you, the J-2 is no *worse* than the average 100-hp two-seater in most of these respects; the shame is that it's not much better. McCulloch is grudgingly admitting that an extra foot of rotor blade and a bigger engine are being "studied"; in aeronautics, however, brute force is not the way. I would think that what is really needed is a much longer two bladed, unarticulated rotor; but that is for Mr. Jovanovich to decide. In the meantime, if I had $20,000 for an unequipped runabout, I think I'd buy a Super Cub and take my change in Bensens.[43]

Kas Thomas, associate editor of *Popular Rotorcraft Flying,* the "Official Publication of the Popular Rotorcraft Association," was far more knowledgeable and, as a result, even less kind to the J-2. He asked:

What is it that has three wheels, two rudders, and looks like a teddy bear with rotorblades?...Step up close to the machine for a better look. What is that you say? A $20,0000 aircraft with a wooden propeller? Yes indeed, the McCulloch has a wooden prop; a Sensenich. And the rotor head? You say the rotor head looks familiar? Why, so 'tis, so 'tis. Straight from Hughes blueprints, 'tis.... The McCulloch J-2 is a cornucopia of unfathomable design quirks. Its posterior end sports an ironing-board of a horizontal stabilizer which, in stark contrast to the rest of the fiberglass-and-metal machine (and for reasons known wholly to the Creator), is *fabric covered.* Also, the J-2 boasts not of one large baggage compartment, but two, separate baggage compartments—one in the nose and one under the pilot's seats—neither of which is overly capacious.

   Speaking of design oddities, what about those stubby wings?...according to McCulloch, they just about lift their own weight in flight...but apart from their cosmetic value, they really seem to be there just to hold up the rudders. True, the fuel is located in the wings, but that might better have been put in the fuselage, certainly increasing the capacity from the current twenty gallons usable (ten in each wing tip). Twenty gallons, after all, is not one heck of a lot for an engine like the Lycoming 0-360-A, which burns nearly ten gallons an hour. But even if the fuel capacity is low, the most unkindest cut of all comes with the knowledge that fuel may be used only from one tank at a time, and must be carefully managed by the pilot.[44]

And famed gyroplane designer and engineer Martin Hollmann commented in late 2001:

When in 1970, the 1,500 lb. gross McCulloch J-2 gyroplane came on the market, I was surprised that Mr. Jovanovich had designed this aircraft around the Hughes 269 helicopter rotor, which had a diameter of 25.3 ft and three blades with a chord of 6.83 inches. Given this size, the disc loading is almost 3.0 lbs/sq. ft. There is no way this gyro could perform well and of course, it did not.[45]

In June, as Thomas' devastating review was published, he participated as a ground crew member in the historic Ken Brock gyroplane cross-country flight. It was only a hint at the great accomplishments that Brock would go on to realize as long-time president of the PRA, but the J-2 would not go on to success—its days were limited. Its manual stated that it could cruise at 105 mph and actually took off at 47 mph in 600 feet—it was not nearly enough to survive.

By 1974 McCulloch had enough and ended production after sinking $30 million in the project. The company was then also under at least one lawsuit, from a student pilot who had been injured in an accident in Michigan and charged the company with product liability.[46] Other than the Rotodyne and Ka-22 government-funded projects, the J-2 was the greatest single investment in gyroplane technology, and it fared no better. A super-version of the J-2 had been produced in the summer of 1972, but it only differed from the original in having a Hartzel three-bladed constant-speed propeller. Production passed to Aero Resources in 1974, which briefly attempted production of the McCulloch J-2 Super Gyroplane powered by 180-horsepower Lycoming 0-360-A2D flat-four engine, but it fared no better than the J-2.[47] Several J-2s survive—for example, one in the Pima Air and Space Museum in Tucson, Arizona, one in the Mid America Air Museum located in Liberal, Kansas, and a Super J-2 in the collection of the Western Aerospace Museum just south of Oakland, California.

AVIAN 2/180

The J-2s may remain primarily underpowered, nonflying museum exhibits, but the Canadian Avian 2/180 has been almost completely forgotten. It was the third certified gyroplane of this period, and justifiably the least successful, although in some ways genuinely innovative and one of the most distinctive gyroplanes produced.[48]

The Avian was produced by Avian Industries, formed in 1958 by one of the most colorful inventors of the twentieth century, Peter Rowland Payne, who had immigrated to Canada from England in 1956 after gaining experience in the British aviation industry. Although he died in 1997, his innovative spirit lives on in such achievements as the nonlethal rubber bullet, the crash-test dummy (anthropomorphic mannequin), groundbreaking mathematical models of how a parachute works, and the stunning SeaKnife hull.

In 1959 Payne joined with colleagues from Avro Canada to found a new company, Avian Aircraft Ltd., located in Georgetown, Ontario, to concentrate on the development of helicopters and autogyros. The company built two prototypes, the first of which, the Avian 2/180A Gyroplane, flew in the spring of 1960. It was a stunning, small two- to three-passenger enclosed-cabin gyroplane with sleek aerodynamic lines and a short, squat, streamlined pylon topped by a three-blade rotor capable of jump takeoffs and a duct-enclosed two-blade pusher propeller in the rear of the fuselage. The ring or cowl that surrounded the propeller featured a vertical stabilizer in back of the propeller, giving the aircraft a distinctive shape. The Avian also featured a nonretractable tricycle landing gear, disc brakes, and a steerable front wheel for ground control.

As the design team had judged jump takeoff ability as crucial for commercial success, the first prototype was mechanically distinguished by rotor tip-nozzles for directing compressed air to achieve a torqueless prerotation. The initial compressed air came from a cylinder attached to the fuselage, which would be filled in flight by a compressor powered by the engine. Although Payne and his associates may have been inspired by the tip-jet success of the Rotodyne, then successfully flying in England, the Avian 2/180A was not a mechanical success. The problems of developing a successful compressed rotor system proved daunting with the prototype; first flying less than a year later, in early 1960,[49] it suffered severe damage in an accident. Avian then opted for a more conventional, proven mechanical prerotator system of a belt drive attached to the engine and engaged by means of a clutch assembly in the second, 2/180B prototype.

Three aircraft were produced and about three hundred hours of flight-testing accomplished by 1964, when the Canadian government announced that it would fund the additional development and testing for certification. The design was significantly improved, with aluminum and fire-resistant fiberglass fairings replacing the 2/180A's heavier steel structure, which resulted in better performance, as cruise speed was increased from 80 mph to 100 mph, with an impressive top speed of 120 mph. Although maximum rotor speed was 263 rpm in forward flight, the three-blade rotor could be prerotated to 360 rpm to achieve jump takeoffs. When the rotor reached the maximum, the engine was declutched and the blades were collectively placed into an eight-degree positive angle, causing the kinetic energy stored in the rotor blades to lift the Avian into the air. It was a convincing demonstration, and certification was granted in 1967[50] or 1968,[51] but the aircraft never entered production. Refining a prototype is a demanding task, and certification can be exhausting, but production is an altogether daunting challenge that has frustrated more than one talented designer, and the Avian fell victim to the estimated cost of production.

Although the company advertised and apparently received 116 advanced orders,[52] that was not enough to actually commence production. By 1972 it had been placed into receivership and was subsequently sold to a group of Listorvel, Ontario, businessmen led by Harvey Krotz that apparently counted on further government support to enter production. It was not to be, and by January 1977 it was reported that the three remaining Avian prototypes, production rights, and plans were available for around $200,00,[53] although it is not known if a buyer was found. The Avian 2/180 then disappeared from view and, never having achieved production, had virtually no impact on subsequent development, but it then resurfaced when a surviving model was sold to Pegasus Rotorcraft Ltd. along with both the certification rights. In 2002 notice appeared that the aircraft had been renamed the Pegasus Mk III.[54] It is a tribute to the vision of Rowland Payne that the Pegasus, virtually identical to the Avian, may yet become a successful commercial venture; the company has announced the intent for future production, a result that would stand in contrast to the only single-passenger certified gyroplane, the Beagle-Wallis 116, which was never intended for the commercial market and which failed to gain acceptance by the British military.

NOTES

1. "TSR2: If Only... ," *Aircraft Illustrated* 34, no. 6 (June 2001): 50–54, 53.

2. "A Fairey Rotodyne Storey," Traplet Video Productions, Worcestershire, England.

3. In 1989 one publication, apparently abandoning the party line of cancellation due to noise, claimed that the Rotodyne was cancelled because "the lumbering Rotodyne was a clumsy aircraft to fly." "The Flat Risers: The Ups and Downs of VTOL (Part 2)," *Take-off* 1, part 5 (1989): 138–14, 139.

4. Jean-Pierre Harrison, "Fairey Rotodyne," *Air Classics* 22, no. 44 (April 1996): 44–47, 60–62, 64–66, 79–80, 61.

5. Frank Anders, "The Problem Solver," *Air Classics* 30, no. 10 (October 1994): 50–58, 53.

6. Michael Taylor, *The World's Strangest Aircraft: A Collection of Weird and Wonderful Flying Machines* (New York: Barnes and Noble Books, 1996), pp. 89–91.

7. John W.R. Taylor and H.F. King, *Milestones of the Air: JANE'S 100 Significant Aircraft* (New York: McGraw-Hill Book Company, 1969), pp. 136–37.

8. Martin Hollmann, *Helicopters* (Monterey, California: Aircraft Designs, Inc., 2000), pp. 130–31.

9. John Everett-Heath, *Soviet Helicopters: Design, Development and Tactics* (London: Jane's Publication Company, 1983), p. 33; Charles Gablehouse, *Helicopters and Autogiros: A Chronicle of Rotating-Wing Aircraft* (Philadelphia: Lippincott, 1967) pp. 108–9.

10. Everett-Heath, pp. 162–63; *World and United States Aviation and Space Records* (Washington, D.C.: National Aeronautic Association, 1987), pp. 89–90.

11. Everett-Heath, p. 33.

12. Ibid., p. 32.

13. For a description of Nagler's life and work, see Don Parham, "The Bruno Nagler Story," *Homebuilt Rotorcraft* 12, no. 3 (March 1999): 8–12. For a specific analysis of the Nagler Vertigyro, see Don Parham, "Nagler's Vertigyro," *Homebuilt Rotorcraft* 14, no. 7 (July 2001): 8–10.

14. George Townson and Howard Levy, "The History of the Autogiro: Part 2," *Air Classics Quarterly Review* 4, no. 3 (Fall 1977): 113 (photograph of Nagler Heli-Giro Aeronca conversion).

15. For a picture and technical description of the Nagler Vertigyro VG-1, see William Green and Gerald Pollinger, *The Aircraft of the World* (New York: Doubleday and Company, Inc., 1965), p. 338; Parham, Don "Nagler's Vertigyro," p. 9.

16. *Popular Rotorcraft Flying* 2, no. 2 (Spring 1964): 15 (photo of Nagler's Vertigyro in flight).

17. Albert G. Fischer, "Germany's VFW Sports a Hollow-Bladed Helicopter-Autogyro—with a McCulloch!" *Popular Rotorcraft Flying* 5, no. 3 (September 1967): 30; "VFW Joins the Parade," *Popular Rotorcraft Flying* 9, no. 4 (July–August 1971): 31.

18. Kas Thomas, "The Umbaugh Story: Rags to Riches (and Back?)," *Popular Rotorcraft Flying* 10, no. 2 (March–April 1972): 10, 23; "'It's All Yours' he said: Kas Thomas Flys the 18A," *Popular Rotorcraft Flying* 10, no. 2 (March-April 1972): 8–9; Glenn Bundy, "A Dream Dies Again, But Is It the Last Time?" *Rotorcraft* 39, no. 5 (August 2001): 41–42; A.C. Bass, "Pilot Report: Umbaugh Gyroplane," *Flying* 70, no. 1 (January 1962): 44–45, 110–12; "Newly Returned to the Rotorcraft Scene: The Air & Space 'Heliplane,'" *Rotor Gazette International,* no. 13 (May–June 1994): 3–4, 6. For photographs and renderings, see covers of *Aero Modeller* 27, no. 320 (September 1962) and *Meccano Magazine* 47, no. 6 (June 1962); "The Queer Birds: Air & Space Model 18-A," *Flying* 76, no. 4 (April 1965): 40–41, 45–46.

19. Igor B. Bensen, *A Dream of Flight* (Indianapolis, Indiana: The Abbott Company, 1992), p. 13, 39–40 (photograph of Umbaugh making his only flight in a cabin-equipped B-7M).

20. Thomas, "Umbaugh Story, p. 10; Bass, p. 110 (Umbaugh interview).

21. Thomas, "Umbaugh Story," p. 10.

22. John T. Potter, letter to author, July 7, 2002.

23. Bass, p. 110.

24. Ibid., p. 112.

25. Potter, letter to author.

26. Thomas, "Umbaugh Story," pp. 10, 23, citing "Queer Birds."

27. There is some confusion as to how many gyroplanes were actually produced by A & S. Thomas maintains that "Air & Space manufactured and delivered nearly 70 finished gyroplanes before their untimely demise. At the time Air & Space closed their doors, they had approximately 50 ships under their roof in various stages of construction, and more back orders than they could have filled in years." Thomas, "Umbaugh Story," p. 23; but see Bundy, p. 41 ("By the end of 1965 a total of 110 machines had been built"); John T. Potter in a draft article entitled "Gyroplane Outlook" claims that at the end of 1965 Air & Space had "over 200 fully committed [orders] in local escrow account."

28. Potter, letter to author.

29. Thomas, p. 23.

30. The advertisement for the two 18-As can be found on page 27 of *Popular Rotorcraft Flying* 11, no. 3 (August 1973).

31. Woody De Saar, "First PPL(G) on Air & Space 18-A in Ireland," *Rotorcraft* 39, no. 1 (February–March 2001): 11, 44–45.

32. Long thought to have vanished, this film is only available in the NTW-WGBH archives in Boston, Massachusetts, and there is a copy that can be viewed in the Museum of Television and Broadcasting in New York City.

33. Don Farrington, "Rotorcraft Training," *Rotorcraft* 28, no. 4 (June–July 1990): 42.

34. For a photograph of the refurbished and improved 18-A, see the cover of *Popular Rotorcraft Flying* 19, no. 3 (June 1981).

35. "Newly Returned," pp. 3–4, 6.

36. For background information on this dual-control two-seat tandem recreational and training autogyro, see Michael J.H. Taylor, (chief ed.), *Brassey's World Aircraft and Systems Directory 1900/2000* (London, UK: Brassey's, 1999), p. 358.

37. For the announcement of the Farrington Aircraft Corporation Air & Space America, Inc. auction, see *Rotorcraft,* 39, no. 2 (April 2001): 35.

38. Peter Garrison, "Everybody Loves an Autogyro," *Flying* 88, no. 2 (February 1971): 66–68, 66.

39. "Return of the Flight of the Phoenix," *Popular Flying,* January–February 1971, 10.

40. For photographs of the McCulloch J-2, see Martin Hollmann, *Flying the Gyroplane* (Monterey, California: Aircraft Designs, Inc., 1986), p. 51; "Return of the Flight of the Phoenix," p. 11; Bill Sanders, "The Rebirth of N4353G and N4364G," *Rotorcraft* 28, no. 4 (June–July 1990): 14–19.

41. Garrison, p. 68.

42. The estimates of J-2 production range from eighty-two to ninety. In part this confusion may be compounded by the fact that the successor company, Aero Resources, apparently completed some unfinished models and constructed one, or

possibly two, Super J-2 models with more powerful engines. See Sanders, p. 14 ("The McCulloch Corporation had built only 85 J-2's, mostly during 1971 and 1972.").

43. Garrison, pp. 66–68.

44. Kas Thomas, "Like a Theodore Bear," *Popular Rotorcraft Flying* 9, no. 3 (May–June 1972): 11–16; the Spanish Aeronautica Industrial SA (AISA), which had built some of the earliest Cierva Autogiros in the late 1920s, also produced an autogyro in a stub-wing, twin boom tail, three-blade rotor configuration very similar to the J-2. David Mondey (ed.), *The Complete Illustrated Encyclopedia of the World's Aircraft* (Secaucus, New Jersey: Chartwell Books, Inc., 1978; updated by Michael Taylor, 2000), p. 98.

45. Martin Hollmann, "Designing Rotor Blades," *Rotorcraft* 39, no. 9 (December 2001–January 2002): 18–23, 18; Hollmann, *Flying the Gyroplane,* p. 50.

46. "Rotary Connection," *Popular Rotorcraft Flying* 13, no. 5 (October 1975): 23.

47. Ron Herron, designer of the Cierva-type "pusher" Little Wing autogyro, owns and flies a 1972 Super J-2.

48. For photographs of the Avian 2/180, see David Mondey, *Complete Illustrated Encyclopedia,* (2000) p. 115; John W. R. Taylor, *Helicopters and VTOL Aircraft* (Garden City, New York: Doubleday, 1968), p. 78; Giorgio Apostolo, *The Illustrated Encyclopedia of Helicopters* (New York: Bonanza Books, 1984), p. 44 (rendering); Harry McDougall, "Avian Gyroplane," *Flying,* 74, no. 4 (April 1964): 44–45, 79–80.

49. John W. R. Taylor, *Helicopters and VTOL Aircraft,* p. 78; Mondey, *Complete Illustrated Encyclopedia,* p. 114; Martin Hollmann, "The Avian Gyroplane," *Gyroplane World,* no. 4 (January 1977): pp. 2–3, where the author asserts that the first prototype never flew.

50. Apostolo, p. 44.

51. Mondey, *Complete Illustrated Encyclopedia,* p. 114.

52. See letter to editor from Ross Bowes in *Popular Rotorcraft Flying* 10, no. 2 (March–April 1971): 24.

53. Hollmann, "The Avian Gyroplane," p. 2.

54. "Avian 2/180 Gyro May Be Revived," *Homebuilt Rotorcraft* 15, no. 11 (November 2002): 4.

Chapter 14

TRANSFORMATION OF THE PRA: KEN BROCK, MARTIN HOLLMANN, AND THE ULTRALIGHT REVOLUTION

Shown in the photo at the controls of her husband's gyrocopter, Mrs. Rose was quoted as saying: "It's thrilling to fly this machine and not at all dangerous. Any woman could use one."

Cy Rose, *Popular Rotorcraft Flying*

By the late 1960s the Vietnam war dominated the nightly news and America was losing an increasing number of personnel as aircraft were downed beyond prudent or possible rescue-helicopter range. The United States Air Force embarked on an experimental program to develop the DDV, or Discretionary Descent Vehicle.[1] This was to be an integral part of an ejection mechanism and provide for a controlled, and hopefully safer, landing in an area not threatened by the enemy. Bensen, having worked with the Rotachute, designed almost a quarter of a century earlier to accomplish a similar task, immediately saw an application for his Gyroglider and Gyrocopter and submitted a proposal.

The air force engaged Bensen Aircraft Company to produce three different experimental (or "X") concept aircraft, with the first being the X-25, a rudimentary rotary glider, in response to the mission requirement of a disposable, "one use" aircraft. It was a basic Bensen frame with an attached seat and a rotor that would deploy upon ejection. As the rotor began to spin, the lift created would allow the pilot to then steer by means of a pedal-controlled rudder. This was even more basic than either the Fa-330 or Rotachute, as each of these rotary kites had direct control of the rotor. The X-25A and B corresponded to the Bensen Gyrocopter and Gyroglider configurations, but with a floor-mounted stick control. This was presumably to make it easier for the downed pilot to fly, as he would be more familiar with that arrangement, which was found in virtually every fighter plane rather than the overhead hanging-stick. The X-25A was a B-8M Gyrocopter powered by a 1,600-cc McCulloch horizontally opposed four-cylinder engine, with which the air force was thoroughly familiar, as it had previously proven its reliability in target drones while the X-25B was a B-8 Gyro-glider.

The X-25A first flew in May 1968 and proved a successful concept aircraft. Even though the X-25A was more complex than called for by the "disposable aircraft" mission profile, it established the feasibility of rotary-wing aircraft for the projected military rescue mission, even at jet-fighter speeds. However, the DDV never entered production, as the government ended funding in the early 1970s. All three models were then sent to the Air Force Museum located at Wright Patterson Air Force Base near Dayton, Ohio, but only the X-25A (68-10770) and B (68-10771) were intended for display.[2] The X-25 was apparently dismantled. It was yet another military opportunity for the Gyrocopter that failed to materialize. But the late 1960s and early 1970s were to see far more dramatic changes to the American rotorcraft movement, with the rise of its most significant influence, Ken Brock of Anaheim, California, and his fifteen years' leadership of the Popular Rotorcraft Association.

Ken Brock, born in Hollis, Oklahoma, in 1932 and subsequently described as perhaps "the most successful person ever involved with homebuilt rotorcraft,"[3] started with Bensen plans in 1957. He had previously served in the United States Air Force and after being released from military service, returned to Vernon, Texas, where he met Marie, who would become his wife after a whirlwind courtship. It was to prove a most productive and long relationship for the American rotorcraft movement. Even though he had opened a small machine shop in Long Beach, California, and had both access and skill with tools, it took him almost two years to complete the project. As there were no Certified Flight Instructors (CFIs), he taught himself how to fly, first with a Gyro-glider and then with a powered machine. Flight training was so unsystematic that Brock and Ed Nielesky destroyed several sets of wooden rotor blades as they unlearned the lessons from the gyrogliding and then mastered powered flight.

There were no PRA Fly-Ins, so the Bensen Gyrocopter was taken to Experimental Aircraft Association gatherings. When Brock read of the first PRA Fly-In at the airport in Raleigh-Durham, North Carolina, home of the Bensen manufacturing facility and the new PRA, he decided to attend. It was his initial contact with the Bensen PRA organization and a revelation—he had been one of the first to *actually* complete an aircraft from plans, and his experience and skills as a machinist served him extraordinarily well. Brock had fabricated parts including his own spindle-type rotor head and wooden blades and had gotten proficient flying at what would become a site he would use throughout the rest of his life, the El Mirage dry lake northeast of Los Angeles. He came to the first North Carolina PRA Fly-In as a skilled, experienced pilot with an intimate understanding of how the Bensen Gyrocopter was built and flown. Given his interest and at Marie's suggestion, it was natural, then, that he should in 1965 become a Bensen dealer.

Ken Brock first began selling Bensen plans, then kits, and he remained the West Coast representative for Bensen Aircraft for many years. And his machine shop, which would become the renowned Ken Brock Manufacturing, Inc., soon began selling Gyrocopter parts and engine mounts for the Bensen craft. But Brock proved to be an effective combination of two contradictory themes, curious visionary and committed pragmatist, and this was to lead him and the gyroplane movement into a new direction. He was often quoted as saying, "There's nothing that's ever been made that can't be improved on," and "Don't tell me, show me." He immediately saw ways of improving the Bensen design, proceeded to make changes, and then proved the effectiveness of the new designs by extensive personal flight-testing. His regular air show performances at the annual EAA Fly-Ins at Oshkosh, Wisconsin, thrilled hundreds of thousands of spectators, who were almost completely unaware that this gyroplane pioneer was born shortly after the start of the Great Depression and had been flying for over four decades! All they saw from the ground were steep turns, moves that caused one's heart to pound and breath to stop, tight spiral descents, and a signature finale—a fall from high altitude with engine off, ending in a dramatic flared landing that always brought a thunderous applause.

It was fortunate that Ken Brock was taking fixed-wing flying lessons at Long Beach in the late 1950s while constructing his Gyrocopter from Bensen plans. As a fixed-wing pilot he had mastered the use of the control-stick mounted on the floor, but Bensen's control system was based on an overhead control-stick, an angle of metal directly connected to the rotor hub ending in a horizontal T-bar. The gyro pilot held the T-bar with two hands and tilted the rotor head to achieve control, as previously developed for the Rotachute. Brock, perceiving that if the Gyrocopter had a familiar control system, it would be easier for experienced pilots to make the tran-

Ken Brock in his KB-2, awaiting takeoff at the Great Western Fly-In at Sacramento, California. (Courtesy of Stu Fields.)

sition to rotary flight, proceeded to innovate a control-stick system that is currently employed in over 90 percent of all gyroplane designs.

Additionally, Brock significantly improved the Bensen "flying lawn chair" design, with "its plywood tail, canvas seat sprung with strips of old inner tubes, heavy steel wheels and an upside-down control stick."[4] Addressing each component, Brock systematically improved the Bensen design, and builders took note. Observing that the Bensen design called for an externally mounted gasoline can (like the one in a garage used to refuel a lawn mower), Brock realized that such an arrangement created aerodynamic drag and negatively impacted performance. His answer, now universally accepted except by those passionately committed to the purity of the original Bensen design, was to incorporate the fuel tank into the seat itself.[5] Sales of the seat-tank, as described by Brock in a 1993 interview, "just took off like a rocket."[6] He also improved the engine mount and throttle and introduced quality manufactured parts and a complete support service for builders. The end result was that by 1970 the Brock KB-2, originally fashioned on the Bensen model, had assumed its own identity and found a following.[7] It was at that point that Bensen and Brock agreed that because the KB-2 had deviated from the specified Bensen plans, it would henceforth be called a *gyroplane*. Two years later Brock would become president of the PRA and lead it in new directions, even as he would himself gain new heights.

In 1971 Brock achieved national recognition when he departed Long Beach, California, on June 11 and arrived at First Flight Airport at the Wright Brothers Memorial monument eleven days later. The official Kitty Hawk, North Carolina, Register read: "June 21, 11:53 Ken Brock, Anaheim, Calif. NONE N2303 gyrocopter, coast-to-coast, Long Beach to Kitty Hawk." His route had followed Interstate Route 40 and then Route 66, had taken eleven days and ten nights with stops at forty-four airports along the way, as well as stops at at least one trailer court in Needles, California, several truck stops and gas stations when he could not find the local airport, and one landing on an incomplete bridge overpass.[8] Accompanied by his volunteer ground crew of John Bruce and University of California student Kas Thomas, then associate editor of *Popular Rotorcraft Flying,* Brock would roar into the sky and down the highway for each flight segment, only to land and wait for the ground crew to catch up. Encountering the possibility of danger at each turn, it was the first gyroplane transcontinental flight since the May 1931 Autogiro flight by Johnny Miller and the June 1931 flight by Amelia Earhart, and when questioned as to why he had done it, Brock stated: "It's something I've wanted to do for a long time.... I guess you could say I just wanted to show that it could be done."[9] Although Brock's achievement would also be duplicated by Howard Merkel in October–November 1989 in a Jerrie Barnett gyroplane,[10] it would not, nor could it, have the same impact.

It is clear that much more was going on in 1971 than a transcontinental flight and that it was to be of paramount significance for the American gyroplane movement. Brock had, deliberately or perhaps inadvertently, focused attention on the capabilities of his KB-2 model, N2303, which is today on permanent exhibit at the EAA Museum in Oshkosh. That recognition was not long in coming. Ken met Marie and his son Terry at the 1971 PRA Fly-In at Edenton, North Carolina, which had begun unofficially on Thursday, June 24, with pilot check-in. Edenton had been the next-to-last stop on his flight to Kitty Hawk, and Brock now returned in triumph. At the concluding Saturday evening banquet FAA official (and PRA member) Juan Croft was the featured speaker, who eloquently praised the PRA, but the highlight of the evening was undoubtedly the awards ceremony conducted by *Popular Rotorcraft Flying* editor and

Bensen associate Ed Trent—Brock received six trophies,[11] but also "won something that no judge could award. Ken Brock won the respect of not only the two hundred people that attended the banquet, but also of every air-minded person in the world that heard of his coast-to-coast flight. Ken received not just one, but three standing ovations!"[12] In honoring Ken the membership was also honoring his innovations, the significance of which was not lost on PRA president Igor Bensen.

In 1972, ten years after founding the PRA, Bensen stepped down as its president, assuming the title of president emeritus and passing leadership responsibility to Ken Brock.[13] In many ways the imperative that Bensen step down was predictable—the seeds of member discontent had been sown in the very foundation of the association. Viewing the directors as primarily advocates of and participants in Bensen marketing (at least half of the directors were Bensen dealers) and *Popular Rotorcraft Flying* as a Bensen in-house newspaper and promotional tool (not surprising given its origins), there was a chorus of criticism regarding the constant PRA emphasis on Bensen designs, the necessity of using only Bensen parts, and the general unwillingness to allow for innovation. But most of all, however unwillingly, Bensen recognized that there would be a revolt of members if he did not step down. Although Bensen and his adherents would continue to advocate his designs and factory parts[14] years after he resigned and chapters would complain of continued Bensen domination of the PRA,[15] it is clear that the association was entering into what might be justly characterized as its most productive and innovative period under the leadership of Ken Brock.

In his farewell address[16] delivered at the concluding banquet on June 24 at the PRA annual Convention and Fly-In held, not in North Carolina, but in Rialto, California, Bensen cited a previous editorial statement made in 1964 that he would "be perfectly happy to yield my job to anyone who can provide PRA with an equal degree of leadership and competence...." The citation of his previous statement was accurate but disingenuous, as it was perfectly clear that there was not anyone competent enough to replace him in 1964, but it had become equally obvious in 1971 that Ken Brock was such a leader. Bensen also cited his 1969 ordination as a priest in the Russian Orthodox Greek Catholic Church of America and Canada[17] and the need to spend weekends in pursuit of his ministry. But he also spoke of a darker and ultimately more significant need for a change of leadership, as he acknowledged that "wearing two hats as president of the PRA and as president of Bensen Aircraft inevitably led to the accusations that I have used my high position in the PRA to promote the interest of Bensen Aircraft Corporation." He had, in fact, anticipated in 1964 that this could happen, when he wrote, "When I accepted the job of PRA president in October, 1963, I anticipated with a heavy heart that my motives would be questioned and my connection to Bensen aircraft Corporation used to throw doubt on my impartiality."[18] This member view was reinforced with the preeminence of Bensen dealers on the PRA board of directors and with the PRA headquarters located in the Bensen North Carolina factory. Then, and until 2001,[19] the PRA directors were elected by and selected from those with lifetime memberships, and twenty-three life members elected the new board, which consisted of Ken Brock and Ed Trent and five new directors: Chuck Beaty, Tampa, Florida; Bob Thomas, El Monte, California; Steve Phaneuf, Burbank, California; Niel Kilchriste, Las Vegas, Nevada; and Jim Duke, Fullerton, California.

With five of the seven from California and Nevada, it is not surprising that at the first meeting convened in Anaheim on Monday, June 26, the new PRA board elected Ken Brock president and voted to relocate the PRA headquarters to southern California. Ed Trent, who had been the fourth member of the PRA and had been a *Popular Rotorcraft Flying* editor since its first issue in the spring of 1963, resigned as editor and PRA secretary and became a vice president. It marked the changing of the guard, and the fifteen years of Ken Brock's presidency would see the PRA go in new and exciting directions.

Under Ken Brock's leadership, there was a new freedom for experimentation and gyroplane development, even as Bensen remained president emeritus and a genuinely revered figure to the PRA membership. The Sunstate Rotor Club, the local chapter of the PRA, honored Bensen by naming its annual Wachula, Florida, fly-in "Bensen Days" and scheduling it to celebrate his February birthday. This event was later shifted to April to coincide with the larger, EAA-sponsored Sun 'n Fun Fly-In at Lakeland, Florida, and it was at Bensen Days in 1993 that the Gyrocopter pioneer took his last flight, in a unique three-seat autogyro flown by famed designer Bill Parsons.[20] Bensen Aircraft had closed its doors in December 1986,[21] having fallen on hard times after the death of Mary Bensen a few years earlier, who had been a major, if unacknowledged, factor in its success, and after liability lawsuits that had drained its funds. Bensen announced at Bensen Days on April 5, 1991, that he was starting a new Canadian company called Bensen International Gyrocopter Transport Company (BIG-T Company and a new association called Universal Flying Organization (UFO). Neither this new company or rival organization got off the ground, and Igor Bensen died in February of 2000, mourned by the thousands who had first glimpsed the vision of flight from the seat of a "flying lawn chair."

It was equally evident in the 1970s that the PRA was largely a white male–dominated organization, as it remains today. Indeed, a member of PRA Chapter 31 from San Diego, California, complained in early 1977, "Looks like the gyro movement has just about hit bottom!—Seems that an article about gyros written about Chapter 5 and Ken Brock appeared in a nudie magazine. What our movement was doing there is anybody's guess. Maybe we've got something we don't know about!!"[22] Although a significant minority presence or influence would not emerge, women were beginning to play important, if often unacknowledged, roles. Mary Bensen had served as a director and treasurer and as chief supporter of her husband, Igor Bensen, for most of the early years of the PRA, and Marie Brock functioned as Ken Brock's business associate in all his endeavors.[23] As observed by noted gyro journalist, author, and then-editor of *Rotorcraft* Paul Bergen Abbott, "No article about Ken Brock is quite complete without including Marie."[24] Mary Bensen and Marie Brock each shared in and very significantly contributed to her husband's accomplishments. Marion Springer, a proud descendent of Choctaw Indians and a member of the Choctaw Nation, became an outstanding pilot, the first female CFI,[25] and a respected journalist. Marion, in her *Blue Angel,* which was later converted to the white *Born Free* Gyrocopter (N2066), was a frequent participant in the regional fly-ins in California, and she and her husband, Al, appeared in the 1982 film *The Great Skycopter Rescue.* In that otherwise eminently forgettable B movie, Marion's husband appears in the credits as Doc Springer, undoubtedly taken from his nickname of "Docko."[26]

Springer was the most visible female Gyrocopter pilot and also wrote prolifically on safety issues.[27] For many years she gathered and analyzed accident reports, a tradition later admirably carried on by Ken Brock, based on a yearly compilation of National Safety Transportation Board. Helen Darvassy, who attempted to solo in the fall of 1964, only to have the rotor come off while taxiing down the strip,[28] became the Editor East of *Popular Rotorcraft Flying* and made a lengthy and outstanding contribution with her thoughtful commentary. Loyal to Bensen, she continually stressed fidelity to his vision and design and the necessity of using only Bensen parts. In a very real sense the rotorcraft movement moved beyond her, but she was still writing in such a manner in 1978, years after Bensen had stepped down. But contributing editors Springer, Darvassy, Mary Van Hoten, and managing editor Arlene Baker (1974–88) carved out editorial roles that others would assume in the future—Stephanie Gremminger and Kathryn Fields would each become editor of the PRA signature publication *Rotorcraft.* And in 2001 Alida "Lisa" De Vries,[29] only the third woman to be designated a gyroplane CFI, following in the footsteps of Marion Springer and Patricia Thomas,[30] was further honored as a Master CFI, a status achieved by fewer than 300 of the then 78,000 CFIs in the United States.[31]

During the years 1972–87 of Brock's leadership, the PRA had held conventions and fly-ins from California to Oklahoma, and "very un-Bensen-like rotorcraft began appearing at these conventions including Jerrie Barnett's gyros, distinguished since 1962 by their ruggedness resulting from the use of welded steel tubing and heavier certified aircraft engines, with a resulting superb safety record, and Martin Hollmann's two-place Sportster gyro."[32] This was the direct result of a freer atmosphere for innovation, the publication of the *Guide to Homebuilt Rotorcraft* by Kas Thomas and Paul Bergen Abbott's stream of flight and construction manuals,[33] the entry of companies such as Vancraft (renamed Sport Copter), Canada's RAF (Rotary Air Force), Dennis Fetter's Air Command, and Ernie Boyette's Rotor Flight Dynamics, Inc. But of greatest significance were the introduction of a two-place Gyrocopter trainer by Bill Parsons[34] and the creation of an entirely new category of gyroplane that made it possible for thousands of new pilots to take to the air, the ultralight.

Bensen had progressed from Gyroglider to Gyrocopter, and his recommended training for new pilots followed a similar and rigidly defined and mandated regimen, which was set forth in his building instructions and flight manuals.[35] One learned starting with a ground-based "point trainer" at the Bensen Plant, boom training (where the Gyroglider is towed into the wind by a vehicle to which the aircraft is attached by a rigid boom), and then flying in back of a towed truck to which a towline is attached, permitting greater altitude and eventual free-flight when the towline is detached after ascent. Only after mastering gliding was the pilot allowed to progress to power Gyrocopter flight, but this mandated sequence has an inherent flaw—although the Gyroglider training could be on a two- (or three-) place machine, powered flight could only be done in a single-seat aircraft with no possibility for dual training other than in the prohibitively expensive certified factory-built McCulloch J-2 or Umbaugh (Air & Space) 18-A machines. This lack of a powered dual trainer resulted in many accidents, as there were significant differences in performance characteristics between the glider and powered flight. This inspired Bill Parsons in 1985 to design a trainer.

Parsons was an experienced flight instructor, and his initial design reflected that familiarity. His Parsons trainer was a Bensen B-8M, with a

Martin Hollmann and the HA-2M Sportster. (Courtesy of Martin Hollmann.)

longer keel to accommodate a second seat, dual controls, and a rotor head attached by an upside-down-U-shaped tandem double mast. Although this was not the first two-place autogyro, as Chuck Vanek had previously flown his two-plane and as Hollmann's Sportster had been flying for many years, it was the first based on the familiar Bensen airframe and available to the average builder. The first Parsons trainer (NWP54) was used to train over six hundred pilots and is on display at the American Helicopter Museum and Education Center at the Brandywine Airport, West Chester, Pennsylvania. The introduction of the Bensen-based Parsons trainer revolutionized flight training and ushered in new models, but the coming of an entirely new category of gyroplane, the ultralight, vastly increased the availability of flight opportunities, and that was the creation of perhaps the most well-educated aeronautic engineer since Cierva and Bensen—Martin Hollmann.

Martin Hollmann, at six feet, seven inches, is probably the tallest person ever to fly in a gyroplane, and he has also been one of the outstanding and outspoken gyroplane designers and theoreticians whose innovations have changed the nature of the industry and the manner in which autorotational technology survived. His interest in gyroplanes began in 1965,[36] and he gained, like all such enthusiasts at the time, a familiarity with but not an enthusiasm for the Bensen design. While an engineering student at San Jose State University, Hollmann was introduced to the work of Cierva and received a student fellowship grant for $400 to build and test a set of fiberglass rotor blades, a design effort that was the beginning of innovative achievements that would produce two models that changed the gyroplane industry in America.

Recognizing in 1969 that there existed a need for a homebuilt two-place gyroplane so that powered training could become the preferred mechanism for flight instruction,[37] Hollmann juggled a new job at Martin Marietta in Orlando, Florida, graduate studies at the University of Central Florida, and designing and constructing first the rotor blades and later the fuselage of his new two-place gyroplane, dubbed the Sportster.[38] Receiving his master's degree in mechanical engineering in 1974, Hollmann returned to a new job in California, driving cross-country in a Volkswagen Beetle towing the completed Sportster. Designed as a side-by-side two-place trainer in 1972,[39] it was the world's first successful two-seat amateur-built gyroplane, and it pointed to Hollmann's goal of revolutionizing gyro flight training. Designed to be constructed by the amateur, it met the FAA requirements of being more than 51 percent self-built and thus qualified for the designation as an "experimental" aircraft. Although it represented a major deviation from the Bensen insistence that two-place Gyro-Glider training was the only way to learn to fly, the Sportster also marked a major departure from the Bensen plans.[40] It immediately acquired advocates and admirers, with Dr. Tom Butler and Hofstra University graduate Walter "Skip" Tyler becoming the first builders to construct a Sportster and at the same time signing on as East Coast distributors. Butler had been the first passenger in the Sportster and was immediately impressed, as others would be, by the thorough preparatory engineering analysis incorporated into the design, which produced outstanding structural integrity in an aircraft capable of carrying two large adults. The first plans-built Sportster was constructed by Tyler and was one of the first gyroplanes to be powered by a Lycoming 0-320 aircraft engine, at the time a stunning innovation for the homebuilt rotary market.

The aircraft was introduced at the PRA 1976 Fly-In at Rockford, Illinois, as an amateur-built experimental aircraft, and although Hollmann had temporarily lost his student pilot license and was unable to fly, Butler soloed the Sportster that Hollmann and Verne Tobin had towed behind the venerable Volkswagen Beetle 2,300 miles from California. The public flight of the Sportster was well-received and generated an enthusiasm that remained undimmed when the craft unexpectedly settled to the ground with two pilots aboard the next day—effectively demonstrating the over-all ruggedness of the airframe and rotor to withstand high-impact loads without failure. Hollmann and his colleagues dubbed this incident, poten-tially catastrophic in other gyroplanes, the "super drop test." Although it takes an estimated one thousand hours to build, this two-place model has remained popular for over a quarter of a century. It has proven a remark-ably safe aircraft and has spawned several copies, including the Glanville Skymaster[41] from Canada, the Marchetti Avenger,[42] and the Tyler Shadow.[43] Hollmann, however, dismissed Kemp Glanville in 1978 as a Sportster dealer because he modified the design and added doors, and Hollmann later warned consumers that Glanville had traced the Sportster blueprints and was offering them for sale. Hollmann claimed that the mod-ified design was unsafe in a power-off flying condition, firmly condemn-ing this "Sportster imitation."[44] Having built a remarkably safe flying machine, it is supremely ironic that Hollmann's next innovation would come while he was recuperating from a broken back as the result of a crash—in a powered motor-glider called the Condor.

In the interim, however, Hollmann, never a fan of either the PRA or its president Ken Brock, whom he has continuously disliked for over three decades,[45] came to the conclusion that the PRA and *Popular Rotorcraft Flying* were deliberately neglecting non-Bensen designs, particularly from foreign countries, based on letters he received in response to his rotor blades and Sportster designs. He founded an alternative organization, enti-tled the International Gyroplane Organization, and commenced a news-letter in October 1976. Written primarily by Hollmann during its few years of existence, *Gyroplane World* contained a mixture of articles dealing with the Sportster, historical views of Autogiros, and a continual direct and indirect criticism of the Bensen and Brock designs. It was clearly a parti-san effort published out of Hollmann's Cupertino, California office. Although he claims to have had "up to 4,900 newsletters to send out each month,"[46] the organization folded in 1982 after Hollmann's crash, with lit-tle or any evidence of having had an impact.

In 1982, surviving the crash in his Condor upon takeoff at Freemont, California, Hollmann dealt with the boredom that accompanied the immo-bility made necessary by a broken back by designing a new, single-place gyroplane that he eventually dubbed the Bumble Bee. It was to be the first ultralight gyroplane—the ultralight movement was popular in the fixed-wing community with the allowable aircraft weight being no more than 245 pounds and friends had previously urged Hollmann to design a gyro-plane that would conform to this category. Now that he had the chance,

Hollmann, always the engineer, characteristically began with stress analysis. Realizing that such a project would be made possible by the new, lighter Kawasaki TA440-A and Rotax 447 engines coming from Japan and Austria, Hollmann knew that the time was right for such a design.

Hollmann found the project convenient to his physical recovery, because having been released from the hospital after a three-month stay, his doctors cautioned him against lifting more than 10 pounds, effectively preventing him from preparing the Sportster for flight, with its 65-pound rotor blades. He felt that he would be able to manage a lighter craft, and his analysis and design efforts resulted in a prototype a year later. The machining was done by his old friend Skip Tyler in New York and shipped west. The Bumble Bee prototype weighed just 190 pounds (the kits that Hollmann eventually sold weighed in at 230 pounds empty and could carry a useful load of 270 pounds, effectively able to lift a gross weight of 500 pounds and accommodate all but the heaviest of pilots). It was 12 feet long and 7.3 feet wide, with a mast extending 7.7 feet in the air, on top of which was a 23-foot diameter rotor. This seemingly fragile aircraft, powered by the dependable Kawasaki TA440-A engine developing thirty-eight horsepower, could fly at speeds as low as five mph and as fast as sixty-four mph. The little flying machine could climb at an impressive 800 feet per minute (fpm), and with its three-gallon fuel tank (ultralight aircraft are limited to five gallons), it had a sixty-five-mile range. The kit version employed a Rotax 447 engine (a change necessitated when the original Kawasaki seized during a demonstration flight), which developed an impressive forty-two horsepower through a gear reduction drive spinning a sixty-inch, two-bladed prop, and it could climb at 1,500 fpm carrying a useful load of 270 pounds.

As an ultralight , the Bumble Bee did not require a pilot's license, and it could be assembled for flight in fifteen minutes and dismantled for storage in a garage or spare room. Although it was still incomplete when displayed at the Hollister, California, air show in the summer of 1983, it was enthusiastically received, attracting a large, interested crowd. One of those who managed to draw Hollmann away for questions was pilot Allan Tatarian, who became the Bumble Bee test pilot.[47] It was subsequently shown and verified at the EAA Oshkosh Airshow as the first ultralight gyroplane. The Bumble Bee was technically sophisticated, with an extremely simple airframe that could initially be constructed for about $5,400. Cost effectiveness was achieved even given that Hollmann had developed a new and stunningly innovative highly laminar-flow rotor system that weighed in at an amazing thirty-five pounds, ready to fly, and was 30 percent more efficient that the best metal rotor. Hollmann's twenty-three-foot rotor blades, with their aluminum leading-edge extrusion, foam core, and multilayered fiberglass construction, made the Bumble Bee's outstanding performance possible and significantly influenced the subsequent ultralight revolution.

The ultralight, a single-passenger aircraft produced in conformance with the Federal Aviation Regulation (FAR) Part 103, weighs no more than 254 pounds, carries a maximum of five gallons of fuel, has a top speed of no more than 63.25 mph (50 knots), and is forbidden from commercial use. In exchange for this modest performance, neither the aircraft nor the pilot need to be licensed (although virtually every authority suggests and Hollmann insisted on ten to fifteen hours of flight training before solo flight).

Although there had been a powered ultralight design as far back as the Piper-Marriott autogyro, which had flown on October 2, 1960,[48] Hollmann was the first to design and offer an ultralight kit—but others soon followed. While Hollmann sold only forty Bumble Bee kits, ceasing because of the intensive labor in the production and inability to obtain liability insurance,[49] he continued selling the plans, and his aircraft led to a series of derivative designs whose names echo their heritage, notably Dr. Ralph Taggert's Gyrobee[50] and GyroTECH, Inc.'s Honey Bee.[51] Several companies specialize in ultralight gyroplanes, and many manufacturers offer an ultralight option, usually a stripped-down, lighter version of their heavier and more powerful experimental model, to capitalize on this entry-level market. Ken Brock even created an ultralight version of his gyroplane, the KB-3, which has achieved a great deal of enduring popularity.[52]

The 1980s were good years for Ken Brock—he flourished as president of the PRA and in his business. Ken and Marie Brock had incorporated a second company, Santa Ana Metal Stamping, and it was doing well with fabrication of high-quality metal parts for the composite planes designed by Ken's good friend Burt Rutan and for other homebuilt kits. He had become a member of the Screen Actors Guild, as he was increasingly involved as a film and television gyroplane pilot, and was featured in commercials for Sears. An Australian gyroplane had even appeared in the 1981 movie *Mad Max 2: The Road Warrior,* in which the "Gyro Captain" emerged as the hero by the end of the movie.[53] But in 1986 it all came to a crashing halt, as Ken was struck in the head by a rotor and almost died.

In June 1986 Brock was flying his gyroplane at El Mirage dry lake bed along with PRA board member and longtime friend Russell W. "Russ" Jansen.[54] Brock approached from the west, passed over some power lines, and safely landed, but Jansen, failing to see the lines, flew into them and dove into the ground, an instantly fatal accident. Brock, standing next to his aircraft, rushed to aid his friend but was struck by his rotor, which was still in motion. The accident report[55] suggested that the usually meticulous and careful Brock had been felled as his KB-2 was turned by the prop wash from a nearby aircraft. He always flew with a helmet, and that now saved his life; friends were amazed when, after weeks of hospitalization, he attended the EAA AirVenture Fly-In six weeks later. It was one of the few EAA gatherings at which he did not fly his trademark exhibitions, but he and Marie were not about to miss their many long-standing friends. He would, after arduous physical therapy, resume flying. He would give up the presidency of the PRA to George J. Charlet Sr. the following year, but the adventure of a lifetime was just around the corner!

In September 1989 gyroplane pilot Beverly Johnson, wife of Academy Award–winner Mike Hoover, asked Ron Menzie to build a tandem gyroplane "capable of performing in the harshest conditions on earth while carrying two persons, survival gear and heavy camera mounts and cameras."[56] Johnson now wanted not only a two-passenger gyroplane, she and her husband wanted Ken Brock to fly with them in Antarctica while filming a *National Geographic* TV special. The Ron Menzie *Ice 90* gyroplane was completed in two months, transported to Brock's machine shop in California, where it was tested and further modified before it was crated, flown to Santiago, Chile, and placed on board a ship for the final southward leg. And it was duly noted that Lan Chile, the name of the domestic commercial airline, had been painted on the tail of the two-passenger gyroplane by local personnel.

Brock spent January through February 1990 in Antarctica, during which he flew both gyroplanes for nearly thirty hours and spent three weeks on the polar ice cap. It was the time of his life, and when asked if he would go again, he stated in typical Ken Brock-fashion, "I'd go in a minute! It was just a lot of fun and good old time."[57] Such photographic survey work had proven hazardous in the past for naturalists, and Philippe Cousteau had crashed on just such an expedition while exploring Easter Island on October 18, 1976, breaking his tibia and fibula and requiring evacuation to Los Angeles for medical treatment,[58] but Brock's performance under harsher conditions had been flawless. However, five days after Brock left for the trip back to California, another pilot, who was inexperienced with the gyroplane, was killed while trying to fly the two-passenger model.

Brock remained a revered figure and helped Igor Bensen whenever he could during the final years of his life as Bensen experienced financial difficulties and a physical decline. Brock's performances at the EAA Air

Venture Fly-Ins at Oshkosh were always crowd-pleasers, earning him a reputation as the finest gyroplane pilot in the world, and he never missed a PRA national convention/fly-in. But 2001 was to prove his last PRA fly-in, as he perished in a crash of his T-18 plane on October 19, 2001, when a malfunction in the tail wheel assembly while landing caused the plane to swerve off the runway and, striking a post, flip over. Ken Brock died instantly from a broken neck, and Marie suffered a minor injury—tributes poured in from around the aviation world. He was a beloved figure, although he would describe himself as just "old Ken Brock." His voice had been stilled, but the innovation wrought by the KB-2, the first gyroplane, would continue wherever and whenever pilots took to the air in his aircraft and in those gyroplanes he inspired.

NOTES

1. Peter Tasker, "Historic Wings X-25A/B," *Autogyro 1/4ly,* no. 6 (2000): 29–31.

2. There is some confusion as to the air force serial numbers of the Bensen X aircraft. Those given in the text are from Ibid., p. 29; but see "Bensen Autogyros," (pamphlet #30) *Aircraft of the World* (International Masters Publishers AB, 1998), which has an illustration of the X-25A with serial number 10772.

3. Paul Bergen Abbott, "Meet Ken Brock," *Rotorcraft* 31, no. 6 (September 1996): 12–17, 13.

4. Ibid., p. 13.

5. Sheldon M. Gallager and Howard Levy, "New Build-Your-Own Copters: More Power, Looks, Convenience," *Popular Science* 140, no. 6 (December 1973): 143–45, 46 (photograph of Brock seat-tank).

6. Abbott, "Meet Ken Brock," p. 14.

7. For Ken Brock's comparison of the Bensen and KB-2, see Abbott, "Meet Ken Brock," p. 15.

8. Kas Thomas, *Guide to Homebuilt Rotorcraft* (New York: Crown Publishers, 1976), p. 59 (photo of N2303 in Jackson, Tennessee).

9. For a description of the transcontinental flight, see Kas Thomas, "From Long Beach to Kitty Hawk…By Gyrocopter," *Popular Rotorcraft Flying* 9, no. 5 (September–October 1971): 10–16; "Staff Interview—Ken Brock—From Long Beach to Kitty Hawk…By Gyrocopter," *Popular Rotorcraft Flying* 9, no. 5 (September–October 1971): 28–31.

10. "Lone Eagle Gyro Pilot Sets Coast to Coast Record," *Rotorcraft* 27, no. 8 (December 1989–January 1990): 25; Erlene Barnett, "What Kind of a Person Flies Coast to Coast," *Rotorcraft* 28, no. 1 (February–March 1990): 16–17; Howard Merkel, "The Long Way Home: Part 2," *Rotorcraft* 29, no. 2 (April 1991): 50–52; Howard Merkel, "The Long Way Home: Part 3," *Rotorcraft* 29, no.

3 (May 1991): 24–31; Howard Merkel, "The Long Way Home: Part 4," *Rotorcraft* 29, no. 4 (June–July 1991): 42–45.

11. Brock's awards included Longest Distance Flown In, Best Gyrocopter–Operational, EAA Trophy, Man & Machine (Over 75 Hours), and the *Mechanix Illustrated* Trophy.

12. "The People Come First: 9th International PRA Fly-In," *Popular Rotorcraft Flying* 9, no. 5 (September–October 1971): 18–21, 20.

13. See Ed Alderfer, "The Designers: 30 Years of Rotorcraft Designers," *Rotorcraft* 27, no. 3 (May Extra 1989): 26–29.

14. Helen R. Darvassy, "Splinters," *Popular Rotorcraft Flying* 16, no. 2 (April 1978): 8–9, 12.

15. Helen Darvassy, editorial complaining of "lost chapters," *Popular Rotorcraft Flying* 15, no. 3 (June 1977): 2; See also Helen Darvassy, "Splinters," *Popular Rotorcraft Flying* 16, no. 2 (April 1978): 8–9, 12: 8. ("Then we have the group that brushes the whole question aside, but avers their reason for not attending the Rockford Fly-in is that it is a Bensen-run outfit anyhow, and they don't have anything to say about it. May as well forget about it and start our own club.")

16. Igor B. Bensen, "President's Farewell Message: Keep 'Em Whirling!" *Popular Rotorcraft Flying* 10, no. 2 (March–April 1972): 2, 28.

17. "The Reverend Igor B. Bensen," *Popular Rotorcraft Flying* 7, no. 4 (July–August 1969): 18.

18. Igor B. Bensen, "Needed: More Leadership," *Popular Rotorcraft Flying* (Spring 1964): 2.

19. For the announcement that the PRA would henceforth select directors by popular vote of all its members, see *Rotorcraft* 39, no. 1 (February–March 2001): 41. It should be noted that directors still will be elected only by those with life memberships.

20. For a video record of that final flight, see Dan Leslie, *Bensen Days 2000* (Macon, Georgia: Rotor/Wings Sports TV, 2000), video.

21. The December 1986 date is from Bensen. Igor B. Bensen, "Where Do We Go from Here?" *Rotorcraft* 29, no. 4 (June–July 1991): 25–26. However, Abbott claims that Bensen Aircraft ceased operations in 1988. Igor B. Bensen, *A Dream of Flight,* (Indianapolis, Indiana: The Abbott Company, 1992), p. 14 (Abbott introduction).

22. "Buzzing Around with Chapters," *Popular Rotorcraft Flying* 9, no. 2 (March–April 1977): 6–8; Richard Ashby, "Come Fly with Me," *Knight* 5, no. 7 (July 1966): 63–65.

23. "I guess Marie might have been somewhat responsible for getting me into the business. I was just a participant and didn't have much time to spend with people who wanted information. But she did. She talked to Igor Bensen and we started selling plans. Then we got into kits and became a West Coast rep for Bensen Aircraft for a long time." Abbott, "Meet Ken Brock," p. 14.

24. Ibid., p. 13.

25. Marion Springer, "In Pursuit of a Gyroplane Rating," *Rotorcraft* 27, no. 3 (May Extra 1989): 16–19.

26. Marion Springer, "Story of a Blue Angel," *Popular Rotorcraft Flying* 8, no. 1 (January–February 1970): 20–21.

27. See, for example, Marion Springer, "Gyrocopter Pilots Are Their Own Worst Enemies," *Popular Rotorcraft Flying* 16, no. 2 (April 1978): 7; "Grant Me Patience, Oh Lord, but Hurry!" *Homebuilt Rotorcraft* 7, no. 2 (February 1994): 4–5; Marion Springer, "No Training Wheels, Please," *Rotorcraft* 28, no. 4 (June–July 1990): 40. Springer was still writing on safety over twenty years later. Marion Springer, "Accident Prone and Macho Personalities," *Homebuilt Rotorcraft* 12, no. 6 (June 1999): 13–14; Marion Springer, "Endorsed for Solo," *Rotorcraft* 29, no. 5 (August 1991): 28–30.

28. Helen R. Darvassy, "Breaking the Woman Barrier!" *Popular Rotorcraft Flying* 2, no. 1 (Winter 1964): 10–11, 20.

29. Stephanie Gremminger, "Flying 'Team' Makes Transition to Gyroplanes," *Rotorcraft* 36, no. 6 (September 1998): 24.

30. Gary Goldsberry, "Along Time Dream Is Coming True," *Rotorcraft* 27, no. 1 (February–March 1989): 15; "First Two PRA CFI Scholarships Awarded," *Rotorcraft* 27, no. 3 (May Extra 1989): 19; "First PRA Scholarship Recipients in the Cockpit," *Rotorcraft* 27, no. 7 (October–November 1989): 5, 9.

31. Alexander "Sandy" Hill, "Alida 'Lisa' De Vries, Master CFI," *Rotorcraft* 39, no. 4 (June–July 2001): 30.

32. Abbott, "Meet Ken Brock," p. 12.

33. Paul Bergen Abbott, *The Gyroplane Flight Manual* (Indianapolis, Indiana: The Abbott Company, 1988, 1992, 1996); Paul Bergen Abbott, *The Gyrocopter Flight Manual* (Indianapolis, Indiana: The Abbott Company, 1977; Indianapolis, Indiana: Cranberry Corners, 1983; Indianapolis, Indiana: The Abbott Company, 1986); Paul Bergen Abbott, *So You Want to Fly a Gyrocopter* (Indianapolis, Indiana: Cranberry Corners, 1977); Paul Bergen Abbott, *Understanding the Gyroplane* (Indianapolis, Indiana: The Abbott Company, 1994).

34. Bill Parsons, "Why I Prefer the Tandem Trainer," *Rotorcraft* 31, no. 5 (August 1993): 21–22.

35. *Bensen Gyro-Glider: Building Instructions with Operating and Flight Manuals* (Raleigh, North Carolina: Bensen Aircraft Corporation, 1967), pp. 28–36.

36. Martin Hollmann, "Gyroplane Designs," *Rotorcraft* 38, no. 5 (August 2000): 32–35, 32; However, it has also been reported that Hollmann's interest in gyroplanes began in 1968 when his instructor at San Jose State University introduced him to the work of Juan de la Cierva. Glenn Brinks, "We Fly the Hollmann Sportster Gyroplane," *Air Progress Aviation Review* 1981.

37. "Rotorcraft Trainers," *Popular Rotorcraft Flying,* (Summer 1963), 12.

38. Helen R. Darvassy, "First Showing of the HA-2M Sportster," *Popular Rotorcraft Flying* 14, no. 5 (October–December 1976): 8–10; Don Dwiggins,

"Hollmann HA-2M Sportster," *Homebuilt Aircraft,* (October 1979); Martin Holl-mann, "HA-2M Sportster Tested," *Popular Rotorcraft Flying* 13, no. 4 (Decem-ber 1975): 34; Martin Hollmann, *Flying the Gyroplane* (Monterey, California: Aircraft Designs, Inc., 1986).

39. Martin Hollmann, "The HA-2 Sportster: An Ultralight Two Place Gyro-plane," *Sport Aviation* 24, no. 1 (January 1975): 16–20.

40. Martin Hollmann, "HA-2M Sportster Tested," p. 34; Martin Hollmann, "A New Wind Is Being Blown," *Gyroplane World,* no. 13 (October 1977): 2–3.

41. The Glanville Skymaster won the Best Two-Place Gyro Award at the July 1988 PRA National Convention. Shelly Goldsberry, "Middletown PRA Convention '88," *Rotorcraft* 26, no. 5 (October–November 1988): 18–21 (photo on p. 20); Martin Hollmann, "Glanville Aircraft," *Gyroplane World,* no. 23 (August 1978): 4.

42. Martin Hollmann wrote, via a telefaxed letter, on March 2, 2001, to the author:

> In 1978, Frank [Marchetti] turned up at Rockford with a VW Sportster type air-frame, two place gyroplane which was incapable of lifting two people adequately. [Walter] Skip Tyler and I talked to Frank for about two hours and convinced him to rebuild his airframe and use a Lycoming 0-320 engine. Skip built and sold to Frank several of the parts for his aircraft such as the control stick assembly and both Skip and I talked to Frank on the telephone numerous times to help Frank with his tech-nical problems. No mention of this help which was given free was made. Needless to say, both Skip and I are a little disappointed that no mention is made of our efforts to support Frank with his Sportster type airframe gyroplane. Frank has a nice aircraft and both Skip and I are proud to have helped Frank.

43. Martin Hollmann, "Gyroplane Designs," *Rotorcraft* 38, no. 5 (August 2000): 32–35, 33.

44. Martin Hollmann, "Sportster Imitation, Buyer Beware," *Gyroplane World* 2, no. 5 (August 1979): 4.

45. Martin Hollmann, letter to author, August 31, 2001.

46. Hollmann, letter to author.

47. Allen Tatarian, "Bumble Bee Gyroplane," *Ultralight Aircraft/Homebuilt Aircraft* (1984): 66–67.

48. William Piper, "The Piper-Marriott Autogyro," *Popular Rotorcraft Flying* 7, no. 6 (November–December 1969): 10–13. Their autogyro weighed just 230 pounds. It was still flying forty years later; "One-Man Rotary Wing Craft," *Air Progress* 16, no. 3 (June/July 1964): 77–79, 77.

49. Hollmann, *Flying the Gyroplane,* p. 105.

50. Mel Morris Jones, "Talkshop: A Conversation with Ralph Taggert," *Fly Gyro!* no. 6 (July 2001): 4–6; Ralph Taggert, "Making the Most of Part 103: The Gyrobee," *Rotorcraft* 28, no. 7 (October–November 1990): 45–49.

51. Steph[anie] Gremminger and Greg Gremminger, "Choices, Choices, Choices: 2000 Rotorcraft Directory," *Rotorcraft* 38, no. 1 (February–March

2000): 17–27; Stephanie Gremminger, "GyroTECH, Inc. Aims for the Entry Level Builder and Pilot," *Rotorcraft* 38, no. 5 (August 2000): 30–31.

52. Jack Cox, "The KB-3: An Ultralight Gyroplane," *Sport Aviation* 35, no. 6 (June 1986): 47–48.

53. Tom Milton, "A Gyroplane Movie Trivia Quiz," *Rotorcraft* 34, no. 3 (May 1996): 39; see also James Brown, "Gyrocopters: The Australian Story," *Rotorcraft* 28, no. 5 (August 1990): 20–24; Bruce H. Charnov, "A Gyroplane Film Trivia Quiz, Part 22," *Rotorcraft* 39, no. 7 (October 2001): 21–23.

54. For a photo of Jansen flying with the Brock seat-tank, see Gallager and Levy, p. 145.

55. Doug O'Connor, "Are Gyroplanes Safe?: Part 2," *Rotorcraft* 34, no. 2 (April 1996): 35–39, 37.

56. Ron Menzie, "Ice 90: The Antarctica Expedition," *Rotorcraft* 28, no. 3 (May Extra 1990): 27–29, 28; Ken Brock, "My Adventure in Antarctica," *Rotorcraft* 29, no. 4 (June–July 1991): 20–23.

57. Brock, p. 22.

58. "Gyrocopter Accident," *Gyroplane World,* no. 4 (January 1977): 5.

Chapter 15

THE EMERGING GYROPLANE FUTURE

We're a community that doesn't come from a land of megabucks, R&D budgets, computer aided design and wind tunnel testing. We come from a land where people just kick over ideas that they've written on scratch pads. Somebody makes a machine in a small shop, and when the time comes for a flight test, someone climbs in and wrings it out a little bit. I think there's a little bit of Orville and Wilbur in everything we do.

Dr. Bill Clem, world altitude record–holder,
as quoted in *Rotorcraft*

Vittorio Magni[1] was born near Milan in 1938 and by 1956 was employed by Agusta Helicopters in the transmission, engine, and airframe departments. In 1962 he joined the helicopter division of Montedison but left for Silvercraft Spa in 1967 to work on ultralight helicopter development. In the process he became a licensed pilot, and in 1967 he purchased, like so many before him, a set of Bensen plans. In 1968 Magni constructed a Gyroglider, and converted it to a modified Bensen Gyrocopter with a 1,600-cc VW engine. Modifications also included a shock-absorber mounted engine, an original rotor head, steerable nosegear, and fiberglass tail surfaces. It was the first Gyrocopter in Italy, and its improvements were also incorporated into a two-passenger dual-control Gyroglider constructed in 1972. In 1973 the Bensen was further modified with a sixty-horsepower Franklin engine, an open fiberglass cockpit pod, a more extensive instrument panel, and installation of an electrical system. Magni was incrementally and cautiously introducing improvements to the basic designs, but his next effort represented significant improvement in technology—in October 1973 Magni purchased the plans, prototype, and manufacturing rights to Jukka Tervamäki's JT-5 autogyro, renaming it the MT-5 Eligyro (MT being derived from Magni Tervamäki). Encouraged by the growing acceptance of the model, Magni created VPM (Vittorio P. Magni) in 1976, with divisions to produce composite parts for the growing aeronautical market and to develop gyroplane models.

In 1980 Magni asked Tervamäki to design a two-seat side-by-side autogyro with a powerful 150-horsepower Lycoming aircraft engine. The resulting aircraft, the MT-7, was a superaerodynamic composite enclosed-cabin autogyro with twin tail surfaces off of a horizontal cross piece and an enclosed pusher engine. It was the most technologically advanced autogyro at the time and certainly one of the most beautiful ever produced, and the sale of its production rights along with the MT-5 to the Spanish Cemenesa in 1986 brought Magni commercial success. Magni then incorporated the MT advances into his own VMP one- and-two-passenger models, which went on to worldwide commercial success. In 1996 the company was renamed MAGNI GYRO, and it went on to produce the existing models, develop new ones, and provide quality composite parts for the aeronautical market. Under the direction of Vittorio Magni, his sons Luca and Pietro, and his daughter-in-law Lisa, the company produces the finest autogyros in the world, incorporating state-of-the-art composite bodies, utilizing Austrian Rotax engines, and exemplify the highest quality of amateur and factory-built aircraft.

The excellence of design and superb quality of workmanship in the kits convinced Greg and Steph Gremminger of Ste. Genevieve, Missouri, to create Magni-USA L.L.C. in 2000 to offer the Italian models to the American market. Gremminger's M16 Tandem Trainer won the award of Best Gyroplane at the 2000 Sun 'n Fun 2000 EAA Fly-In at Lakeland, Florida, and three awards at that year's PRA convention, including Best Commercial Ship and the prestigious Person and Machine award. It was an auspicious beginning for Magni in America.

The 1980s also saw the development and introduction of Jim McCutchen's composite Wind Ryder single-place enclosed autogyro. McCutchen, of Broomfield, Colorado, had achieved a notable niche with the formation of McCutchen SkyWheels Corporation in 1984. At a time when American rotor blades were made of wood, McCutchen had pioneered in the introduction of composite construction of rotor blades, successfully introducing fiberglass-reinforced vinyl ester plastic molded around an aluminum spar, which produced blades that were flexible but robust, could be molded in a variety of efficient shapes, and were exceptionally strong.[2] McCutchen then joined with longtime friend Kurt Shaw to create Wind Ryder Engineering in 1985 for "the purpose of developing and manufacturing the finest autogyros,"[3] and the two friends succeeded with the production of the Wind Ryder Hurricane, which was first offered as a kit in 1989. The kit contained all parts but the paint and required only 150 hours to complete, as all the necessary machining and finishing of the metal frame and composite fuselage and fairing had been done at the factory. It was a streamlined, enclosed-cabin craft with a rotor pylon built around a central keel with the landing wheels on either side to the rear of the cabin enclosed in fiberglass "pants" and a horizontal composite tail with three vertical vanes to the rear of the faired Rotax 532 engine.[4] The aircraft weighed only 450 pounds the result of being 99 percent composite construction, and could be licensed as an "experimental" aircraft, as the FAA deemed that assembly constituted at least 51 percent of its construction. The gyroplane achieved a top speed of 115 mph and comfortably cruised at 90 mph, with only a student pilot's license required. It was, however, at $13,500 (increased in 1972 to $17,000), the most expensive single-place machine, and its cost contributed to its end.[5] Also contributing to the demise of the Wind Ryder was a fatal accident in 1992 in Arkansas when Harry Cordon, an experienced gyroplane pilot and CFI, plunged to the ground in a friend's Wind Ryder while flying at 400 feet.[6] Additionally, McCutchen discovered in February 1992 that a good friend, to whom he had entrusted the manufacturing of rotor blades while he was involved with the Wind Ryder, had been embezzling money for a long time by accepting funds but not filling orders. McCutchen wrote to all PRA members informing them of the scam and pledged to make good on all unfilled orders. The end result was, however, that the Wind Ryder, having been a successful technological achievement, failed to achieve a comparable market success.

Although the composite entry of Richard Bentley of Cottonwood, Arizona, in 1991 was even more stunning than the Wind Ryder, it achieved less success and, as it never entered production, has been all-but-forgotten. The Richard Bentley–designed Mängoos Stealth Gyroplane was a two-passenger cabin gyroplane with a dark black Stilts Poly Fiber covering attached to a chromium-molybdenum alloy steel frame. It was a unique design, and the tooling for producing kits had been constructed at the same time as the prototype. The fully enclosed cockpit, a tandem arrangement that allowed for both two passengers and an extremely narrow, aerodynamic compound fuselage, featured opaque tinted Lexan windshields and side windows, which helped the all-black aircraft achieve "the sophisticated looks of a high tech military gunship."[7] The aircraft featured a pair of short rear wings sweeping back from the fuselage in back of the landing gear, one on top of the other but attached to common vertical fins on either side, allowing the claim that the model had "the most horizontal surface of any gyroplane."[8] Bentley claimed a top speed in excess of 100 mph, a cruising speed of 80 mph, and outstanding performance, and the model had received a favorable reception at the 28th International PRA Fly-In/Convention held at Hearne, Texas, July 20–23, 1990, where Bentley had won the award for Outstanding New Design.

The Mängoos subsequently appeared on the cover of the April 1991 issue of *Rotorcraft,* but Bentley, making a belated decision to take the prototype to Bensen Days in Florida, arrived too late. Being in Florida, however, he took the prototype to Sun 'n Fun, the EAA Fly-In at Lakeland, but he left early after being informed that he would not be allowed to fly.[9] He then took the prototype to the 1991 PRA annual convention during July 19–22 in Brookville, Ohio, where it did not achieve the acclaim of a year earlier. Editor Paul Bergen Abbott wrote that an "unfinished example of Rich Bentley's Mängoos, better known as the 'Stealth Gyro,' is said to hold two people, although nobody can see into that enclosed black cockpit to count the occupants."[10]

Bentley subsequently perished along with a passenger in the summer of 1991 when the Mängoos crashed while he was giving demonstration flights in Las Vegas. The accident investigation found that the motor mounts had loosened and the engine swung around, then the prop fatally injured both occupants.[11] Thus the Mängoos passed from view—no one ever pursued its development, although it is rumored that a second aircraft had been built and subsequently placed in storage—but it did achieve a curious celebrity, as it was featured in the 1992 Andy Sidaris movie *Hard Hunted.* In the movie the Mängoos is flown by one of the villains, and it is inspired by *You Only Live Twice,* complete with machine guns and rockets. The Mängoos flying scenes are quite well done, and, had this movie received wider distribution, it would have impressed many. However, the movie, starring *Playboy* Playmates, was never widely seen, as a legal dispute apparently kept it from all but the video rental market. The film is dedicated to Bentley's memory, and the stunning flying scenes are all that remain of the Mängoos.

The gyroplane performances in the 1988 *New Adventures of Pippi Longstocking* and the Disney 1991 feature *The Rocketeer* were less impressive. In the former, Steve Pitcairn flew *Miss Champion* to rescue Pippi and her friends. The PCA-2 also furnished the model in the *Rocketeer,* which was partially filmed at the Ken Brock hangar at El Mirage, California. In this film the hero and his girlfriend are rescued from a burning Nazi blimp by an Autogiro flown by the legendary aviator/industrialist /filmmaker Howard Hughes. In the 1991 *"Rocketeer" Official Movie Souvenir* magazine, which presents original storyboards, it is clear that the Autogiro was modeled on *Miss Champion.* However, in the 1989 *"Rocketeer" Adventure* magazine, the Autogiro is clearly labeled *Missing Link,* the PCA-2 (NC10781) flown by Johnny Miller.

The 1980–90s also saw the introduction of new trends in America. George D. Pate III's Pate Autogyro,[12] Ron Herron's outstanding series of Little Wing tractor-autogyros,[13] and John VanVoorhees's Pitbull continued a tradition of almost seven decades and found devoted followings, but perhaps the most dramatic "tractor" configuration was that created by Afro-American artist-designer David Gittens, who would design and build the Ikenga, a strikingly original, award-winning gyroplane.[14]

Gittens, born in Brooklyn, earned a New York Institute of Visual Art degree in communications, advertising, and graphics.[15] Relocating to Europe in the mid 1960s, after years as staff photographer with *Car & Driver* magazine, he began shooting for *Vogue and Harper's Bazaar.* It was while on a freelance assignment for *Queen* magazine that he met Ken Wallis while using Wallis's Gyrocopter as visual background a fashion shoot. This meeting fired Gitten's interest; he subsequently received a commission in 1966 to develop the concept/styling of a four-place autogyro being developed in Blackpool and began research on the historical background of the Autogiro. Returning to America in 1979, he received a grant to develop a unique gyroplane, but it was six years later, while residing in Santa Fe, New Mexico, that he began designing and formed a firm funded by Canadian investors to build a flying prototype called Wind Dancer. The completed vehicle was displayed at the 1986 EAA AirVenture Fly-In, to great interest. The "fuselage was long and racy looking, with what looked like a composite sailplane nose section that had been married to a pusher gyroplane with triple vertical tails...a tinted sailplane canopy gave it an artistic touché of real class."[16]

The Wind Dancer looked stunning, and its Mazda engine promised great thrust but was not cost effective for the target kit-builder market; so it was back to the drawing board in 1987. While Gittens found the PRA to be a conservative, Bensen-oriented, mostly white organization, he later stated that "it was also a gift that a few members of the PRA stepped out of the

David Gittens standing next to his award-winning Ikenga 530Z Gyroplane, which is currently in the collection of the National Air and Space Museum of the Smithsonian Institution. *Ikenga,* meaning "man's creative life force" in the Ibo language of eastern Nigeria, had first been used on his GT automobile designs of the mid-1960s in London. The Ikenga 530Z received an FAA Airworthiness Certificate at Santa Fe, New Mexico on March 29, 1988, and although Gittens had taken flying lessons with Bill Parsons, the craft was first flown by experienced pilot Mark Hallett. (Courtesy of David Gittens.)

norm, were my mentors and supported the building of the very unique Ikenga. Jim Eich, Jerrie Barnett, Bill Parsons and Martin Hollmann gave heartfelt support to this project and Martin's *Gyroplane Design* book was the foundation piece of technical information that paved the way."[17] One of these members was "Helicopter Ed" Alderfer, who had constructed a two-place autogyro called the Gyrochopper III, "a tiny functional medical tractor gyro for missionary work in the Far East."[18] Gittens, remembering his study of the early Autogiros, turned to their design in 1987, intending to produce an aircraft both for the local kit builder and for use by those in less-developed countries, through his newly created company Gyro 2000, which was funded by a group of Santa Fe businessmen. He named his revolutionary design the Ikenga, meaning "man's creative life force" in the Ibo language of eastern Nigeria, a name first used on his GT automobile designs of the mid-1960s in London. In an interview printed in the *Dallas Morning News* on February 26, 1989, Gittens stated, "The name [Ikenga] resonates with what I'm trying to do, to use my creativity on the planet…It also reflects the creative potential of the Black community; there's been a great lacking there."

The Ikenga 530Z received an FAA Airworthiness Certificate at Santa Fe on March 29, 1988, and although Gittens had taken flying lessons with Bill Parsons, the gyroplane was first flown by experienced pilot Mark Hallett. Gittens's goal had been to produce a rugged design from available components and that would cost less than $10,000, be readily expandable to a planned two-place configuration for training, was easy to fly, and could provide cheap transportation for people in poor countries. What he had produced was different from any other design—the keel and cross member to which the landing wheels were attached appeared Bensen-like, but that is where the similarity ended. The seat, which was also the fuel tank, was formed of foam-filled molded polypropelene, and the pilot straddled it like a motorcycle seat. The mast consisted of an inverted *U,* an innovation of chromoly tubing that served as a roll bar, stiffened and strengthened the autogyro frame, and protected the pilot and on which it was topped by cheek plates that secured the Parsons rotor head. For control, a horizontal bar was attached to the rotor head, which had push rods descending on both sides of the fuselage to a lower horizontal control bar, on which the control stick was mounted. The ninety-horsepower Suzuki 5300Z engine had a ground-adjustable three-blade propeller and control was enhanced by twin all-flying composite flying rudders. The Ikenga had a top speed of 120 mph, could cruise at 90 mph, had a 350-mile range on fifteen gallons of gasoline, and had a ceiling of 14,500 feet—it was an excellent performer and wowed the gyroplane community in 1988.

It received the Grand Champion award at the Albuquerque International Air Show, the Best New Rotorcraft Idea award at the PRA Annual Convention/Fly-In, and was honored as the Reserve Grand Champion at the EAA Convention/Fly-In at Oshkosh, Wisconsin. This critical acclaim, however, was not matched by economic success, and a lack of further development capital dissolved the production hopes of Gittens and the original investors. The Ikenga 530Z was accepted into the National Air and Space Museum of the Smithsonian Institution on June 8, 1994. It remains the only aircraft designed by an Afro-American in that museum, and as such, and also distinguished by its lyric lines embodying an artistic vision of flight, it occupies a unique and treasured place in American aviation history.

Gittens then turned to a design better adapted to cost-effective manufacture, and the Ikenga Cygnus 21 series was developed as a more conservative/user-friendly aircraft for kit production in 1988. Utilizing the basic Ikenga/Bensen keel-and-cross-member structure, Gittens now utilized off-the-shelf components that reduced production costs. Unlike the inverted *U* masts in the Ikenga, which went from side-to-side, the Parsons masts were now attached to the keel in front and back of the seat with the Parson's rotor directly above the pilot's head, with a metal fuselage frame that could be bolted together. The pilot sat on the 15.5 gallon seat-tank just behind a semicircular, dark-tinted windshield. In front of the windshield the aircraft had a downward-sloping composite fairing, and it was powered by a tractor engine with an eye-catching five-bladed prop[19] mounted on top of the fairing.

Gittens claimed the weight was only 246 pounds, obviously intending it for the ultralight market, and it had an anticipated kit price of $10,000. The Cygnus 21 was also named Grand Champion Rotorcraft at the 1989 Albuquerque International Air Show, but also like the Ikenga, it failed to achieve commercial success as, the tractor configuration did not achieve acceptance. After the 1989 air show the Ikenga Cygnus 21 was redesigned with an all-new composite rudder and stabilizer and designated the Cygnus 21T. The Cygnus 21T offered as an option an SWS Intelligence Gathering Platform, video downlink equipment for attachment of a camera for law-enforcement use, and the MicroAg crop-spraying system, but it too would fail to find a market.

In late 1988 and 1989 Gittens and his investors were negotiating with a number of Japanese companies interested in producing a gyroplane for the Japanese recreational market. Mitsui and Sumitomo were the most competitive for an Ikenga to appear with their corporate name at the Sky Sport Japan Air Show in November of 1989. Sumitomo offered a more favorable business relationship and financial option, and a contract was made between the parties. Sumitomo was interested in exploring sports recreation, with a special interest in using Sumitomo real estate as a basis for flying clubs. Baca Development, an ulralight aircraft company in Albuquerque, was contracted to fabricate the frame and control components for the two Japanese gyroplane kits, designated Ikenga Cygnus 21P, a "pusher" configuration readily adaptable to the Bensen/Brock-style Gyrocopter/gyroplane.

In mid-September 1989 the Sumitomo representatives arrived to finalize and accept the unassembled kits. The day the Japanese signed off on the contract and returned to Japan to await arrival of the kits, Gittens found himself in intensive care, having suffered what doctors interpreted as a heart attack. He remained in the hospital for four days, suffering from extreme stress and exhaustion, and the resulting convalescence made it impossible for the two kits to be assembled, test-flown, and disassembled in time for shipment to Japan. Colleagues then arranged for all Cygnus 21P components to be shipped to Richard Bentley in Cottonwood, Arizona, during Gittens's recovery. Bentley, with the help of Dennis Renner, prepared the models for shipment to Japan. But complications in Arizona resulted in the model that was exhibited in the Sky Sport Japan Air Show being different from Gittens's original design, and the project faltered.

The Ikenga Cygnus 21TX was the final gyroplane in this series, incorporating many of the parts salvaged from the destroyed Cygnus 21T. It looked similar but featured a redesigned composite stabilizer and all-flying rudder. It was built with First Arrival Trauma Pack for responding to medical emergencies and, having received the Grand Champion Rotorcraft award at the 1990 Albuquerque International Air Show, was featured in the January 1991 issue of *Popular Science*.[20] It was not to be, however, and Gyro 2000 closed its doors on February 29, 1992. In a letter to Paul Bergen Abbot, then *Rotorcraft* editor and publisher, Gittens stated, "As an artist, it has been a great blessing to have had the gyroplane as my canvas,"[21] and the words of Rabindranath Tagore that he selected to include on his Smithsonian donation also applied to his gyro venture: "While leaving no trace of wings in the air I am glad I have had my flight." Gittens had virtually no impact on the commercial course of the American rotorcraft movement, yet his models still stand out and may yet, seen by museum-goers, inspire others to dream and design.

Although David Gittens was the most artistic of designers, Air Command's Dennis Fetters[22] stands out as perhaps the most enigmatic. He had, through trial and error, finally succeeded in mounting an Austrian Rotax 447 engine on a Bensen frame—an accomplishment that had also been achieved by Larry Ramal, who was equally credited by Paul Bergen Abbott with adapting the Rotax engine to the gyroplane[23]—and introduced an entirely new aircraft, the Air Command 447, into the kit-built market in 1984. It was not the first ultralight, nor was it the first kit, as Hollmann was offering Bumble Bee kits in 1983; but the Air Command 447 was the first widely distributed ultralight kit. It was innovative, with weight-saving use of composite plastic components in the rudder, propeller, and rotor blades; shock-mounted landing gear made necessary by the lightness of the round-tubing frame; and a refined quality finish for the kit parts.[24]

Demand was immediately forthcoming, as the kit was a truly complete but unassembled aircraft, not merely a Bensen-type collection of parts, and Fetters established a factory to meet the large number of resulting orders. An early fatality of Air Command 447 pilot, Gary Gibler in Liberty, Missouri, led Fetters to build a two-place trainer, modifying the Air Command 447 by placing side-by-side seats and installing a more powerful, sixty-five-horsepower Rotax 532 water-cooled engine. Subsequent models followed, and composite body pods and aerodynamic fairings were introduced even as the heavier and more powerful Air Command models became best-sellers in the kit-built market. Fetters was also market-savvy in building in an upgrade ability so that the owner of an older model could retrofit newer engines, enclosures, and even a two-place seat.[25] The company and Fetters were riding high,[26] but it was not to last.

By August ownership of Air Command had passed to a new company, called Venture Industries, which had purchased it from the government after seizure for failure to pay employee withholding. Fetters had fallen on hard times after defaulting a lawsuit, resulting in a settlement of $500,000.[27] But he had a great deal of his ego invested in the ultralight Air Command 447, and to the amazement of some, and surprise of a few, the president of the new company was none other than—Dennis Fetters! The Air Command was his baby, and he was not about to lose it if it could be kept going, by whatever possible legal maneuvers. But the troubles facing Air Command did not go away; by early fall of 1990 it had been sold to Mark and Jeff Pearson of Daytona Beach, Florida, but the brothers were not able to make a go of the enterprise, given that there were thousands of dollars' worth of unfilled orders with deposited funds unaccounted for. In early May of 1992, as announced in *Rotorcraft*,[28] the Air Command operation was acquired by R & D Aeronautical Engineering, Inc. of Plano, Texas.

R & D was headed by Harold F. "Red" Smith, an aeronautical engineer, who stated that the company would honor all outstanding orders, restoring confidence in the company. Smith proceeded to make some improvements to the Air Command design and then to do what marks him as one of the most honorable and far-sighted gyroplane manufacturers in the world— Air Command issued an Emergency Bulletin that informed owners of the changes necessary to insure safer operations. It was an unprecedented move and, garnering rave reviews, has continued to mark Air Command as an industry leader.

Smith's acquisition was a daring act, as the Air Command, introduced into England in 1988 in one- and two-seat configurations, had subsequently suffered a number of accidents, leading the British Civil Aeronautics Authority (CAA) in 1991 to suspend the Air Command models' Permits-to-Fly and to ground the fleet. That suspension would last over ten

Jim Vanek flying a sportcopter Vortex. (Courtesy of Jim Vanek.)

years and lead to a University of Glasgow study of gyroplane safety headed by Dr. Stewart S. Houston of the department of aerospace engineering.[29] A solution was proposed, related to changing the configuration to a more stable format, and an upgraded machine entered flight-testing in December 2001. That machine featured a raised seat, based on a kit introduced by Air Command and originally developed by Larry Neal, who gave the rights to the company in exchange for a new engine, which transformed the aircraft into a center thrust configuration originally developed by Chuck Beaty and Ernie Boyette.

Dennis Fetters then turned his attention to the development of the Revolution Helicopter Corporation, Inc. (RHCI) Mini-500 Rotax-powered kit-built helicopter. Subsequently, Dennis and his wife, Laura González Fetters, who had served as RHCI's operations manager, disappeared in 2000, owing a "substantial loan to the Small Business Administration."[30] RHCI, after all, had been located in Excelsior Springs, Missouri, on W. Jesse James Road!

Gyrocopter pioneer Chuck Vanek's company, Vancraft, changed its name in 1991 to Sport Copters and established a state-of-the art production and flight training facility at Scappoose, Oregon, and marketed both ultralight and experimental kits and gyroplanes. His son, Jim Vanek, who assumed leadership of the company in 1988, dedicated himself to the development of the "elite of single-place gyroplanes."[31] Known for his commitment to perfection, Jim established a reputation as one of the top pilots and innovators in the industry. With his wife and business partner Kelly (also a gyroplane pilot), Jim has established an enviable international business reputation for producing and supporting an outstanding aviation product. In 1997 he achieved everlasting fame as the first pilot to loop a gyrocopter, the Sport Copter Vortex.

As a result of his experiences and demonstrated competencies, Vanek holds a membership in the ICAS (International Council of Air Shows) and the FAA Statement of Acrobatic Competency. It is a singular achievement that places him with the late Ken Brock among the greatest of gyroplane pilots.

At the same time RAF in Canada emerged as an international source of larger, enclosed "experimental" two-place all-year gyroplanes and quickly carved out a share of that market. RAF of Canada also established an international network of CFIs to insure quality flight training, even requiring at least ten hours of flight instruction as a condition of purchase.

Significant contributions were also made on an ongoing basis by two men renowned in the American and, increasingly, the worldwide rotorcraft movement, longtime Florida collaborators Chuck Beaty and Ernie Boyette. Boyette is most well-known for the creation of Dragon Wing rotor blades and the Dominator gyrocopter, both produced by his company Rotor Flight Dynamics, Inc. (RFDI). As many, he began with a Bensen kit but soon went his own way.[32] After experiencing numerous engine shutdowns with his Bensen McCulloch engine, Boyette mounted a heavier, more reliable Continental engine but found that he needed more efficient rotor blades than were then available. Consulting with Beaty as to rotor blade construction, he built his first composite rotor blades out of foam, wood, steel, and fiberglass. The blades performed well, and his move to Tampa, Florida, in the mid-1980s led to a serious collaboration with Beaty, the result being the independent creation of an effective airfoil, although it also started a friendly competition with Beaty in blade design. Each set about the task of besting the other, Boyette by pragmatic experimentation with different blade shapes and Beaty by "number crunching." Boyette carved his blades out of aluminum and took two of his blades to the 1990 Bensen Days, where Carl Schneider and Jim Smith prepaid for blades, and as word spread, Boyette received four additional deposits. He now had to deliver.

After consulting with Beaty, Boyette constructed a building, secured the necessary extrusion die to create the blades, and acquired a suitable aerospace glue to bond the composite to the extruded aluminum spar, the "spine" of the blade. He called his product Dragon Wings, and they have achieved great success; he also sells kits and Dominator gyrocopters, for which he is equally known. The Dominator began as an experimental autogyro called Big Bird.[33] It received that name when Boyette raised the seat so that its top was almost level with the engine, with the thrust line passing horizontally through the engine and the pilot. Boyette and Beaty had sensed that when the pilot sat below the engine thrust line, as in a Bensen/Brock design, the center of gravity was always below the line of thrust, which threatened to push the aircraft over while in flight. But when the center of gravity was in line with the center of thrust, greater stability was achieved.

Boyette now sought to incorporate this stability into the Dominator, as it was common knowledge that a pilot's reactions in controlling the gyroplane, delayed by the control system that tilted the rotor head, could accentuate aircraft oscillations in increasing ups and downs, a condition known as PIO (pilot induced oscillation), or porpoising, and when this caused the aircraft to flip or "bunt" over, a condition known as the power pushover, the results were always fatal. The Boyette Dominator reduced the possibility of a power pushover by placing the center of gravity in line or slightly above the center of thrust by raising the pilot—the product was an ungainly aircraft that some considered downright ugly, but it flew well and attracted a devoted international following. Carl Schneider, with his High Seater, also constructed a centerline-thrust machine and experimented with various configurations, helping to refine the design.[34] Subsequent analysis by the French aeronautical engineer Jean Fourcade has borne out Boyette's contention[35] that the Dominator is a stable aircraft, and it is surely no accident that Fourcade himself flies a Dominator. By 2001 the Dominator kit was being manufactured in conjunction with Boyette's RFDI by Modus Verticraft Inc. (MVI) in Taiwan.[36]

Ernie Boyette married Connie Watterson on September 2, 1989, while flying at 300 feet above the airport at Clewiston, Florida, in the prototype Big Bird gyrocopter—the world's first such wedding. He was heard to declare: "To all my friends on the ground and to the Great Spirit in the sky who gave man the wisdom to fly, let it be known that on this day, I, Ernie Boyette, and Connie Watterson were wed forever more—to fly the skies as one until our life is done."[37] That aircraft, which had first flown in March 1988, led to the claim by Chuck Beaty of having invented the "tall tail" and that claim provoked a firestorm of adverse reaction. The tall tail was a vertical symmetrical airfoil surface placed directly behind the engine, which spanned the diameter of the prop and improved control by reducing the torque roll and yaw.

In 1988 Paul Bergen Abbott stated, "Tall tails were originally designed by Chuck Beaty, who also contributed to the final development of Big Bird,"[38] and Beaty himself then claimed the invention in October of that year.[39] But it was not so, and although Beaty, the longtime technical editor of the PRA publications and guru to several generations of gyro pilots, had made an undeniable contribution to the rotorcraft movement with his numerous technical analyses and autogyro configurations, he did not invent the tall tail, and famed designer Jerrie Barnett's wife Erlene let him know it in the December issue of *Rotorcraft!* She wrote, "Tall tails a 'new thing'? I think not. Since Jerrie Barnett first started flying and designing his own gyroplanes in 1962 they all had 'tall tails'...Let's get the facts straight, gang, before we start giving credit where credit is not due."[40] And, in reply, Beaty backed down, giving credit to Barnett for the "full span tail," attempting to still somehow stake out the tall tail, but he was taken to task a month later by no less than Wing Commander Ken Wallis, who pointed out that he had flown the WA-117 on March 24, 1965, with a "very tall all-moving tail, supported at the top and bottom,"[41] a claim to which Beaty did not reply. And indeed, Sport Copter advances the credible claim, complete with dated photographs, that Chuck Vanek was flying with a tall tail as early as 1959![42]

There is little doubt, however, of the contributions of Boyette and Beaty to the rotorcraft movement. As summed up by Russ King in 1998, Dominator purchaser and pilot, "Ernie possesses a rare combination of talent—the abilities to design, to build, and to fly. The existence of his machine and the fact that he's still alive after all of his research and development says it all. He is also a man of integrity."[43] Although it is far more difficult to quantify Beaty's impact, few would deny that he is perhaps the premier technical writer in the American gyroplane community, and in his capacity as technical editor of *Rotorcraft,* he has been able to introduce complex aeronautical concepts and applications to thousands of readers around the world with a sensibility and clarity unavailable from virtually any other source. Together, these two longtime friends have helped to shape the future of Bensen's vision and helped it down roads that the founder never anticipated.

The PRA, under the direction of Gary Goldsberry, who had become president in 1989, created a separate entity called PRA Mentone that purchased a local airport in Mentone, Indiana, in 1995 for $151,403.22, of which $51,100 was donated by the PRA for the down payment. Although

Rick Marshall flying Snobird at the PRA Fly-In, Mentone, Indiana 2000. (Courtesy of Stu Fields.)

the PRA board, which had made the decision to purchase the airport largely without consulting the membership, had been criticized,[44] PRA members made, and continue to make, individual contributions toward support of PRA Mentone.

In September 1995 LeRoy Hardee advertised an auction of donated used "aviation items, rotor blades, props, all types of engines, airframes, instruments, parts, wheels, parts, parts, parts!" Hardee advised purchasers to "[b]ring cash because we do not take American Express, Master Card, Visa or Discover." The stated purpose was to "aid in the purchase of the new Popular Rotorcraft Association home airport. If you have ever wanted to give, now is the time. We are accepting anything aviation related from helmets to complete aircraft."[45]

PRA Mentone pays for the airport from a variety of sources, including monthly rent from the PRA for its office space, hangar rents, sale of aviation gas, and shares of the profits from grain or hay sales that a local farmer grows on the unused part of the airport property. This has served as the focus of PRA activity, its national offices, the nearby Archimedes Rotorcraft Museum, and in 1996 it saw the first national Fly-In/PRA Convention at Mentone. That convention, now held at various other locations in addition to Mentone, has emerged as the most important gathering of the gyroplane community in America and is the focus of this international industry, along with Wallis Days in England (created in 1997 by noted English *Rotor Gazette International* editor Rowland Parsons), Bensen Days in Florida, and Magni's Day in Italy.

The gyroplane movement had truly gone international by the end of the century with PRA-type national associations in Australia, Britain, France, New Zealand, and South Africa and the publication of the long-standing PRA's *Rotorcraft* (formerly *Popular Rotorcraft Flying*) (1963–), Don Parham's *Homebuilt Rotorcraft* (1989–), Rowland Parsons' *Rotor Gazette International* (1992–2002), Ron Bartlett's *Autogyro 1/4ly* (1999–), and Mel Morris Jones's *FlyGyro!* (2000–2002). Jones quickly gained an international readership as he published a far-reaching mix of technical and historical articles. He had also been successful in getting the British aeronautical authorities to change the medical certification requirements from a helicopter pilot–like full physical to a self-certification for gyroplane pilots, allowing a large number of pilots (including Ken Wallis, then in his eighties) to fly. Although his publication ceased in 2002 due to Jones's medical condition, it had gained a devoted readership and was distinguished by its quality and photographs.

The 1995 PRA Convention at Greencastle, Indiana, and 1998 Bensen Days at Wachula, Florida, saw significant technological achievements that, while having virtually no economic impact on the industry, fired the imaginations of observers and served notice that innovation was thriving. In 1993 the PRA had announced a competition for an Autogyro Performance Award, initially suggested by Jay Carter Sr., for a homebuilt autogyro that could accomplish a jump takeoff and win a race. No homebuilt aircraft had accomplished this feat, and Carter offered $5,000 to fund the prize, later increased to $20,000 by the PRA, the Hurst Foundation, Dan Haseloh of RAF, Red Smith of Air Command International, Duane Hunn, and longtime PRA member and *Rotorcraft* contributor Art Evans.[46] No one successfully accomplished the required jump takeoff in 1994, but two pilots did the following year: Dick DeGraw and Johnny Hay.

DeGraw, an engineer and helicopter pilot from Michigan, constructed a one-of-a-kind aircraft, and it stunned all who saw it as perhaps the most unusual autogyro to ever fly. He had encased a rebuilt EA-82 eighty-five-horsepower Subaru engine within a metal tube framework topped by a rotor hub with three Dragon Wing rotor blades, and a Warp Drive propeller—all encased in a clear Lexan polycarbonate skin, which allowed for all parts to be inspected and provided a sleek aerodynamic shape. DeGraw's wife, Karol, had named the autogyro after the picture of a rhinoceros on the paper that initially covered the plastic skin—she called it the GyRhino. DeGraw, in the transparent cockpit, prespun his rotor to 470 rpm at zero pitch, snapped the blades into positive angle by means of a collective pitch control, and leaped into the air, easily clearing the ten-foot barrier. He had, by means of a power splitter, diverted 10 percent of the engine's power to the rotor, which allowed him to fly with a flatter pitch, resulting in less drag, and he completed the required course in 15 minutes, 50.32 seconds, and then waited for Johnny Hay to make his run.

Hay had modified a Bensen-type gyrocopter with an aerodynamic composite Air Command body powered by a turbocharged Volkswagen engine. He employed a specially designed rotor head that allowed him, by means of a prerotator utilizing a cogbelt system, to prespin the Hughes two-blade rotor prior to takeoff. When the rotor had achieved 440+ rpm at a zero pitch, Hay could snap the blades into a positive seven-degree pitch, causing the little craft to jump six to seven feet before commencing forward flight. He cleared the ten-foot barrier placed fifty feet from the takeoff point and raced to the nearby airport, returning in 15 minutes, and 56.32 seconds, an excellent performance for what was essentially an off-the-shelf homebuilt aircraft, but it was not enough to win the $20,000 Autogyro Performance Award—DeGraw had won, but his GyRhino never entered production and remained a unique attraction at national fly-ins.

The other achievement took place at the 1998 Bensen Days, where Denver, Colorado, physician Bill Clem established a new autogyro altitude record of an astounding 23,438 feet (4.43 miles) in a modified Boyette Dominator (N36MR).[47] Clem, board certified in emergency medicine and hyperbaric medicine, had attempted the feat on Thursday, April 16, 1998,[48] with an ascent to 22,500 feet, only to discover that the battery of the video camera made available by the National Aeronautic Association (NAA) to record the achievement had frozen at 15,000 feet. Clem then made an eight-hour round trip to Gainesville, Florida, to secure a recording barograph, a device approved by the Fédération Aéronautique Internationale to certify the flight, which was mounted in his Dominator Friday morning. He suited up and, although tired from the drive, took off for a new record. He rose steadily and seventy minutes after taking off radioed the ground that he had reached 23,400 feet, at which point he began the eighty-minute descent. He landed completely exhausted, with evidence of frostbite on his exposed cheeks, but the computer display showing the recording device data proved he had reached 24,438 feet—a new world's altitude record.

Of equal organizational significance, in 1998 the PRA was a strong, vibrant organization with over five thousand registered members, triple what it had been ten years before, and more than fourteen times the first year's membership. New models were appearing yearly, and the home-built market, often unorganized and seemingly chaotic, was strong in its diversity and characterized by opportunity ranging from retro tractor models to handcrafted jump takeoff machines, and, with the application of new materials technology, the introduction of New Zealand's Mac Gillespie's futuristic U.F.O. Helithruster, with its alloy frame, fiberglass body, and gray-tinted double bubble wrap-around windscreen, looking like something out of a science fiction movie.[49] By the end of the twentieth century it was obvious that the gyroplane industry had, far from disappearing, grown into a multifaceted, truly international enterprise. That which had begun in 1923 in Cierva's Spain was about to enter into the new millennium with the creations of Groen Brothers Aviation (particularly the Hawk 4 series), with a revived opportunity for international commercial success, and with perhaps the most innovative new development since the flight of the C.4, the CarterCopter (CC).

GBA, headed by brothers David and Jay Groen, a Salt Lake, Utah–based company,[50] has developed a family of larger Hawk 4 gyroplanes targeted to the agricultural market (e.g., crop spraying, field inspection), law enforcement, package delivery, and passenger shuttle service. The Hawk 4 story really begins, like so many others, when a fifteen-year-old David Groen saw a Bensen Gyrocopter ad in a 1966 *Popular Mechanics* and was hooked,[51] but failing to convince his high school shop teacher to let him build a Bensen as a class project (the teacher was afraid Groen would get killed in the flying lawn chair), the dream of flight had to wait until he graduated high school, joined the army, and became a helicopter pilot, flying combat missions in Vietnam in 1969. The helicopter experience, eventually thousands of hours, featured a great deal of experience practicing autorotation, what pilots do when they have an emergency, and led him back to the gyroplane and a March 1986 partnership with his brother Jay. H. Jay Groen, a successful businessman, had extensive experience in the Far East, particularly the People's Republic of China, as an economist for the Central Intelligence Agency (CIA) and was a proficient Chinese linguist. The brothers had coauthored *Huey,* a best-selling novel dealing with the Vietnam War, and now were about to collaborate to "build the world's greatest homebuilt gyroplane."[52] The company produced and flew proof-of-concept aircraft, technology demonstrators, and production prototypes in its developmental program. It intended from the beginning to produce an FAA-certified aircraft,[53] but of greater importance, it had designed a sophisticated and effective business infrastructure to support the commercial venture. Jay Groen stated in 2000 that "we spent a lot of time just designing how to be successful at the business of producing these aircraft long before we ever got the right aircraft."[54]

Making its first public appearance at the 2000 EAA Fun 'n Sun Fly-In in April 2000, the Hawk 4 four-passenger gyroplane produced great interest, but it had also flown for potential customers privately, and the company had already received cash deposits for 160 aircraft. Even though the aircraft was priced at a significant $295,000, its performance (useful load of 960 pounds, maximum speed at 12,000 feet of 150 mph, cruising speed of 130 mph, cruise range of 420 miles) and anticipated direct operating expenses of $80 per hour (with an additional $20 for insurance and $36 for aircraft amortization) were only a fraction of those of a helicopter, with the added safety of a less mechanical complexity on an aircraft that could not stall.

The Hawk 4 was augmented by the Hawk 4T, a Rolls-Royce-powered turbine version, which initially flew on July 12, 2000. It was the first turbine-powered gyroplane and, selling at $749,000, the most expensive GBA gyroplane; yet, when compared to helicopters of comparable capacity, it was readily received by the marketplace. Market response was enthusiastic and GBA decided to shift its focus from the piston-engine-powered Hawk 4 to the 4T. Anticipating certification in the last quarter of 2001 and commencement of deliveries in 2002, GBA announced in August 2001 that a production facility would be built in Phoenix, Arizona, representing an optimism in America unknown since the days of Harold Pitcairn's PCA-2.[55] By 2001, China had ordered 200 Hawk 4 series gyroplanes and taken options on another 300, stating that it planned to use them for 100- to 200-mile air taxi operations, and *Time* magazine, in its November 19, 2001, issue, named the Hawk 4 as one of the best Inventions of the Year.

The September 11, 2001, terrorist attack on America, in which civilian airliners were hijacked and crashed by religious fanatics into New York City's World Trade Center and the Pentagon in Washington, D.C., killing thousands of innocent people, led directly to severe restrictions in civilian aviation as the result of a heightened concern for national security. This made it difficult for the company to obtain additional developmental funding in the OTC (Over-the-Counter) Bulletin Board stock market, and GBA announced in early 2002 that it was slowing the laborious and costly FAA certification process, consolidating its production facilities in Buckeye,

Arizona, and temporarily laying off personnel whose efforts were directed toward the civilian market in an attempt to conserve resources until the capital markets were more favorable to aviation companies. GBA had decided to concentrate on the military and law-enforcement markets, areas that do not require certification and for which the company maintained the Hawk 4 was well-suited. The Utah Olympic Public Safety Command (UOPSC) made use of a Hawk 4 during the 2002 Olympics for observation and to provide security surrounding the Salt Lake International Airport. Cierva, Pitcairn, Kellett, the French, the English, the Germans, the Russians, the Japanese, and even Ken Wallis had attempted to convince the military as to the suitability of gyroplane technology for aerial reconnaissance—now GBA succeeded in equipping its aircraft with a Flir Systems, Inc. day/night observation system, a Spectrolab Inc. SX-5 search light, an Avalex Technologies flat panel display, a Broadcast Microwave Services realtime video downlink system, and a law-enforcement communications radio stack. James Bond would've been proud!

Additionally, in a return to the earliest tractor Autogiros, GBA exhibited a converted Cessna C337 Skymaster as the Hawk 6 at the 2001 EAA AirVenture.[56] The fixed wings had been replaced by short stub wings, a twin-boom tail was added, a front-mounted Rolls-Royce 250-B17F2 engine was utilized, and outward-opening clamshell doors had been fitted to the rear of the aircraft to allow for cargo loading. The aircraft made use of the Hawk 4 rotor system added to the top of the cabin. The prototype was destined for a Russian company that intended to use it for oil pipeline maintenance, where it could operate more economically than a comparable helicopter, carry personnel and equipment, and land on dirt roads along a pipeline. GBA had also received a letter of interest from an overnight package-delivery company. The model was dubbed the Hawk 6G (the *G* for "government use"), as GBA intended to aggressively market it for official use.[57] Given the enthusiastic reception of the Hawk series of gyroplanes and the business acumen of the Groen brothers and their associates, it is likely that they will be successful and that the Autogiro, in its newest gyroplane configurations, will achieve an acceptance that has been elusive since the PCA-2 and C.30A flew over American and European skies. But the greatest success may yet be achieved by a technological breakthrough now flying in experimental form—the CarterCopter.

CarterCopters L.L.C., subsequently Carter Aviation Technologies (CAT), the creation of talented designer Jay Carter Jr., has created both the CarterCopter (a proof-of-concept testing platform),[58] and the Heliplane (a VTOL design system).[59] The prototype CarterCopter, which began flying in September 1998, is a five-passenger composite gyroplane with both a two-blade rotor *and* wings. It is capable of jump takeoffs and flying as a fixed-wing aircraft, in effect a convertiplane. The jump takeoff capacity is the product of an ultra-high-inertia rotor in which each blade tip is weighted with sixty-two pounds of depleted uranium. The rotor is pre-rotated between 365 rpm and 425 rpm at a zero pitch, and when snapped into a positive angle, the kinetic energy stored in the blades enables it to jump into the air. Jay Carter and his associates hope that the unique weighted rotor blades will enable the sleek aircraft to accomplish something never done before, to enter a realm only imagined—which they have trademarked as Extreme-Mu Flight.

One of the ironies of rotary flight is that the faster the rotor spins, the greater drag it produces and because of this drag, the modern rotary aircraft speed record is only 249 mph.[60] The weights in the tips of the CC's rotor blades allow them to be slowed during flight and yet still remain rigid, which should, paradoxically, allow the CC to fly faster and farther than any other rotorcraft in history. The CC's ultra-high-inertia weighted rotor blades are designed to allow them to be slowed and yet retain their rigid shape, even as lift is shifted to the wings. This slowing of the rotor blade resulting in a decrease of drag had first been disclosed in U.S. patent No. 3,155,341, issued to the Ryan Aeronautical Company on November 3, 1964, but it had never been achieved, in part because as the rotor slowed, it would in fact droop and *increase* drag. But the CC's patented rotor blade system and tip weights will allow a reduction in drag to occur where slowing of the rotor blades will allow the CC to enter the realm of Extreme-Mu flight as lift shifts to the CC's wings.

Mu is the ratio between (1) the forward speed of the aircraft and (2) the speed of the rotor blade tip in relation to the aircraft. Thus if the rotor blade tip is moving at 300 mph and the aircraft is flying at 300 mph, Mu = 1. But if aircraft in this example could slow its rotor tip speed to 100 mph while flying at 300 mph, Mu = 3. Because rotational drag is the product of rotor tip speed, when the rotor is slowed, drag will be reduced, and the CC will not only be able to fly faster, it will also become much more efficient! In fact, "a rotor turning at 300 rpm produces 27 times more rotational drag than a rotor slowed to 100 rpm."[61] Carter anticipates that the CC, with its ability to slow the rotor and reduce drag while shifting lift to its wings, will be able to establish a new rotary speed record and fly with an efficiency that is as good or better than fixed-wing aircraft. Carter has speculated that when Mu ratios of 4 or 5 are reached, the reduced drag will allow speeds approaching a truly revolutionary 500 mph. Although the CC is a proof-of-concept vehicle, the company plans to license the technology, in a manner similar to Cierva and Pitcairn. But it is also proceeding with applying the technology to the next generation, the Heliplane.

Utilizing the latest in technological advancements, including engine design, control mechanisms, proprietary rotor development, and extensive use of composite materials, the Heliplane will have the capacity to take off, hover, and land as a helicopter but will cruise with maximum efficiency as a gyroplane, with a rotor that is both unloaded and slowed, along with efficient high-aspect ratio wings. Of particular potential interest to the military, seeking such a craft for battlefield insertion, the Heliplane's designer anticipates using autorotation for high-speed descents, thus avoiding "settling under power" in a vortex ring state. This latter characteristic should prove of great interest to the United States Marine Corps, which has invested heavily in the Bell-Boeing V-22 Osprey, a competing "tiltrotor" technology that suffered from such vortex ring states. The $44 million transport, under development for eighteen years, crashed twice, killing all onboard. In the wake of the crash of April 8, 2000, one of the criticisms voiced about the Osprey was that it could not autorotate, precisely the projected strength of the Heliplane.

Carter estimates that the largest configuration Heliplane would be the biggest gyroplane ever built, with an empty weight of 90,000 pounds, a length of 106 feet, and height of 43 feet. Its VTOL-design takeoff weight is projected at 160,000 pounds at 7,000 feet density altitude. The Heliplane, a combination of helicopter, gyroplane, and fixed-wing aircraft, may yet emerge in the twenty-first century to soar in the skies about our cities, just as its Autogiro ancestor did eight decades ago. It is not surprising, then, that the Heliplane has already been called the "21st Century Rotodyne."[62]

NOTES

1. See Howard Levy, "Italian Import: The Magni Gyroplane Makes Its U.S. Debut," *Kitplanes* 18, no. 2 (February 2001): 41–44; Don Parham, "Magni Gyros: Quality Gyroplanes from Europe," *Homebuilt Rotorcraft* 14, no. 6 (June 2001): 6–7.

2. See Paul Bergen Abbott, "The Outstanding New Rotorcraft Equipment of the 1980's," *Rotorcraft* 27, no. 8 (December 1989–January 1990): 17–19.

3. Jim McCutchen, "An Open Letter to All PRA Members," *Rotorcraft* 30, no. 3 (May 1992): 19.

4. *Rotorcraft* 27, no. 1 (February–March 1989): 30; *Rotorcraft* 27, no. 8 (December 1989–January 1990): cover and 17–18; *Rotorcraft* 28, no. 1 (February–March

1990): 34; *Rotorcraft* 31, no. 6 (September 1993): 21, 25; *Popular Mechanics* 165, no. 11 (November 1988): cover.

5. Don Parham, "Sport Pilot/Light Sport Plane," *Homebuilt Rotorcraft* 14, no. 9 (September 2001): 14–16, 16.

6. Art Evans, "Tragedy in Arkansas," *Rotorcraft* 30, no. 3 (May 1992): 25.

7. Richard Bentley, "Mängoos 'Stealth' Gyroplane," *Rotorcraft* 29, no. 2 (April 1991): cover 42–43.

8. Ibid., p. 42.

9. Jim Eich, "'Sun 'n Fun' at Lakeland Florida," *Rotorcraft* 29, no. 44 (June–July 1991): 39.

10. Paul Bergen Abbott, "An Amazing National Convention," *Rotorcraft* 29, no. 6 (September 1991): 20, 30–31, 20.

11. Conversation between author and Gary Goldsberry, PRA president, who was one of the first to reach the Mängoos wreckage, May 25, 2002.

12. Kerry Cartier, "Dallas A&P Designs Nostalgic Tractor Autogyro," *Rotorcraft* 37, no. 4 (June–July 1999): 22–25.

13. Ron Herron, "First Flight of a New Tractor Autogyro," *Rotorcraft* 34, no. 8 (November 1996): 6–7; also see Ron Herron, "Flying Backwards in a Tractor Autogyro," *Rotorcraft* 33, no. 5 (August 1995): 32–34; Ron Herron, "History of Little Wing Autogyros," *Fly Gyro!* no. 4 (March–April 2001): 12–14; Ron Herron, "Little Wing 'Roto-Pup,'" *Homebuilt Rotorcraft* 8, no. 11 (November 1995): 12–13; Ron Herron, "Bringing Back the Autogiro," 33, no. 1 (February–March 1995): 12–13 (citing inspiration by the [David] Kay Gyroplane of the 1930s).

14. David Gittens and Kia Woods, "Ikenga, An Artist's Approach to Gyroplane Design," *Rotorcraft* 27, no. 1 (February–March 1989): 44, back cover (article and photographs of Ikenga).

15. Dick Cavin, "Ikenga...Reserve Grand Champion," *Sport Aviation* 38, no. 3 (March 1989): 46–49.

16. Ibid., p. 47.

17. Email to the author from David Gittens dated May 2, 2001; subsequent email comment of February 16, 2002.

18. Gittens, email to author, May 2, 2001; *Rotorcraft* 26, no. 5 (October–November 1988): 20 (photograph of Alderfer tractor autogyro); Ed Alderfer, "A Study of the Tractor Gyroplane," *Rotorcraft* 26, no. 6 (December 1988–January 1989): 17–18.

19. Paul Bergen Abbott, "New Shape from Santa Fe," *Rotorcraft* 27, no. 7 (October–November 1989): 33.

20. "What's New: Cygnus 21TX," *Popular Science,* January 1991.

21. "Gyro 2000 Closes Doors," *Rotorcraft* 30, no. 2 (May 1992): 7.

22. Charlie Yaw, "The Air Command Story," *Rotorcraft* 26, no. 4 (August–September 1988): 32–33.

23. Abbott, "Outstanding New Rotorcraft Equipment," p. 18.

24. Paul Bergen Abbott, "The Best of the 1980's," *Rotorcraft* 27, no. 8 (December 1989–January 1990): 14–17, 16; Ed Alderfer, "The Designers: 30 Years of Rotorcraft Designers," *Rotorcraft* 27, no. 3 (May Extra 1989): 26–29, 28.

25. Paul Bergen Abbott, "Rotorcraft Choices 1989," *Rotorcraft* 27, no. 1 (February–March 1989): 22–30, 24.

26. Paul Bergen Abbott, "Air Command Fly-In," *Rotorcraft* 26, no. 4 (August–September 1988): 16–17; Betty Jo Charlet, "Air Command Does It Again," *Rotorcraft* 27, no. 7 (October–November 1989): 38–39.

27. Paul Bergen Abbott, "A New Name at Air Command," *Rotorcraft* 28, no. 5 (August 1990): 19.

28. "A New Owner for Air Command," *Rotorcraft* 30, no. 4 (June–July 1992): 6.

29. Dr. Stewart S. Houston, "Eight Years of Gyroplane Research in the UK: A *Very* Personal Reflection," *Fly Gyro!* no. 3 (January–February 2001): 12–16.

30. "Rotorcraft in the News," *Homebuilt Rotorcraft* 14, no. 9 (September 2001): 4.

31. Email to author from Kelly Vanek of February 28, 2002; see also Dave Martin, "Evolving While Revolving: Jim Vanek's New Sport Copters Continues a Family Tradition," *Kitplanes* 12, no. 4 (June 1995): 36–41.

32. "PRA Interview: Ernie Boyette," *Rotorcraft* 37, no. 3 (May 1999): 4–6; Ernie Boyette, "Ernie Finds the Lost Chord," *Rotorcraft* 29, no. 2 (April 1991): 15–21.

33. Paul Bergen Abbott, "Big Bird Takes Off," *Rotorcraft* 26, no. 4 (August–September 1988): 12–13.

34. "Meet the Board: Carl Schneider," *Rotorcraft* 36, no. 9 (December 1998–January 1999): 10; *Rotorcraft* 36, no. 4 (June–July 1998): 25 (Schneider photograph).

35. Jean Fourcade, "Longitudinal Stability of Gyroplanes," *Rotorcraft* 37, no. 4 (June–July 1999): 14–19.

36. *Rotorcraft* 40, no. 1 (February 2002): 7.

37. Paul Bergen Abbott, "Marriage Made in Heaven," *Rotorcraft* 27, no. 8 (December 1989–January 1990): 12–13.

38. Abbott, "Big Bird," p. 13.

39. Chuck Beaty, "Tails: Tall and Otherwise," *Rotorcraft* 26, no. 5 (October–November 1988): 15.

40. Erlene Barnett, "A Twice Told Tall Tail," *Rotorcraft* 26, no. 6 (December 1988–January 1989): 6.

41. Wallis letter to the editor, *Rotorcraft* 27, no. 1 (February–March 1989): 4–5.

42. Jim Vanek, *Sport Copter* (Scappose, Oregon: Sport Copter, Inc., 2000), p. 7.

43. Russ King, "From Lessons to Solo," *Rotorcraft* 36, no. 4 (June–July 1998): 7.

44. Don Parham, "From the Editor's Desk," *Homebuilt Rotorcraft* 8, no. 11 (November 1995) and Buck Buchanan and Jim DiGateano, "Letters to the Editor" *Homebuilt Rotorcraft* 8, no. 11 (November 1995) CFI pp. 18–20.

45. "Auction! Auction! Auction!" *Homebuilt Rotorcraft* 8, no. 9 (September 1995): 13.

46. Paul Bergen Abbott, "The $20,000 Takeoff!" *Rotorcraft* 33, no. 6 (September 1995): 44–47; Paul Bergen Abbott, "How Did Dick Do It?" *Rotorcraft* 33, no. 6 (September 1995): 31–32.

47. Sandy Love, "The Sky's No Limit," *Rotorcraft* 36, no. 4 (June–July 1998): 34–35; Stephanie Gremminger, "Over Four Miles High in a Gyrocopter!" *Rotorcraft* 36, no. 3 (May 1998): 18–19.

48. Love, p. 35, where the author places the first flight with the video camera on Thursday; but see Mike Stinnett, "Bensen Days 1998," *Rotorcraft* 36, no. 4 (June–July 1998): 21–23, at p. 23, where the author claims that the camera flight was on Wednesday, April 15. Gremminger would agree with the Thursday dating, as she places the record flight on Friday, April 24. Gremminger, pp. 18–19.

49. Ron Bartlett, "The UFO-Helithruster: An Update," *Autogyro Quarterly,* not 11 (January 2002): 26–27; Mel Morris Jones, "Talkshop: The UFO Helithruster," *Fly Gyro!* no. 2 (November–December 2000): 8–11 (photograph on the front cover).

50. Don Parham, "High Tech Rotorcraft," *Homebuilt Rotorcraft* 12, no. 10 (October 1999): 6–11.

51. Stephanie Gremminger, "The Hawk 4 Gyroplane Aims to Bring 'Gyroplanes to Their Logical Progression,'" *Rotorcraft* 38, no. 5 (August 2000): 8–9, 35, 8.

52. Ibid., p. 9.

53. "Hawk Gyro Goes to China," *Rotorcraft* 33, no. 1 (February–March 1995): 9.

54. Gremminger, "Hawk 4," p. 9.

55. William B. Scott, "Hawk 4T Breathes New Life into Gyroplanes," *Aviation Week & Space Technology* 153, no. 19 (November 6, 2000): 54–56.

56. Bruce H. Charnov, "Groen Brothers at Oshkosh," *Fly Gyro!* no. 7 (September 2001): 16.

57. Ron Bartlett, "Groen Brothers Hawk 6G Gyroplane," *Autogyro 1/4ly,* no. 8, 8 (text and photograph).

58. Parham, "High Tech Rotorcraft," pp. 6–11; Rod Anderson, "CarterCopter (CC) R&D: Something for Everyone," *Rotorcraft* 38, no. 7 (October 2000): 32–35; Jay Carter Jr., "CarterCopter: A High Technology Gyroplane" (paper presented at the American Helicopter Society Vertical Lift Aircraft Design Conference, San Francisco, California, 2000).

59. Jay Carter Jr., "The CarterCopter (CC) Heliplane" (paper presented at the Precision Strike Technology Symposium 2000, Johns Hopkins University, Laurel, Maryland, October 2000).

60. Paul Bergen Abbott, "Will This New Autogyro Go 400 Miles per Hour?" *Rotorcraft* 34, no. 6 (September 1996): 31–32.

61. "Update: The CarterCopter (CC) Heliplane." Handed out at EAA Air Venture Fly-In, 2002. (CarterCopters Information Sheet 2001); Abbott, "Will This New Autogyro Go 400?" p. 10.

62. "21st Century Rotodyne," *Wingspan International,* no. 8 (September–October 2001): 65.

THE ESSENTIAL AUTOGIRO/ AUTOGYRO/GYROCOPTER/ GYROPLANE HISTORY LIBRARY: AN ANNOTATED LIST

The following constitutes an essential library of books and articles dealing with the history of the Autogiro (Cierva and Cierva-licensed autorotation aircraft), autogyro (non-Cierva machines), Gyrocopters/gyrocopters (capitalized when applied to Bensen machines prior to the end of the Bensen Company in 1988), and gyroplanes (generic term and applied to post-1970 Ken Brock gyrocopters). Many are little known to readers but *frequently* cited by writers who have obviously not read the originals. I have indicated sources for each book, but it should be noted that copies may often be found at www.abebooks.com, which is a superb worldwide rare-book network touted by the *New York Times* as an Internet business that works. Additionally, your local college or university library may be able to locate a copy and request an interlibrary loan. Finally, do not neglect Internet auction sites—these books often come on at reasonable rates.

Cierva, Juan de la, and Don Rose. *Wings of Tomorrow: The Story of the Autogiro.* New York: Brewer, Warren & Putnam, 1931. The start of it all—Autogiro history begins with this book. Fairly rare but copies available.

Sanders, C.J., and A.H. Rawson. *The Book of the C.19 Autogiro.* London: Sir Isaac Pitman & Sons, 1931. Charles J. Saunders headed the design team for Avro in the manufacture of the C.19 Autogiro. Arthur H.C.A. "Dizzy" Rawson was a Cierva test pilot *who accompanied the C.8W to America and who made the first flight—prior to Harold Pitcairn.* Book generally unavailable, but occasionally through www.abebooks.com a copy becomes available.

The Autogiro. Philadelphia: Pitcairn-Cierva Autogiro Company of America, 1930; *The Autogiro.* Philadelphia: Pitcairn-Cierva Autogiro Company of America, 1933; *Some Facts of Interest about Rotating-Wing Aircraft and the Autogiro Company of America.* Philadelphia: The Autogiro Company of America, 1944; Pitcairn, Harold F. "Juan de la Cierva: In Memoriam." Philadelphia: Autogiro Company of America, January 9, 1939. These four publications by the Autogiro Company of America constitute an essential record of Autogiro development in America, with an invaluable photographic record. The text is sometimes contradictory but always informative. The final slim publication is Harold F. Pitcairn's tribute to his friend Juan de la Cierva on the third anniversary of his death. These books are quite rare and only occasionally come up at auction.

Brie, R.A.C. *The Autogiro and How to Fly It.* London: Sir Isaac Pitman & Sons, 1933. 2nd ed., 1934. "Reggie" Brie was chief test pilot for Cierva Autogyro Company Ltd., and the foreword is by Juan de la Cierva. Book generally unavailable, but occasionally through www.abebooks.com a copy becomes available.

Cierva, Juan de la. "The Autogiro." *Journal of Royal Aeronautical Society* 34, no. 239 (November 1930): 902–21; Cierva, Juan de la. "New Developments of the Autogiro." *Journal of the Royal Aeronautical Society* (December 1935): 1125–43. Cierva greatly valued his membership in The Royal Aeronautical Society (TRAS), and his invited lectures are clear, concise statements of Autogiro history. These lectures are generally not quoted and will only be found in aviation libraries that maintain TRAS journals. Obviously, more available in England.

Hafner, Raoul. *The Hafner Gyroplane.* Reprinted from the *Journal of the Royal Aeronautical Society* (February 1938). An Austrian active in helicopter development, created a series of autogyros in England in 1930s and later, as British citizen during World War II, developed the Hafner "Rotachute" and Malcolm "Rotabuggy." The Rotachute was one of the major influences on Igor Bensen's Gyro-Glider. Read this book along with Clouston, Air Commodore A. E. *The Dangerous Skies.* London, England: Cassell & Company Limited, 1954. Clouston was Hafner's test pilot and has personal observations on the Hafner aircraft. Book generally unavailable, but occasionally through www.abebooks.com a copy becomes available.

Haugen, Lieutenant Victor. "Principles of Rotating Wing Aircraft." *Aeronautics* 2, no. 7 (October 16, 1940): 420–38; Lunde, Professor Otto H. "Development and Operation of the Autogiro." *Aeronautics* 2, no. 9 (October 30, 1940): 548–62; Miller, John M. "Civil Uses of the Autogiro." *Aeronautics* 2, no. 10 (November 6, 1940): 611–624. Three installments in the 1939–40 series of lectures on aeronautics. Each presents Autogiro history and photographs. At the time of their publication they were the most comprehensive early Autogiro history compilation. John M. "Johnny" Miller was the Autogiro pilot who beat Amelia Earhart cross-country in May 1931, and he had just completed flying the Eastern Air Lines Experimental Autogiro Air Mail flights between the roof of the Philadelphia 30th Street Post Office and nearby Camden, New Jersey, airport between July 6, 1939, and July 5, 1940. This lecture series is still available in aviation libraries and occasionally comes up for sale at auction.

Gregory, Hollingsworth Franklin. *Anything a Horse Can Do: The Story of the Helicopter.* Introduction by Igor Sikorsky. New York: Reynal & Hitchcock, 1944. Revised editions were published as *The Helicopter; or, Anything a Horse Can Do.* New York: Reynal & Hitchcock, 1948 and London: George Allen & Unwin 1948; and as *The Helicopter.* South Brunswick, New Jersey: A. S. Barnes, 1976. H. Franklin Gregory became the United States Army Air Force expert on rotary-wing aircraft, gained great experience with the Kellett Autogiro, was commanding officer of the U.S. Army Autogiro School at Dayton, Ohio, and was instrumental, through his administration of the Dorsey-Logan Bill funds, in channeling government money away from Autogiro development to the Sikorsky and Platt-LePage helicopters. His insider account of the military attitude toward the Autogiro helps greatly in understanding why authorities turned away from the Pitcairn and Kellett models. Fairly rare but can be found in some libraries and thus secured through interlibrary loan.

Gablehouse, Charles. *Helicopters and Autogiros: A History of Rotating-Wing and V/STOL Aviation.* Rev. ed. Philadelphia: Lippincott, 1969. Previous edition published as *Helicopters and Autogiros: A Chronicle of Rotating-Wing Aircraft.* Philadelphia: Lippincott, 1967. Has a good generalized account of the development of the Autogiro in relationship to the helicopter. Still generally available in libraries.

Courtney, Frank T. *Flight Path.* London: William Kimber, 1972. Also published as *The Eighth Sea.* New York: Doubleday & Co., 1972. Frank Courtney was Cierva's first test pilot after Cierva relocated from Spain to England in 1925. Courtney has sharp, self-serving observations of Cierva. (See also Hannon, Bill. "Those Infuriating 'Palm Trees.' " *Popular Rotorcraft Flying* 7, no. 6 (November–December 1969): 30, which is an account of an address Courtney gave before a local PRA chapter, describing the flight of a Cierva C.6 at Farnborough in 1925. Courtney describes the chance meeting and dinner with Cierva on the evening of December 8, 1936, the night before Cierva died in a plane crash at Croydon). Book still found in some libraries and available in used book stores and on the internet.

Smith, Frank Kingston. *Legacy of Wings: The Story of Harold F. Pitcairn.* New York: Jason Aronson, 1981. Smith was a lawyer and admirer of Harold F. Pitcairn. This is a devotional biography covering Pitcairn's aviation achievements from the Mailwing, Eastern Air Transport (eventually Eastern Airlines, after being acquired in the fall of 1929 by a group headed by Chandler Keys and Glen Curtiss), and the Autogiro. In his admiration, Smith often fails to make accurate judgments regarding his subject, always choosing the flattering interpretation; but this book is *essential* to understanding the man and his Autogiros, with many photographs and diagrams. (Helpful to read with George Townson's book and especially with Peter Brooks's *Cierva's Autogiros,* which presents the same events from a different perspective.) Privately printed in an edition underwritten by the Pitcairn family, it can regularly be found at auction and at www.abebooks.com. Sometimes available for sale at Historic Aviation. The book also served as the basis for the EAA film *Legacy of Wings,* which is available from EAA and at its museum at Oshkosh, Wisconsin, and available at the Museum of the American Helicopter Museum & Education Center, Brandywine Airport, West Chester, Pennsylvania. It should be noted that the book is extremely handicapped by having an exceptionally limited index. (See also Smith, Frank Kingston. "Mr. Pitcairn's Autogiros." *Airpower* 12, no. 2 (March 1983): 28–49; Aellen, Richard. "The Autogiro and Its Legacy." *Air & Space Smithsonian,* December 1989/January 1990, 52–59; Anders, Frank. "The Forgotten Rotorcraft Pioneer: Harold F. Pitcairn." *Rotor & Wing International,* May 1990, 34–37. Reprinted as "The Forgotten Rotor." *Rotorcraft* 28, no. 7 (October–November 1990): 30–34.

LePage, Wynn Laurence. *Growing Up with Aviation.* Ardmore, Pennsylvania: Dorrance, 1981. W. Laurence LePage worked with Pitcairn and the Kellett brothers and has a brief but insightful insider account of the historical period. He later joined with Haviland H. Platt to create the Platt-LePage helicopter, which was allocated funds from the Dorsey-Logan Bill; the allocation that gave rise to the unfounded rumors of a conspiracy to deny financial support for Autogiro development. This book is frequently overlooked in Autogiro research because of its title, the association of its author with the helicopter, and the 1981 publication, almost forty years after the last Pitcairn Autogiros. It is sometimes available from aviation book dealers and found in libraries.

Townson, George. *Autogiro: The Story of "the Windmill Plane."* Fallbrook, California: Aero Publishers, 1985. Townson wrote an idiosyncratic account of the Autogiro, with many photographs and diagrams. Generally available for sale at Historic Aviation and at the Museum of the American Helicopter Museum & Education Center, Brandywine Airport, West Chester, Pennsylvania. See also Townson, George, and Howard Levy. "The History of the Autogiro: Part 1." *Air Classics Quarterly Review* 4, no. 2 (Summer 1977): 4–18; Townson, George, and Howard Levy. "The History of the Autogiro: Part 2." *Air Classics Quarterly Review* 4, no. 3 (Fall 1977): 4–19, 110–14.

Brooks, Peter W. *Cierva's Autogiros: The Development of Rotary-Wing Flight.* Washington, D.C.: Smithsonian Institution Press, 1988. This is *the* most comprehensive account of the Autogiro and indispensable reading—if you could read only one book from this list, this is it. It has an exceptionally detailed index and comprehensive listing by registration number of all the Autogiros produced, with good accounts of England, America, Germany, France, Japan, and Russia and with abbreviated references to the Fairey Rotodyne and the autogyros of Wing Commander Ken Wallis. Generally available at www.abebooks.com, occasionally online, and in a fair number of libraries. See also Capon, P. T. "Cierva's First Autogiros: Part 1." *Aeroplane Monthly* 7, no. 4 (April 1979): 200–205; Capon, P. T. "Cierva's First Autogiros: Part 2." *Aeroplane Monthly* 7, no. 4 (May 1979): 234–40; Brooks, Peter W. "Rotary Wing Pioneer." *Aeroplane,* December 9 and 16, 1955.

Bensen, Igor B. *A Dream of Flight.* Indianapolis, Indiana: The Abbott Company, 1992. Nominally authored by Dr. Igor Bensen, it was prepared by Paul Bergen Abbott, who had purchased the rights (see Abbott, Paul Bergen. "The Story of Dr. Bensen's Fabulous Book." *Rotorcraft* 37, no. 5 [August 1999]: 8–9). Went out of print in early 1999 when the supply ran out and it proved too expensive to reprint. This was apparently put together from Bensen articles and does not name Bensen's parents or even mention his wife, Mary, who made a significant but here unacknowledged contribution to the American rotorcraft movement. There are four copies listed nationally in libraries, and it can be secured on an interlibrary loan—if your local facility cannot do this, consult a university reference librarian. See also Gilley, Rick. "Dr. Igor Bensen The Man and His Machine." *Fly Gyro!* September/October 2001, 20–22, which contains details, such as Bensen's parents' names, not found in *A Dream of Flight;* "Design Classroom." *Collected Works of Design Classroom.* California: Popular Rotorcraft Association, 1974. Introduced by Kas Thomas, this was a forty-five-page compilation of the mostly Bensen-written *Popular Rotorcraft Flying* columns on rotorcraft design and history.

Hollmann, Martin. *Helicopters.* Monterey, California: Aircraft Designs, Inc., 2000; *Flying the Gyroplane.* Monterey, California: Aircraft Designs, Inc., 1986. Martin Hollmann, creator of the HA-2A Sportster and later the ultralight Bumble Bee, has a great deal of historical information contained in these two books available directly from the author. With numerous diagrams and photographs, valuable for scope (but not depth) of its coverage of the Autogiro, with references to the Focke-Achgelis FA-330 rotary-wing kite (see also *Aircraft of the National Air and Space Museum.* 4th ed. Washington, D.C.: Smithsonian Institution Press, 1991), the Bensen Gyrocopter, the Air & Space (Umbaugh) 18A, the McCulloch J-2, and the autorotational work of Anton Flettner and Frederich von Doblhoff. Reprints the Fifth Cierva Memorial Lecture from the *Journal of the Royal Aeronautical Society* by Heinrich K.J. Focke—see especially the report of comments made at the lecture by Rauol Hafner and O.L.L. Fitzwilliams, designers of the Hafner Rotachute that inspired Igor Bensen, by August Stepan, associate of Von Doblhof in World War II and later jet-tip rotor expert on the Fairey Rotodyne, and by the great J.A.J. Bennett, successor at Cierva Autogiro Company Ltd. after Cierva's December 1936 death and codesigner (with Captain A. Graham Forsyth) of the Fairey Rotodyne.

Thomas, Kas. *Guide to Homebuilt Rotorcraft.* New York: Crown Publishers, 1976; Abbott, Paul Bergen. *The Gyroplane Flight Manual.* Indianapolis, Indiana: The Abbott Company, 1988, 1992, 1996; *The Gyrocopter Flight Manual.* Indianapolis, Indiana: The Abbott Company, 1977; Indianapolis, Indiana: Cranberry Corners, 1983; Indianapolis, Indiana: The Abbott Company 1986; *Want to Fly a Gyrocopter.* Indianapolis, Indiana: Cranberry Corners, 1977; *Understanding the Gyroplane.* Indianapolis, Indiana: The Abbott Company, 1994. These books are generally oriented toward the Bensen Gyrocopter and, as such, are obsolete *but of continuing interest for Autogiro/Gyrocopter /autogyro/gyroplane history.* The later books remain available from Paul Bergen Abbott.

Wood, Derek. *Project Cancelled: British Aircraft That Never Flew.* Indianapolis/New York: The Bobbs-Merrill Company Inc., 1975. The most comprehensive description of development of the Fairey Rotodyne, arguably the most impressive autogyro ever to fly, and almost completely unknown due to its cancellation and destruction by the British government in 1962. See also Anders, Frank. "The Problem Solver." *Air Classics* 30, no. 10 (October 1994): 50–58; "The Flat Risers: The Ups and Downs of VTOL (Part 2)." *TakeOff* 1, part 5 (1989): 138–43; Harrison, Jean-Pierre. "Fairey Rotodyne." *Air Classics* 22, no. 44 (April 1996): 44–47, 60–62, 64–66, 79–80; "The Fairey Rotodyne: Nearly the Answer." *West Coast Aviator,* September/October 1995, 35–37; Hislop, George S. "The Fairey Rotodyne." Paper presented before the Helicopter Association of Great Britain and The Royal Aeronautical Society, London, England, November 7, 1958; and for a film of the Rotodyne flying, see "A Fairey Rotodyne Storey," Traplet Video Productions, Worcestershire, England). Wood's book found in many libraries and readily available via interlibrary loan. Journal articles fairly unavailable—film available from Traplet. For more recent summary, see Charnov, Bruce. "The Fairey Rotodyne: An Idea Whose Time Has Come—Again: Part 3." *Fly Gyro!* March/April 2002, 4–8; "Part 2." *Fly Gyro!* November/December 2001, 4–7; "Part 1" *Fly Gyro!* September/October 2001, 4–7.

Spenser, Jay P. *Whirlybirds: A History of the U.S. Helicopter Pioneers.* Seattle and London: University of Washington Press, 1998. Brief but informative description of American Autogiro development and involvement of the military. Also deals with American developers Buhl Aircraft and E. Burke Wilford. Main strength is availability. In a similar vein, see Beard, Barrett Thomas. *Wonderful Flying Machines.* Annapolis, Maryland: Naval Institute Press, 1996 (describes American Autogiro pilots) and Young, Warren R. *The Helicopters.* Alexandria, Virginia: Time-Life Books, 1982 (description of Autogiro development and copies of the 1930s Pitcairn Autogiro advertisements).

BIBLIOGRAPHY

BOOKS

Abbot, Charles Greeley. *Great Inventions.* Vol. 12. Originally published as *Smithsonian Scientific Series.* Washington, D.C.: Smithsonian Institution Press, 1944.

Abbott, Paul Bergen. *The Gyroplane Flight Manual.* Indianapolis, Indiana: The Abbott Company, 1988, 1996.

————. *The Gyrocopter Flight Manual.* Introduction by Dr. Igor Bensen. Indianapolis, Indiana: The Abbott Company, 1977, 1986. Revised edition, Indianapolis, Indiana: Cranberry Corners, 1983.

————. *So You Want To Fly a Gyrocopter.* Indianapolis, Indiana: Cranberry Corners, 1977.

————. *Understanding the Gyroplane.* Indianapolis, Indiana: The Abbott Company, 1994.

Adams, Eustace L. *The Flying Windmill.* New York: Grosset and Dunlap, 1930.

Aircraft of the National Air and Space Museum, 4th ed. Washington, D.C.: Smithsonian Institution Press, 1991.

Almond, Peter. *Aviation: The Early Years (The Hulton Getty Picture Collection).* Germany: Könemann Verlagsgesellschaft mbH, 1997.

Andersson, Lennart. *Soviet Aircraft and Aviation 1917–1941.* Annapolis, Maryland: Naval Institute Press; London: Putnam Aeronautical Books, 1994.

Andrews, Allen. *The Flying Machine: Its Evolution Through the Ages.* New York: G. P. Putnam's Sons, 1977.

Apostolo, Giorgio. *The Illustrated Encyclopedia of Helicopters.* New York: Bonanza Books, 1984.

The Autogiro. Philadelphia: Pitcairn-Cierva Autogiro Company of America, 1930.

The Autogiro. Philadelphia: Pitcairn-Cierva Autogiro Company of America, 1933

Bakus, Jean L. *Letters from Amelia: An Intimate Portrait of Amelia Earhart.* Boston, Massachusetts: Beacon Press, 1982.

Barker, Ralph. *The RAF At War.* Alexandria, Virginia: Time-Life Books, 1981.

Beard, Barrett Thomas. *Wonderful Flying Machines.* Annapolis, Maryland: Naval Institute Press, 1996.

Bensen, Igor B. *A Dream of Flight.* Indianapolis, Indiana: The Abbott Company, 1992.

Bensen Gyro-Glider: Building Instructions with Operating and Flight Manuals. Raleigh, North Carolina: Bensen Aircraft Corporation, 1967.

Boyne, Walter J. *The Aircraft Treasures of Silver Hill.* New York: Rawson Associates, 1982.

Boyne, Walter J., and Donald S. Lopez. *Vertical Flight.* Washington, D.C.: Smithsonian Institution Press, 1984.

Bratukhin, A. G., ed. *Russian Aircraft.* Moscow: Mashinostroenie, 1995.

Brie, R. A. C. *The Autogiro and How to Fly It.* London: Sir Isaac Pitman and Sons, 1933, 2d. reprint edition, 1934.

Brooks, Peter W. *Cierva Autogiros: The Development of Rotary-Wing Flight.* Washington, D.C.: Smithsonian Institution Press, 1988.

Butler, Susan. *East to the Dawn: The Life of Amelia Earhart.* New York: Da Capo Press, 1999.

Caidin, Martin. *Golden Wings: A Pictorial History of the United States Navy and Marine Corps in the Air.* New York: Random House, 1960. Reprint, 1974.

Carney, Ray. *American Vision: The Films of Frank Capra.* Hanover, New Hampshire: Wesleyan University Press, 1986.

Cierva y Codorníu, Juan de la. *Engineering Theory of the Autogiro.* Cierva Autogiro Co. Ltd., 1929.

Cierva, Juan de la, and Don Rose. *Wings of Tomorrow: The Story of the Autogiro.* New York: Brewer, Warren and Putnam, 1931.

Clouston, A. E. *The Dangerous Skies.* London: Cassell and Company, 1954.

Courtney, Frank T. *Flight Path.* London: William Kimber, 1972. Also published as *The Eighth Sea.* New York: Doubleday, 1972.

Craddock, Jim, ed. *VideoHound's Golden Movie Retriever.* Farmington Hills, Michigan: Visible Ink Press, 2001.

Croome, Angela. *Hover Craft.* Williamsburg, Virginia: Astor Book, 1962; Norwich, Great Britain: Jarrold and Sons, 1960.

"Design Classroom." *Collected Works of Design Classroom.* Anaheim, California: Popular Rotorcraft Association, 1974.

Donald, David, ed. *The Complete Encyclopedia of World Aircraft.* New York: Barnes & Noble Books, 1997.

Dougall, Alastair. *James Bond: The Secret World of 007.* London: Dorling Kindersley, 2000.

Everett-Heath, John. *Soviet Helicopters: Design, Development and Tactics.* London: Jane's Publication Company, 1983.

Eves, Edward. *The Schneider Trophy Story.* St. Paul, Minnesota: MBI Publishing Company, 2001.

Eyermann, Karl-Heinz. *Die Luftfahrt der UdSSR 1917–1977.* Berlin: transpress VEB Verlag für Verkehrswesen, 1977.

Fay, John. *The Helicopter,* 4th ed. New York: Hippocrene Books, 1987.

Ford, Brian. *German Secret Weapons: Blueprint for Mars.* New York: Ballantine Books, 1969.

Ford, Roger. *Germany's Secret Weapons In World War II.* Osceola, Wisconsin: MBI Publishing Company, 2000.

Francillon, R. J. *Japanese Aircraft of the Pacific War.* London: Putnam, 1970.

Francis, Devon Earl. *The Story of the Helicopter.* New York: Coward-McCann, Inc., 1946.

Gablehouse, Charles. *Helicopters and Autogiros: A History of Rotating-Wing and V/STOL Aviation.* Revised edition. Philadelphia: Lippincott, 1969. Previous edition published as *Helicopters and Autogiros: A Chronicle of Rotating-Wing Aircraft.* Philadelphia: Lippincott, 1967.

García, Albors, E. *Juan de la Cierva y el Autogiro.* Madrid: Editiones Cid, 1965.

German Submarine Rotary Wing Kite. London: Combined Intelligence Objectives Sub-Committee, 1945.

Green, William, and Gerald Pollinger. *The Aircraft of the World.* New York: Doubleday, 1965.

Gregory, Hollingsworth Franklin. *Anything A Horse Can Do: The Story of the Helicopter.* Introduction by Igor Sikorsky. New York: Reynal and Hitchcock, 1944. Revised editions were published as: *The Helicopter; or, Anything a Horse Can Do.* New York: Reynal and Hitchcock, 1948; London: George Allen and Unwin, 1948 and *The Helicopter.* South Brunswick, New Jersey: A. S. Barnes, 1976.

Gunston, Bill. *Aircraft of the Soviet Union.* London: Osprey, 1983.

———— *Aviation: The First 100 Years.* New York: Barrons Educational Series, 2002.

————. *Classic World War II Aircraft Cutaways.* New York: Barnes & Noble Books, 1999; London: Osprey Publishing, 1995.

————. *Helicopters at War.* London: Hamlyn, 1977.

————. *History of Military Aviation.* London: Hamlyn, Octopus Publishing Group Limited, 2000.

————. *The Osprey Encyclopedia of Russian Aircraft.* Oxford, England: Osprey Publishing Limited, 1995, 2000.

————. *World Encyclopedia of Aero Engines,* 4th ed. Sparkford. Nr. Yeovil, Somerset: Patrick Stephens Limited, 1998.

Hafner, Raoul. *The Hafner Gyroplane.* London: Raoul Hafner, 1938.

Hallion, Richard P. *Test Pilots: The Frontiersmen of Flight.* Garden City, New York: Doubleday, 1981.

Hancock, Ian. *The Lives of Ken Wallis—Engineer and Aviator Extraordinaire.* Suffolk, England: Norfolk and Suffolk Aviation Museum, 2001.

Hislop, George S. "The Fairey Rotodyne." Paper presented to The Helicopter Association of Great Britain and The Royal Aeronautical Society, London, November 7, 1958.

Hollmann, Martin. *Flying the Gyroplane.* Monterey, California: Aircraft Designs, Inc., 1986.

———. *Helicopters.* Monterey, California: Aircraft Designs, Inc., 2000.

Holmes, Donald B. *Air Mail: An Illustrated History 1793–1981.* New York: Clarkson N. Potter, Publishers, 1981.

Hubler, Richard G. *Straight Up: The Story of Vertical Flight.* New York: Duell, Sloan and Pearce, 1961.

Hunt, William E. *Heelicopter: Pioneering with Igor Sikorsky.* London: Airlife Publishing Ltd., 1998.

Jablonski, Edward. *Man With Wings.* Garden City, New York: Doubleday, 1980.

Jackson, A. J. *AVRO Aircraft Since 1908.* London: Putnam, 1962.

———. *De Havilland Aircraft Since 1915.* London: Putnam, 1962.

Jackson, Donald Dale. *Flying the Mail.* Alexandria, Virginia: Time-Life Books, 1982.

Jackson, Robert. *The Dragonflies: The Story of Helicopters and Autogiros.* London: Barker, 1971.

James, Derek N. *Westland Aircraft Since 1915.* London: Putnam Aeronautical Books; Annapolis, Maryland: Naval Institute Press, 1991.

Jarrett, Philip. *Ultimate Aircraft.* London: Dorling Kindersley, 2000.

Johnson, Brian. *Classic Aircraft: A Century of Powered Flight.* London: Channel 4 Books, 1998.

Johnson, Wayne. *Helicopter Theory.* Mineola, New York: Dover Publications, 1980.

Johnston, S. Paul. *Horizons Unlimited: A Graphic History of Aviation.* New York: Duell, Sloan and Pearce, 1941.

Kaman, Charles H. "Kaman Helicopters and the Evolution of Vertical Flight." Thirty-third Wings Club General Harold R. Harris 'Sight' Lecture, New York, May 15, 1996.

Katz, Ephraim. *The Film Encyclopedia,* 4th revised ed. by Fred Klein and Ronald Dean Nolen. New York: HarperCollins Publishers, 2001.

Kurland, Michael. *Too Soon Dead.* New York: St. Martins Press, 1997.

Lambermont, Paul M., and Anthony Pirie. *Helicopters and Autogyros of the World.* Introduction by Igor Sikorsky. Revised ed. New York: A. S. Barnes, 1970; London: Cassell, 1958, 1970. Previous edition, New York: A. S. Barnes, 1959.

Liptrot, Roger N., and J. D. Woods. *Rotorcraft.* London: Butterworth's Scientific Publications, 1955.

Longyard, William H. *Who's Who in Aviation History.* Shrewsbury, England: Airlife Publishing Ltd., 1994.

Lovell, Mary S. *The Sound of Wings: The Life of Amelia Earhart.* New York: St. Martins Press, 1989.

March, Daniel J., ed. *British Warplanes of World War II: Combat Aircraft of the RAF and Fleet Air Arm 1939–1945.* New York: Barnes & Noble Books, 1998.

Mellen, Joan. *Big Bad Wolves: Masculinity in the American Film.* New York: Pantheon Books, 1977.

Mitchell, William. *Skyways: A Book On Modern Aeronautics.* Philadelphia: J.B. Lippincott Company, 1930.

Mondey, David, ed. *The Complete Illustrated Encyclopedia of the World's Aircraft.* Secaucus, New Jersey: Chartwell Books, Inc. 1978. Reprint updated and expanded by Michael Taylor, 2000.

———. *The Concise Guide to Axis Aircraft of World War II.* London: Chancellor Press, 1984. Reprint, 1996.

Munson, K. *Civil Aircraft of Yesteryear.* New York: Arco Publishing Company, Inc., 1968.

———. *Helicopters and Other Rotorcraft Since 1907.* New York: Macmillan, 1969. Revised ed., London: Blandford Press, 1973.

Nahum, Andrew. *Eyewitness Books: Flying Machine.* London: Dorling Kindersley Books, 1990.

Nayler, J.L., and E. Ower. *Flight To-day.* London: Oxford University Press, 1936. Revised edition, 1942.

Newman, Simon. *The Foundation of Helicopter Flight.* London: Edwin Arnold, Division of Hodder Headline PLC, 1994.

O'Brien, Kathryn E. *The Great and the Gracious on Millionaires' Row.* Utica, New York: North Country Books, Inc., 1978.

Ogden, Bob. *Aircraft Museums and Collections of the World, 9: Eastern and South Eastern Europe and the C.I.S.* Woodley, England: Bob Ogden Publications, no date.

———. *Aircraft Museums and Collections of the World: USA The Western States,* 2d ed. Stamford, England: Key Publishing Ltd., no date.

———. *British Aviation Museums and Collections,* 2d ed. Stamford, England: Key Publishing Ltd., 1986.

———. *British Aviation Museums.* Stamford, England: Key Publishing Ltd., 1983.

———. *Great Aircraft Collections of the World.* New York: Gallery Books, 1988.

Philpott, Bryan. *The Encyclopedia of German Military Aircraft.* New York: Park South Books, 1981.

Pitcairn, Harold F. *Juan de la Cierva.* New York: Autogiro Company of America, 1939.

Pollinger, Gerald. *Strange But They Flew.* New York: G.P. Putnam's Sons, 1967.

Prewitt, Richard H. *Report on Helicopter Development in Germany.* New York, 1945.

Proceedings of Rotating WingAircraft Meeting, October 28 and 29, 1938. Philadelphia: Institute of Aeronautical Sciences, 1938.

Rickenbacker, Edward V. *Rickenbacker.* London: Hutchinson & Co. Ltd., 1967.

Riddle, Donald H. *The Truman Committee: A Study in Congressional Responsibility.* New Brunswick, New Jersey: Rutgers University Press, 1964.

Roseberry, C.R. *The Challenging Skies: The Colorful Story of Aviation's Most Exciting Years 1919–39.* Garden City, New York: Doubleday, 1966.

Rubin, Steven Jay. *The Complete James Bond Movie Encyclopedia.* Chicago, Illinois: Contemporary Books, 1995.

Sanders, C.J., and A.H. Rawson. *The Book of the C.19 Autogiro.* London: Sir Isaac Pitman and Sons, 1931.

Serling, Robert J. *From the Captain to the Colonel: An Informal History of Eastern Airlines.* New York: The Dial Press, 1980.

Shamburger, Page, and Joe Christy. *Command the Horizon: A Pictorial History of Aviation.* New York: A.S. Barnes and Co., Inc., 1968.

Sims, C.A. *British Aeroplanes Illustrated.* London: A. & C. Black Ltd., 1934.

Smith, Frank Kingston. *Legacy of Wings: The Story of Harold F. Pitcairn.* New York: Jason Aronson, 1981.

Smith, J.R., and Antony L. Kay. *German Aircraft of the Second World War.* London: Putnam, 1972.

Some Facts of Interest About Rotating-Wing Aircraft and the Autogiro Company of America. Philadelphia: The Autogiro Company of America, 1944.

Spenser, Jay P. *Whirlybirds: A History of the U.S. Helicopter Pioneers.* Seattle: University of Washington Press, 1998.

Spick, Mike. *Milestones of Manned Flight.* London: Salamander Books Ltd., 1994.

Stanley, P.H. *Historical Outline.* Glenside, Pennsylvania: Engineering Department, Autogiro Co. of America, 1952.

TAKEOFF! How Long Island Inspired America to Fly. Foreword by Nelson DeMille. Melville, New York: Newsday, Inc., 2000.

Tamate, Eiji. *Imperial Japanese Army Ka Go Autogiro* (in Japanese). Tokyo: Kojin Sha, 2002.

Taylor, H.A. *Fairey Aircraft Since 1915.* London: Putnam Aeronautical Books, 1974, 1988; Annapolis, Maryland: Naval Institute Press, 1988.

Taylor, John W.R. *Helicopters and VTOL Aircraft.* Garden City, New York: Doubleday, 1968.

Taylor, John W.R., and H.F. King. *Milestones of the Air: JANE'S 100 Significant Aircraft.* New York: McGraw-Hill, 1969.

Taylor, John W.R., and Kenneth Munson. *History of Aviation.* New York: Crown Publishers, Inc., 1972.

Taylor, Michael J.H., ed. *Brassey's World Aircraft & Systems Directory 1900/2000.* London: Brassey's, 1999.

Taylor, Michael J. H. *Jane's American Fighting Aircraft of the 20th Century.* New York: Modern Publishing, 1988.

Taylor, Michael J. H., and John W. R. Taylor. *Encyclopedia of Aircraft.* New York: G. P. Putnam's Sons, 1978.

Thomas, Kas. *Guide to Homebuilt Rotorcraft.* New York: Crown Publishers, 1976.

Tobin, James. *Ernie Pyle's War: America's Eyewitness to World War II.* Lawrence, Kansas: University Press of Kansas, 1997.

Townson, George. *Autogiro: The Story of 'the Windmill Plane.'* Fallbrook, California: Aero Publishers, 1st ed.; Trenton, New Jersey: Townson, 2d printing, 1985.

Trimble, William F. *High Frontier: A History of Aeronautics in Pennsylvania.* University of Pittsburgh Press, 1982.

Vanek, Jim. *Sport Copter.* Scappose, Oregon: Sport Copter, Inc., 2000.

Villard, Henry Serrano, and Willis M. Allen Jr. *Looping the Loop: Posters of Flight.* Hong Kong: Palace Press, International, n.d.

Walker, John, ed. *Halliwell's Film and Video Guide, 2001,* 16th ed. Great Britain: HarperCollinsPublishers, 2000.

Warleta Carrillo, José. *Autogiro: Juan de la Cierva y su Obra.* Madrid: Instituto de Espana, 1978.

Warleta Carrillo, José. *Autogyro.* Madrid: Instituto de Espana, 1977.

Winkowski, Fredric, and Frank D. Sullivan. *100 Planes, 100 Years: The First Century of Aviation.* New York: Smithmark Publishers, 1998.

Wood, Derek. *Project Cancelled: British Aircraft That Never Flew.* Indianapolis: The Bobbs-Merrill Company, Inc., 1975.

Wood, Tony, and Bill Gunston. *Hitler's Luftwaffe.* New York: Crescent Books, 1978.

World and United States Aviation and Space Records. Washington, D.C.: National Aeronautic Association, 1987.

Yenne, Bill. *Legends of Flight.* Lincolnwood, Illinois: Publications International, 1999.

Young, Warren R. *The Helicopters.* Alexandria, Virginia: Time-Life Books, 1982.

Zazas, James B. *Visions of Luscombe: The Early Years.* Terre Haute, Indiana: SunShine House, Inc. 1993.

ARTICLES

Abbott, Paul Bergen. "The $20,000 Takeoff!" *Rotorcraft* 33, no. 6 (September 1995): 44–47.

———. "Actions of the PRA Board of Directors." *Rotorcraft* 30, no. 4 (June–July 1992): 32.

———. "Air Command Fly-In." *Rotorcraft* 26, no. 4 (August–September 1988): 16–17.

———. "An Amazing National Convention." *Rotorcraft* 29, no. 6 (September 1991): 20, 30–31.

———. "The Best of the 1980's." *Rotorcraft* (December 1989–January 1990): 14–17.

———. "Big Bird Takes Off." *Rotorcraft* 26, no. 4 (August–September 1988): 12–13.

———. "An Evening With Ken Wallis." *Rotorcraft* 34, no. 6 (September 1996): 22–23, 26–27, 29, 30.

———. "FAA APPROVED!" *Rotorcraft* 29, no. 6 (September 1991): 8–9.

———. "From the First PRA Convention to the Latest: A Conversation with Ken Brock." *Rotorcraft* 34, no. 3 (May 1996): 8–10.

———. "Gyroplane Forum at Bensen Days." *Rotorcraft* 29, no. 4 (June–July 1991): 17–26.

———. "How Did Dick Do It?" *Rotorcraft* 33, no. 6 (September 1995): 31–32.

———. "Ken Brock—A Full and Wonderful Life." *Rotorcraft* 39, no. 9 (December 2001–January 2002): 5.

———. "Making History at Mentone: The 34th PRA Convention." *Rotorcraft* 34, no. 6 (September 1996): 5–7, 10–21.

———. "Marriage Made in Heaven." *Rotorcraft* 27, no. 8 (December 1989–January 1990): 12–13.

———. "Meet Ken Brock." *Rotorcraft* 31, no. 6 (September 1993): 12–17.

———. "A New Name at Air Command." *Rotorcraft* 28, no. 5 (August 1990): 19.

———. "New Shape from Santa Fe." *Rotorcraft* 27, no. 7 (October–November 1989): 33.

———. "The Outstanding New Rotorcraft Equipment of the 1980's." *Rotorcraft* 27, no. 8 (December 1989–January 1990): 17–19.

———. "Rotorcraft Choices 1989." *Rotorcraft* 27, no. 1 (February–March 1989): 22–30.

———. "A Standard Biennial Flight Review for Gyroplane Pilots." *Rotorcraft* 28, no. 6 (September Extra 1990): 20–21.

———. "The Story of Dr. Bensen's Fabulous Book." *Rotorcraft* 37, no. 5 (August 1999): 8–9.

———. "The Subaru Story." *Rotorcraft* 29, no. 7 (October–November 1991): 26–28.

———. "Two-Place Gyroplane Training Is Now Legal." *Rotorcraft* 28, no. 6 (September Extra 1990): 11–13.

———. "What Went Wrong at RotorWay?" *Rotorcraft* 28, no. 5 (August 1990): 18–19.

———. "Will this New Autogyro Go 400 Miles per Hour?" *Rotorcraft* 34, no. 6 (September 1996): 31–32.

Aellen, Richard. "The Autogiro and Its Legacy." *Air & Space Smithsonian* 4, no. 5 (December 1989–January 1990): 52–59.

Alderfer, Ed. "30 Years of Rotorcraft Designers." *Official Newsletter of Air Comand Mfg., Inc.* 3, no. 6 (Spring 1988): 2.

———. "The Designers: 30 Years of Rotorcraft Designers." *Rotorcraft* 27, no. 3 (May Extra 1989): 26–29.

————. "Federal Aviation Regulations Part 103—Ultralight Vehicles." *Rotor-craft* 27, no. 5 (August 1989): 28.

————. "Flying the Windmill Plane with Johnny Miller." *Popular Rotorcraft Flying* 24, no. 6 (December 1986): 14–16.

————. "A Home for PRA?" *Popular Rotorcraft Flying* 26, no. 5 (October–November 1988): 6.

————. "A Study of the Tractor Gyroplane." *Rotorcraft* 26, no. 6 (December 1988–January 1989): 17–18.

Alvares, Antonio Angulo. "Juan de la Cierva." *Rotorcraft* 37, no. 9 (December 1999–January 2000): 20–23.

Anders, Frank. "The Forgotten Rotorcraft Pioneer: Harold F. Pitcairn." *Rotor & Wing International* (May 1990): 34–37; reprinted as "The Forgotten Rotor." *Rotorcraft* 28, no. 7 (October–November 1990): 30–34.

————. "The Problem Solver." *Air Classics* 30, no. 10 (October 1994): 50–58.

Anderson, Rod. "CarterCopter (CC) R&D: Something for Everyone." *Rotorcraft* 38, no. 7 (October 2000): 32–35.

Ashby, Richard. "Come Fly With Me." *Knight* 5, no. 7 (July 1966): 63–65.

Attrill, Mark. "Photo Album: Cierva C.30." *Scale Aviation Modeller International* 6, no. 10 (December 2000): 818–819.

"Auction! Auction! Auction!" *Homebuilt Rotorcraft* 8, no. 9 (September 1995): 13.

"Autogiro." *Avion,* January 1963.

"Autogiro." *The New Yorker.* November 1, 1930.

"Autogiro in 1936." *Fortune* 13, no. 3 (March 1936): 88–93, 130–131, 134, 137.

"Autogiro Gives Air Stability." *United States Naval Institute Proceedings* 51, no. 4 (April 1925): 852–854.

"Autogiro News." *Autogiro Company of America,* October 1931, 1–6.

"Autogiros of 1931–1932." *Fortune* 9, no. 3 (March 1932) 48–52.

"'Autogyro' Flies from London to Paris." *United States Naval Institute Proceedings* 54, no. 11 (November 1928): 1010.

"Avian 1/180 Gyro May Be Revived." *Homebuilt Rotorcraft* 15, no. 11 (November 2002): 4.

"Aviatrix of the Autogiro." *Westchester Home Life,* August 1931, 15.

Bairstowe, L. "The Cierva-Auto-Gyro." *Nature,* October 31, 1925.

Ballentine, Commander J.J. "Aircraft Carriers." *Aeronautics* 3, no. 16 (December 18, 1940): 1011–1024.

Bartlett, Ron. "Groen Brothers Hawk 6G Gyroplane." *AutoGyro 1/4 ly* no. 8 (2001): 8.

————. "The UFO-Helithruster—An Update." *Autogyro 1/4ly* no. 11 (January 2002): 26–27.

Bartlett, Ron, and Kathy Fields. "Introducing Ron Bartlett and his Autogyro 1/4ly Magazine." *Rotorcraft* 40, no. 5 (August 2002): 10–11.

Barnett, Erlene. "A Twice Told Tall Tail." *Rotorcraft* 26, no. 6 (December 1989–January 1990): 6.

———. "What Kind of a Person Flies Coast to Coast." *Rotorcraft* 28, no. 1 (February–March 1990): 16–17.

Barnett, Jerrie. "A Gyroplane if an Aircraft." *Homebuilt Rotorcraft* 8, no. 11 (November 1995): 16–17.

———. "To Porpoise or Not to Porpoise, That Is." *Rotorcraft* 28, no. 6 (September Extra 1990): 17–18.

Bass, A. C. "Pilot Report: Umbaugh Gyroplane." *Flying* 70, no. 1 (January 1962): 44–45, 110–112.

"Beagle Wallis WA.116 Autogyro: Origin and Development." *Rotor Gazette International* no. 13 (May–June 1994): 7.

Beaty, Chuck. "Gyro Stability: Understanding PIO, Buntover, and How Gyroplane Rotors Work." *Rotorcraft* 33, no. 5 (August 1995): 18–23.

———. "Tails—Tall and Otherwise." *Popular Rotorcraft Flying* 26, no. 5 (October–November 1988): 15.

Beggs, Bob. "Sport Rotorcraft or A Tribute to the Other Igor." *American Helicopter Museum & Education Center Newsletter* 4, no. 1 (First Quarter 1997): 5.

Bennett, J. A. J. "The Era of the Autogiro (First Cierva Memorial Lecture)." *The Journal of the Royal Aeronautical Society* 65, no. 610 (October 1961): unpaginated.

———. "The Fairey Gyrodyne." *The Journal of the Royal Aeronautical Society* 53, no. 469 (April 1949).

———. "Rotary-Wing Aircraft." *Aircraft Engineering* (January/August 1940).

"Bensen Autogyros." *Aircraft of the World,* pamphlet no. 30. International Masters Publishers AB, 1998, 1–4.

"Bensen B-8M Gyrocopter Design Information." Sales brochure, Raleigh, North Carolina, Bensen Aircraft Company, 1969.

"The Bensen Gyrocopter." *AOPA Pilot* 23, no. 12 (December 1980): 48–50.

Bensen, Igor B. "Buying a Used Copter—Buyer Beware!" *Popular Rotorcraft Flying* 16, no. 2 (April 1978): 9, 12.

———. "Design Classroom: Bensen Model B-1." *Rotorcraft* 31, no. 4 (June–July 1993): 22.

———. "Needed: A Fresh Approach." *Popular Rotorcraft Flying* (Fall 1963): 2; Reprinted in *Sport Aviation* 14, no. 2 (February 1965): 44.

———. "Needed: More Leadership." *Popular Rotorcraft Flying* (Spring 1964): 2.

———. "President's Farewell Message: KEEP 'EM WHIRLING!" *Popular Rotorcraft Flying* 10, no. 2 (March–April 1972): 2, 28.

———. "Rotachute, Rotary Wing Glider-Kite." General Electric Co., Report No. 33200. Schenectady, New York, 1946.

———. "Wanted: Teamwork." *Popular Rotorcraft Flying* (Summer 1963): 2.

———. "What Will It Be." *Rotorcraft* 29, no. 4 (June–July 1991): 27–31.

———. "Where Do We Go From Here." *Rotorcraft* 29, no. 4 (June–July 1991): 25–26.

Bentley, Richard. "Mängoos 'Stealth' Gyroplane." *Rotorcraft* 29, no. 2 (April 1991): 42–43.

Blier, Mike. "Cross-Country in a Gyrocopter? You Must Be Kidding." *Popular Rotorcraft Flying* 13, no. 5 (October 1975): 21–22.

"Bois de la Pierre." *Fly Gyro!* no. 6 (July/August 2001): 19–21.

Bouchard, Daurent "Don." "The Subaru Story." *Rotorcraft* 29, no. 7 (October–November 1991): 26–28.

———. "A Visit to the RAF Gyro Factory." *Rotorcraft* 30, no. 2 (April 1992): 8–13.

Boyette, Ernie. "Ernie Finds the Lost Chord." *Rotorcraft* 29, no. 2 (April 1991): 15–21.

———. "A Tail of Tails." *Rotorcraft* 29, no. 4 (June–July 1991): 37–38.

Boyette, Mike. "What's the Advantage of a 4-Blade Rotor?" *Rotorcraft* 35, no. 9 (December 1997–January 1998): 44.

Bradley, Greg. "What's Wrong With Our Training?" *Homebuilt Rotorcraft* 14, no. 10 (October 2001): 11–12.

"Brave New Aircraft." *LIFE Magazine,* June 14, 1954.

Brie, Reginald A. C. "Practical Notes on the Autogiro." *Journal of the Royal Aeronautical Society* 4, (March 1939); Reprinted as "Pilot's Notes on Flying the Direct-Control Autogyro in 1939." *Rotorcraft* 34, no. 5 (August 1996): 19–21.

Brock, Ken. "My Adventure in Antarctica." *Rotorcraft* 29, no. 4 (June–July 1991): 20–23.

Brooks, Peter W. "Rotary Wing Pioneer." *Aeroplane* (December 9 and 16, 1955): 910–913, 940–943.

Brown, James. "Gyrocopters—The Australian Story." *Rotorcraft* 28, no. 5 (August 1990): 20–24.

Brownridge, David. "A Sycamore Seed for Grownups." *Western People* (December 5, 1996): 5–6; Reprinted in "RAF 2000…a 2 Place Cross Country Gyroplane!" *Rotary Air Force Marketing Inc.* (2000): 32–33.

Bruegger, Bruce. "Building the Dominator Gyro." *Rotorcraft* 28, no. 4 (June–July 1990): 37–39.

Brunak, Bolek, P. E. "Visions." *Popular Rotorcraft Flying* 14, no. 3 (June 1976): 25–28.

Bruty, Paul. "My China Training Trip." *Homebuilt Rotorcraft* 14, no. 9 (September 2001): 5–6.

Bundy, Glenn. "A Dream Dies Again, But Is It the Last Time?" *Rotorcraft* 39, no. 5 (August 2001): 41–42.

Buzzing Around With Chapters. *Popular Rotorcraft Flying* 8, no. 5 (September–October 1970): 8–9.

Buzzing Around With Chapters. *Popular Rotorcraft Flying* 9, no. 2 (March–April 1971): 6–8.

Capon, P. T. "Cierva's First Autogiros—Part 1." *Aeroplane Monthly* 7, no. 4 (April 1979): 200–205.

―――. "Cierva's First Autogiros—Part 2." *Aeroplane Monthly* 7, no. 4 (May 1979): 234–240.

"Captain Yancey Explores Mayan Ruins by Autogiro." *Autogiro News,* February 1932, 1–3.

Carroll, Thomas. "Relative Flight Safety of the Autogiro." *Aero Digest* 17, no. 7 (December 1930): 72.

Carter, Jay Jr. "CarterCopter—A High Technology Gyroplane." Paper presented at the American Helicopter Society Vertical Lift Aircraft Design Conference, San Francisco, California, 2000.

―――. "The CarterCopter (CC) Heliplane." Paper presented at the Precision Strike Technology Symposium 2000, Johns Hopkins University, Laurel, Maryland, October 2000.

Cartier, Kerry. "Dallas A&P Designs Nostalgic Tractor Autogyro." *Rotorcraft* 37, no. 4 (June–July 1999): 22–25.

―――. "A Real Texas Party—PRA 2002 Convention in Texas." *Rotorcraft* 40, no. 5 (August 2002): 4–9.

Cavin, Dick. "Here Come the Two-Cycles." *Popular Rotorcraft Flying* 26, no. 5 (October–November 1988): 26–29.

―――. "IKENGA…Reserve Grand Champion." *Sport Aviation* 38, no. 3 (March 1989): 49.

Charlet, Betty Jo. "Air Command Does It Again." *Rotorcraft* 27, no. 7 (October–November 1989): 38–39.

Charnov, Bruce H. "Amelia Earhart and the PCA-2: A Re-evaluation of the First Woman Autogiro Pilot." *FlyGyro!* (January/February 2001): 4–6.

―――. "Autogiros in the Soviet Union During WWII." *Homebuilt Rotorcraft* 15, no. 11 (November 2002): 8–10.

―――. "AUTOGIRO vs Focke-Wulf 109: A Forgotten Story of WWII Air Combat." *Homebuilt Rotorcraft* 15, no. 7 (July 2002): 10–11.

―――. "Back to the Future: The Rotec R 2800 Radial Engine." *Rotorcraft* 40, no. 8 (November 2002): 4–8.

―――. "Bensen Days, 2001." *FlyGyro!* (May/June 2001): 15–16.

―――. "Bensen Days 2001—First Impressions." *Rotorcraft* 39, no. 4 (June/July 2001): 4–7.

―――. "David Gittens: Afro-American Gyroplane Pioneer." *Rotorcraft* 40, no. 4 (June–July 2002): 18–21.

―――. "An Essential Autogiro/Autogyro/Gryrocopter/Gyroplane History Library." *Rotorcraft* 40, no. 6 (September 2002): 26–28.

―――. "The Fairey Rotodyne: An Idea Whose Time Has Come—Again, Part I." *FlyGyro!* (September/October 2001): 4–7.

―――. "The Fairey Rotodyne: An Idea Whose Time Has Come—Again, Part II." *FlyGyro!* (November/December 2001): 4–7.

―――. "The Fairey Rotodyne: An Idea Whose Time Has Come—Again, Part III." *FlyGyro!* (March/April 2002): 4–8.

———. "From Autogiro to Gyroplane: The Past, Present and Future of an Aviation Industry." *Rotorcraft* 40, no. 3 (May 2002): 6–7.

———. "Groen Brothers at Oshkosh." *FlyGyro!* no. 7 (September/October 2001): 16.

———. "A Gyroplane Film Trivia Quiz, Part II." *Rotorcraft* 39, no. 7 (October 2001): 21–23.

———. "Ken Brock—In Tribute." *Rotorcraft* 39, no. 9 (December 2001–January 2002): 9–11.

———. "Mel Morris Jones: A Tribute." *Rotorcraft* 40, no. 7 (October 2002): 22–23.

———. "More on Gyro Films and the Redisccovery of Richard Bentley." *Rotorcraft* 40, no. 5 (August 2002): 18–19.

———. "Rediscovering the Autogiro: Cierva, Pitcairn and the Legacy of Rotary-Wing Flight." *Hofstra Horizons* (Fall 2002): 3–7.

———. "Rotorfest 2000." *FlyGyro!* (November/December 2000): 19.

"The Cierva Autogiro Type C.30." Cierva Autogiro Co. Ltd. pamphlet, 1933.

"Cierva C.6 Autogiro." *Rotorcraft* 28, no. 4 (June–July 1990): 53.

Cierva y Codorníu, J. de la. "The Autogiro." *Journal of the Royal Aeronautical Society* 34, no. 239 (November 1930): 902–921.

———. "The Autogiro." Lecture at Cambridge University, England, November 8, 1928.

———. "The Autogiro—Its Future as a Service Aeroplane." *Journal of the Royal United Service Institution* (May 1928); Reprinted in *United States Naval Institute Proceedings* 54, no. 8 (August 1928): 696–701.

———. "The C-19 MK IV Type Autogiro—1931." reprinted in *Gyroplane World* 2, no. 2 (February 1979): 2–4.

———. "New Developments of the Autogiro." *Journal of the Royal Aeronautical Society* 39, no. 300 (December 1935): 1125–1143.

———. "A New Way to Fly." *Saturday Evening Post,* November 2, 1929.

———. "Rotary-Wing Aircraft." *Aircraft Engineering,* June 1934.

———. "Rotary-Wing Aircraft." *Cambridge University Engineering and Aeronautical Society's Journal* (1934).

———. "Uses and Possibilities of the Autogiro." *Aero Digest* 17, no. 7 (December 1930): 35.

Cox, Jack. "The KB-3, An Ultralight Gyroplane." *Sport Aviation* 35, no. 6 (June 1986): 47–48.

Curboy, Bob. "In Search of the Golden Rivet." *Rotorcraft* 28, no. 5 (August 1990): 27–29.

Darvassy, Helen R. "Breaking the Woman Barrier!" *Popular Rotorcraft Flying* 2, no. 1 (Winter 1964): 10–11, 20.

———. "First Showing of the HA-2M Sportster." *Popular Rotorcraft Flying* 14, no. 5 (October 1976): 8–10.

———. "Splinters." *Popular Rotorcraft Flying* 16, no. 2 (April 1978): 8–9, 12.

Darvassy, Louis. "'Lady-Bug' Is Born." *Sport Aviation* 14, no. 3 (March 1965): 16–18.

Davies, R.E.G. "Development of the Transcontinental Air Service." *American Aviation Historical Society Journal* 23, no. 1 (Spring 1978): 60–71.

"De La Cierva Autogiro Achieves More Success." *United States Naval Institute Proceedings* 52, no. 275 (January 1926): 142–143.

De Saar, Woody. "First PPL(G) on Air & Space 18-A in Ireland." *Rotorcraft* 39, no. 1 (February–March 2001): 11, 44–45.

Directory of PRA Members 1966. Raleigh, North Carolina, Popular Rotorcraft Association, Inc., 1966.

"Double Gyroplane Has Speed and Power." *Popular Science* 125, no. 2 (August 1934): 47.

Downie, Don. "Little Wing, Plenty of Disc." *Kitplanes* 17, no. 2 (February 2000): 10–13.

Dusek, Josef T. "Bohemian Report." *Rotorcraft* 38, no. 8 (November 2000): 8.

Duval, G.R. "Cierva C.30A." *Aero Modeller* 27, no. 320 (September 1962): 446–448.

Dwiggins, Don. "Hollmann HA-2M Sportster." *Homebuilt Aircraft* (October 1979).

Earhart, Amelia. "Your Next Garage May House An Autogiro." *Hearst's International* (combined with *Cosmopolitan*) 91, no. 2 (August 1931): 58–59, 160–161.

Eich, Jim. "Arliss Riggs and Twenty Years of Gyroplanes." *Popular Rotorcraft Flying* 15, no. 6 (December 1977): 14–16.

———. "The Eich JE-2 Two Seat Gyroplane." *Popular Rotorcraft Flying* 15, no. 2 (April 1977): 18–21.

———. "Jim Eich Travels to Paris, Montardoise, Avignon and Bois de la Pierre." *Rotorcraft* 32, no. 7 (October 1994): 31–35.

———. "GYROPLANE DESIGN." *Popular Rotorcraft Flying* 26, no. 2 (April 1988): 18–21.

———. "P.S. To 'Arliss Riggs and 20 Years of Gyroplanes.'" *Popular Rotorcraft Flying* 16, no. 2 (April 1978): 25.

———. "Putting a French Spin on Your Rotor." *Rotorcraft* 32, no. 7 (October 1994): 36–37.

———. "Sun 'n Fun at Lakeland Florida." *Rotorcraft* 29, no. 44 (June–July 1991): 39.

———. "The XNJ 790 Autogyro." *Rotorcraft* 29, no. 3 (May 1991): 10–13.

Etienne, Philippe. "A Vintage Report from Bordeaux." *Rotorcraft* 29, no. 3 (May 1991): 33.

Evans, Art. "Tail Feathers and Other Stuff." *Rotorcraft* 30, no. 7 (October–November 1992): 33.

———. "Tragedy in Arkansas." *Rotorcraft* 30, no. 3 (May 1992): 25.

"Exhibit Spotlight: Pitcairn PCA-1A." *Vertika* 7, no. 2 (October 2000): 5.

"Fairey Story: The Sad Saga of the Fairey Rotodyne." *WINGSPAN International* no. 8 (September/October 2001): 60–64.

Farrell, Morgan. "No Wings, No Rudders, No Ailerons: A Comparison of the Kellett and Pitcairn Autogiros." *Town and Country* 90, no. 4148 (March 15, 1935): 44–45.

Farrington, Don. "Rotorcraft Training." *Rotorcraft* 28, no. 4 (June–July 1990): 42.

Fetters, Linda González. "Mini-500 Accident Analysis." *Homebuilt Rotorcraft* 12, no. 4 (April 1999): 4–13.

"The First Fatal Autogiro Accident." *Flight* 47, no. 2234 (December 29, 1932).

"The First Fly-In—PRA International Fly-In: Raleigh-Durham Airport, June 15–16, 1963. (We Did It!)." *Popular Rotorcraft Flying* (Summer 1963): 8–11.

"First PRA Scholarship Recipients in the Cockpit." *Rotorcraft* 27, no. 7 (October–November 1989): 5, 9.

"First Two PRA CFI Scholarships Awarded." *Rotorcraft* 27, no. 3 (May Extra 1989): 19.

"First UK Lady Gyro Flyer For 50 Years." *Rotor Gazette International* no. 13 (May–June 1994): 5.

Fischer, Albert G. "Germany's VFW Sports a Hollow-Bladed Helicopter-Autogyro—with a McCulloch!" *Popular Rotorcraft Flying* 5, no. 3 (September 1967): 31.

"The Flat Risers: The Ups & downs of VTOL (Part 2)." *TakeOff* 1, no. 5 (1989): 138–143.

"Flying A Kite: The Focke-Achgelis Fa-330 Rotary Wing Kite." *Rotor Gazette International* 1, no. 2 (July–August 1992): 1–2, 4.

"Focke Achgelis FA-330 Bachstelze (Wagtail)." *FlyGyro!* no. 3 (February 2001): 8–9.

Focke, Heinrich K.J. "German Thinking on Rotary-Wing Development: (5th Cierva Memorial Lecture)." *Journal of the Royal Aeronautical Society* 69 (May 1965): 293–305.

Fourcade, Jean. "Longitudinal Stability of Gyroplanes." *Rotorcraft* 37, no. 4 (June–July 1999): 14–19.

Franks, Richard A. "French Model! Lioré-et-Olivier C.30." *Scale Aviation Modeller International* 6, no. 10 (October 2000): 814–817.

"A Fun Vehicle That Flies." *Mechanix Illustrated* 66, no. 502 (March 1970): 60–61, 145–146.

Gallager, Sheldon M., and Howard Levy. "New Build-Your-Own Copters: More Power, Looks, Convenience." *Popular Mechanics* 140, no. 6 (December 1973): 143–145.

Garrison, Peter. "Everybody Loves an Autogyro." *Flying* 88, no. 2 (February 1971): 66–68.

Gibbings, David. "Rotodyne, the Airliner Whose Hour Came Too Soon." *Proceedings: Society of Flight Engineers,* 26th Annual Symposium, Berlin, June 1995.

Gilley, Rick. "Dr. Igor Bensen: The Man and His Machine." *FlyGyro!* (September/October 2001): 20–22.

Gittens, David, and Kia Woods. "Ikenga, An Artist's Approach to Gyroplane Design." *Rotorcraaft* 27, no. 1 (February–March 1989): 44.

"Glimpses of History." *Popular Rotorcraft Flying* 2, no. 4 (Fall 1964): 13.

Goldsberry, Gary. "Along Time Dream is Coming True." *Rotorcraaft* 27, no. 1 (February–March 1989): 15.

Goldsberry, Shelly. "Middletown PRA Convention '88." *Rotorcraft* 26, no. 5 (October–November 1988): 18–21.

Gunther, Carl R. "Autogiro: The World's First Commercially Successful Rotary-wing Aircraft." *Popular Rotorcraft Flying* 17, no. 5, 7 (October–December 1979).

Graves, Steve. "Where Do We Go From Here?" *Rotorcraft* 30, no. 1 (February–March 1992): 45–46.

"The Great Silver Fleet News." *Eastern Air Lines* 10, no. 4 (July–August 1946).

Gregg, E. Stuart. "Jump Ship." *Smithsonian Air and Space* 15, no. 6 (March 2001): 14–15.

Gremminger, Stephanie. "Flying 'Team' Makes Transition to Gyroplanes." *Rotorcraft* 36, no. 6 (September 1998): 24.

———. "GyroTECH, Inc. Aims for the Entry Level Builder & Pilot." *Rotorcraft* 38, no. 5 (August 2000): 30–31.

———. "The Hawk 4 Gyroplane Aims to Bring 'Gyroplanes to Their Logical Progression.'" *Rotorcraft* 38, no. 5 (August 2000): 8–9, 35.

Gremminger, Steph[anie], and Greg Gremminger. "Choices, Choices, Choices: 2000 Rotorcraft Directory." *Rotorcraft* 38, no. 1 (February–March 2000): 17–27.

Grey, C.C., and Harold F. Pitcairn. "Autogiro or Airplane—Which?" *Everyday Science and Mechanics* 2, no. 12 (November 1931): 658.

"Gyro 2000 Closes Doors." *Rotorcraft* 30, no. 2 (May 1992): 7.

"Gyrocopter Accident." *Gyroplane World* no. 4 (January 1977): 5.

"Gyroglider In Smithsonian." *Popular Rotorcraft Flying* 3, no. 3 (Summer 1965): 19.

Haaskarl Jr., Robert A. "Early Military Uses of Rotary-Wing Aircraft." *The Air Power Historian* 12, no. 3 (July 1965).

Hafner, R. "British Rotorcraft." *Journal of the Royal Aeronautical Society* 70, no. 661 (January 1966).

———. "Domain of the Convertible Rotor." *Journal of Aircraft* 1, no. 6 (November–December 1964).

"The Hafner Rotachute." *FlyGyro!* no. 3 (February 2001): 18–19.

Hager, Uwe P. "Gyros in Germany." *FlyGyro!* no. 6 (July/August 2001): 9.

Hannon, Bill. "Those Infuriating "Palm Trees." *Popular Rotorcraft Flying* 7, no. 6 (November–December 1969): 30.

Hardee, Cindy. "The Coning Angle in Your Flower Arrangement and the Blades Behind Your Sofa." *Rotorcraft* 32, no. 7 (October 1994): 24.

Hardee, LeRoy. "The 1995 PRA Gyro Convention." *Homebuilt Rotorcraft* 8, no. 9 (September 1995): 4–8.

———. "Snowbird Gyroplanes Have New Owners." *Homebuilt Rotorcraft* 8, no. 11 (November 1995): 8–10.

———. "Thoughts on Our New PRA Home." *Rotorcraft* 33, no. 5 (August 1995): 11.

Harrison, Jean-Pierre. "Fairey Rotodyne." *Air Classics* 22, no. 44 (April 1996): 44–47, 60–62, 64–66, 79–80.

Haugen, Victor Lieutenant. "Principles of Rotating Wing Aircraft." *Aeronautics* 2, no. 7 (October 16, 1940): 420–438.

"Hawk 4T Breathes New Life Into Gyroplanes." *Aviation Week and Space Technology* 153, no. 19 (November 2000): 54–56.

"HAWK 4 HOMELAND DEFENDER: A HIT ON CAPITOL HILL." *Rotorcraft* 40, no. 6 (September 2002): 13.

"Helicopter or Autogiro?" *Flying Cadet* 2, no. 2 (February 1944): 46.

Hengel, Paul. "Portrait of a Pioneer Rotary-Wing Pilot." *American Helicopter Museum and Education Center Newsletter* 2, no. 3 (Summer 1995): 3.

Herron, Ron. "Bringing Back the Autogiro." *Rotorcraft* 33, no. 1 (February–March 1995): 12–13.

———. "First Flight of a New Tractor Autogyro." *Rotorcraft* 34, no. 8 (November 1996): 6–7.

———. "Flying Backwards in a Tractor Autogyro." *Rotorcraft* 33, no. 5 (August 1995): 32–34.

———. "A Gyrocopter That Looks Different?" *Homebuilt Rotorcraft* 8, no. 6 (June 1995): 13–16.

———. "History of Little Wing Autogyros." *FlyGyro!* no. 4 (March–April 2001): 12–14.

———. "Little Wing 'Roto-Pup.'" *Homebuilt Rotorcraft* 8, no. 11 (November 1995): 12–13.

———. "Maiden Voyage for LW-3." *Homebuilt Rotorcraft* 9, no. 11 (November 1966): 13.

Hessenaur, Donald P. "Avian Vibration Encounters." *Homebuilt Rotorcraft* 15, no. 11 (November 2002): 5.

Hill, Alexander. "'Sandy' 'Alida' 'Lisa' De Vries, Master CFI." *Rotorcraft* 39, no. 4 (June–July 2001): 30.

Hill, Norman. "Wingless Combat." *Royal Air Force Flying Review* 18, no. 4 (January 1963): 24–25, 57.

Hilton, Maj. R. "The Alleged Vulnerability of the Autogiro." *The Fighting Forces* 11, no. 3 (August 1934).

Hodgess, F.L. "The Weir Autogiros." *Helicopter World* (July–November 1964).

Hollmann, Martin. "The Avian Gyroplane." *Gyroplane World* no. 4 (January 1977): 2–3.

———. "Designing Rotor Blades." *Rotorcraft* 39, no. 9 (December 2001–January 2002): 18–23.

———. "The Focke-Achgelis Fa 330 Gyroplane Kite." *Gyroplane World* no. 14 (November 1977): 1.

———. "Glanville Aircraft." *Gyroplane World* no. 23 (August 1978): 4.

———. "Gyroplane Designs." *Rotorcraft* 38, no. 5 (August 2000): 32–35.

———. "The HA-2 Sportster, An Ultralight Two Place Gyroplane." *Sport Aviation* 24, no. 1 (January 1975): 16–20.

———. "HA-2M Sportster Tested." *Popular Rotorcraft Flying* 13, no. 4 (December 1975): 34.

———. "A New Wind is Being Blown." *Gyroplane World* no. 13 (October 1977): 2–3.

———. "Notes on Gyroplane Rotor Design." *Popular Rotorcraft Flying* 15, no. 2 (April 1977): 27–31.

———. "One of the Last C.30A Autogiros Found in Australia." *Gyroplane World* no. 26 (November 1978): 2–3.

———. "The Pitcairn AC-35." *Gyroplane World* no. 27 (December 1978): 3–4.

———. "Pusher Gyroplanes, Increasing Interest." *Gyroplane World* no. 26 (November 1978): 1.

———. "The Rotachute." *Gyroplane World* no. 7 (April 1977): 1–3.

———. "Sportster Imitation, Buyer Beware." *Gyroplane World* 2, no. 8 (August 1979): 4.

Houston, Stewart S. "Eight Years of Gyroplane Research in the UK—a *Very* Personal Reflection." *FlyGyro!* (January–February 2001): 12–16.

Hovgard, Paul E. "Safety—With Performance." *Aviation Engineering,* September 1931, 12.

Howe, Richard. "Kellett KD-1/YG-1 Autogyro." *American Aviation Historical Society Journal* 23, no. 1 (First Quarter 1978): 49–50.

Hufton, P. A., and A. E. Nutt. "General Investigation into the Characteristics of the Cierva C.30 Autogiro." Air Ministry Aeronautical Research Committee Report and Memoranda No. 1859, London, 1939.

Iaconis, Ron. "Museum Endowment Program." *Rotorcraft* 27, no. 6 (September Extra 1989): 34.

"Icarus II." *The Dude* 3, no. 2 (November 1958): 22–23.

"Igor Bensen Announces Plans for New Aircraft, New Organization." *Rotorcraft* 30, no. 1 (February–March 1992): 15.

Ingalls, David S. "Autogiros—Missing Link." *Fortune* 8, no. 3 (March 1931): 77–83, 103–104, 106, 108, 110.

Ishikawa, Akira. "First Japanese Homebuilt Helicopter Fly-In." *Homebuilt Rotorcraft* 7, no. 11 (November 1994): 18–20.

———. "Japan Jumps into Rotorcraft." *Rotorcraft* 29, no. 4 (June–July 1991): 32–33.

———. "Japan PRA." *Rotorcraft* 33, no. 9 (December 1995–January 1996): 18.

———. "Third Annual Japanese Gyro Fly-In." *Homebuilt Rotorcraft* 15, no. 9 (September 2002): 6–7.

————. "World Altitude Record Set in Dominator Gyro." *Homebuilt Rotorcraft* 11, no. 6 (June 1998): 12–14.

————. "World's 1st Flying Subaru!: A Report on Very Early Flying Use of the Subaru Engine in Japan." *Rotorcraft* 30, no. 3 (May 1992): 8–10.

Issacs, Keith. "Project Skywards." *Rotorcraft* 32, no. 4 (June–July 1994): 6–9; Originally in *Air International* (July 1975) and later reprinted in *Army Motors* Military Vehicle Preservation Club (Summer 1981) and again reprinted in *International AUTOGYRO 1/4ly* no. 10 (October 2001).

"The Jim Montgomerie Story: A Man and His Gyros." *Rotor Gazette International* no. 17 (January–February 1995): 3–5, 10–11.

Jones, Mel Morris. "Talkshop—A Conversation With Ralph Taggart." *FlyGyro!* no. 6 (July 2001): 4–6.

————. "Talkshop: A Conversation With Jukka Tervamäki." *FlyGyro!* no. 1 (September–October 2000): 4–10, 18.

————. "Talkshop: The UFO Helithruster." *FlyGyro!* no. 2 (November–December 2000): 8–11.

Jordanoff, Assen. "Will Autogiros Banish Present Planes?" *Popular Science Monthly* 118, no. 3 (March 1931): 28–30, 146.

"Kamov & Skrzhinsky—Russian Gyroplane Pioneers." *International AUTO-GYRO 1/4ly* no. 10 (October 2001): 18–23.

"The Kay Gyroplane Type 331." *Aeroplane* 24, no. 12 (December 26, 1996): 32–34.

"Ken Wallis Keeps Busy." *Rotorcraft* 27, no. 4 (June–July 1989): 4.

King, Russ. "From Lessons to Solo." *Rotorcraft* 36, no. 4 (June–July 1998): 7.

Kirk, Joe. "Gizmo." *Gyroplane World* no. 10 (July 1977): 1–3.

Klemen, Alexander. "A Debate About the Autogiro." *Scientific American,* November 1931, 337.

————. "A Rival of the Autogiro?" *Scientific American,* November 1931, 336–337.

Klimt, Claudius. "Roland Stagl: A Rising Star, Gyro Designer from Austria." *Rotorcraft* 39, no. 6 (September 2001): 30–33.

Kohn, Leo J. "Mr. Cirva and His Autogiros." *Air Classics* 15, no. 6 (June 1979): 87–93.

Koman, Victor. "PRA Computer Bulletin Board." *Rotorcraft* 29, no. 6 (September 1991): 9.

Kurylenko, George. "Autogyros of Present and Past." *Sport Aviation* 10, no. 6 (June 1961).

"Lady Bugs, United." *Popular Rotorcraft Flying* 2, no. 2 (Spring 1964): 17.

Larsen, Agnew E. "Autogiro Development." *Aviation Engineering* (November 1932): 15.

"Latest Merlin GTS." *FlyGyro!* no. 4 (March–April 2001): 16–17.

Lawton, Peter. "Flying a Wallis Autogyro." *Pilot,* January 1998, 26–28.

Levy, Howard. "Italian Import: The Magni Gyroplane Makes Its U.S. Debut." *Kitplanes* 18, no. 2 (February 2001): 41–44.

————. "Kellett Gyrations." *Aeroplane* 24, no. 1 (January 1996): 32–34.

————. "Lift Without Wings?" *Kitplanes* 15, no. 5 (May 1998): 61–63.

Liptrot, R. N. "Historical Development of Helicopters." *American Helicopter,* March 1947.

————. "Rotating Wing Activities in Germany During the Period 1939–1945." British Intelligence Objectives Subcommittee Overall Report no. 8, London, 1948.

"Lone Eagle Gyro Pilot Sets Coast to Coast Record." *Rotorcraft* 27, no. 8 (December 1989–January 1990): 25.

Love, Sandy. "The Sky's No Limit." *Rotorcraft* 36, no. 4 (June–July 1998): 34–35.

Lunde, Otto H. "Development and Operation of the Autogiro." *Aeronautics* 2, no. 9 (October 30, 1940): 548–562.

MacKay, Hal. "Bug Fighters." *Popular Aviation* 24 (June 1939): 48–50, 82.

Magni, Vittorio. "Machine Was a Dream Come True." *Rotorcraft* 38, no. 3 (May 2000): 12.

Marsh, Alton K. "Son of a Pioneer." *AOPA Pilott* 44, no. 4 (April 2001): 138–143.

Martin, Dave. "Evolving While Revolving: Jim Vanek's New Sport Copters Continue a Family Tradition." *Kitplanes* 12, no. 4 (June 1995): 36–41.

————. "Investing in Rotorcraft Safety." *Kitplanes* 5, no. 4 (April 1988): 40–47.

Mayfield, Jim. "Convereting the Mazda RX-7 Rotary Powerplant." *CONTACT! Experimental Aircraft and Powerplant Newsforum for Designers and Builders* 4, no. 6, issue 23 (November–December 1994): 2–10.

McCutchen, Jim. "An Open Letter to All PRA Members." *Rotorcraft* 30, no. 3 (May 1992): 19.

McDougall, Harry. "Avian Gyroplane." *Flying* 74, no. 4 (April 1964): 44–45, 79–80.

"Meet the Board: Carl Schneider." *Rotorcraft* 36, no. 9 (December 1998–January 1999): 10.

Menzie, Ron. "From England: Gyroplane Training." *Rotorcraft* 28, no. 2 (April 1990): 22–24.

————. "Ice 90—The Antarctica Expedition." *Rotorcraft* 28, no. 3 (May Extra 1990): 27–29.

Merkel, Howard. "The Long Way Home—Part 2." *Rotorcraft* 29, no. 2 (April 1991): 50–52.

————. "The Long Way Home—Part 3." *Rotorcraft* 29, no. 3 (May 1991): 24–31.

————. "The Long Way Home—Part 4." *Rotorcraft* 29, no. 4 (June–July 1991): 42–45.

Meyer, Shirley. "Amelia Earhart, Autogiro Pilot." *Popular Rotorcraft* 16, no. 66 (December 1978): 16–17.

Miller, Guy. "Autogiro Flight Instruction." *Aviation Engineering* (September 1932): 24.

Miller, John M. "Civil Uses of the Autogiro." *Aeronautics* 2, no. 10 (November 6, 1940): 611–624.

———. "The First Scheduled Rooftop Flying Operation in Aviation (Autogiro Air Mail Service at Philadelphia, 1939–40)." *Rotorcraft* 30, no. 6 (September 1992): 24–33.

———. "The First Transcontinental Flights with a Rotary-Wing Aircraft 1931." *Rotorcraft* 30, no. 5 (August 1992): 11–19.

———. "The First Transcontinental Rotary-Wing Flight." *Vertika: The Newsletter of the American Helicopter Museum and Education Center* 7, no. 2 (October 2000): 4.

———. "The First Transcontinental Rotary-Wing Flight—Part 3." *Vertika: The Newsletter of the American Helicopter Museum and Education Center* 8, no. 1 (February 2001): 4.

———. "The Missing Link in Aviation." *Popular Mechanics Magazine* 70, no. 3 (September 1938): 346–351, 134A–135A; Reprinted in *Homebuilt Rotorcraft* 9, no. 1 (January 1996): 4–6; *Homebuilt Rotorcraft* 15, no. 11 (November 2002): 14–16.

———. "Test Flying for Kellett Autogiro Corporation." *Rotorcraft* 30, no. 7 (October–November 1992): 22–28.

———. "UFO Recollections—The Death of Charlie Otto." *American Helicopter Museum and Education Center Newsletter* 8, no. 1 (Spring 1996): 3.

Milton, Tom. "A Gyroplane Movie Trivia Quiz." *Rotorcraft* 34, no. 3 (May 1996): 39.

Mitchell, William. "The Automobile of the Air." *Women's Home Companion,* May 1932.

Montgomerie, Jim. "All Ok in the UK." *Rotorcraft* 28, no. 2 (April 1990): 15.

———. "Autogyro Basics and World Record Flights." *Rotorcraft* 27, no. 2 (April 1989): 12–13.

"The National Advisory Committee for Aeronautics." included in Charles Greeley Abbot, *Great Inventions.* Washington, D.C.: The Smithsonian Institution Series (formerly Smithsonian Scientific Series) 12 (1949): 233–238.

"The Navy's Autogiro." *United States Naval Institute Proceedings* 57, no. 8 (August 1931): 1118–1119.

Neal, Larry. "Pilot Report: CarterCopters® Technology Demonstrator Part 1." *Rotorcraft* 40, no. 2 (March–April 2002): 44–45.

———. "Pilot Report: CarterCopters® Technology Demonstrator Part 2." *Rotorcraft* 40, no. 5 (August 2002): 14–16.

"Negative Gravity: An Accident Analyzed." *Gyroplane News and Small Helicopter* no. 1 (Spring 1990).

Nelson, Lieutenant Commander (CC) William. "The Autogiro as a Military Craft." *United States Naval Institute Proceedings* 57, no. 8 (August 1931): 1092–1095.

"A New Force Behind Air Command." *Rotorcraft* 28, no. 7 (October–November 1990): 9.

"A New Owner for Air Command." *Rotorcraft* 30, no. 4 (June–July 1992): 6.

"Newly Returned to the Rotorcraft Scene: The Air & Space 'HELIPLANE.'" *Rotor Gazette International* no. 13 (May–June 1994): 3–4, 6.

Nix, Maria. "The Hoosier Heliplane—Alive and Well." *Rotorcraft* 34, no. 9 (December 1996–January 1997): 14–16.

Noorduyn, Robert B.C. "Pitcairn PA-19 Cabin Autogiro." *Aero Digest* 22, no. 2 (February 1933): 48–50.

Nye, Willis N. "Pitcairn Cierva C-8." *American Aviation Historical Society Journal* 11, no. 4 (Winter 1966): 278–279.

O'Connor, Doug. "Are Gyroplanes Safe?—Part 2." *Rotorcraft* 34, no. 2 (April 1996): 35–39.

———. "Mixing Rotorcraft and Fixed Wingers." *Rotorcraft* 32, no. 7 (October 1994): 27–28.

O'Connor, Eva. "So Your Husband Wants to Build a Gyroplane." *Rotorcraft* 32, no. 7 (October 1994): 25.

O'Leary, Michael. "It's a Kellett!" *Air Classics* 38, no. 6 (June 2002): 68–72.

"One-Man Rotary Wing Craft." *Air Progress* 16, no. 3 (June/July 1964): 77–79.

Organ, Dave. "Roland Parsons, A Tribute." *International Autogyro 1/4ly* no. 11 (January 2002): 3.

"Oskar Westermayer—Designer Extraordinary." *Autogyro 1/4ly* no. 8: 16–18; Reprinted in *Rotorcraft* 40, no. 5 (August 2002): 24–26.

Parham, Don. "Bensen Days & Sun 'N Fun 2001." *Homebuilt Rotorcraft* 14, no. 5 (May 2001): 9–13.

———. "The Bruno Nagler Story." *Homebuilt Rotorcraft* 12, no. 3 (March 1999): 8–12.

———. "Certification of Amateur-Built Rotorcraft." *Homebuilt Rotorcraft* 7, no. 4 (April 1994): 20.

———. "Details of Bill Parsons' Accident." *Homebuilt Rotorcraft* 10, no. 5 (May 1997): 4.

———. "From the Editor's Desk." *Homebuilt Rotorcraft* 12, no. 1 (January 1999): 2.

———. "High Tech Rotorcraft." *Homebuilt Rotorcraft* 12, no. 10 (October 1999): 6–11.

———. "Magni Gyros—Quality Gyroplanes From Europe." *Homebuilt Rotorcraft* 14, no. 6 (June 2001): 6–7.

———. "Nagler'sVertigyro." *Homebuilt Rotorcraft* 14, no. 7 (July 2001): 8–10.

———. "Pilot's Notes on Flying the Direct-Control Autogyro in 1939." *Rotorcraft* 34, no. 5 (August 1996): 19–21.

———. "Sport Pilot/Light Sport Plane." *Homebuilt Rotorcraft* 14, no. 9 (September 2001): 14–16.

Parker, David. "THE FAIREY ROTODYNE—Nearly The Answer." *West Coast Aviator* 5, no. 1 (September/October 1995): 35–37.

Parsons, Bill. "Astronaut Jim Irwin Learns to Fly Gyro." *Popular Rotorcraft Flying* 26, no. 1 (February 1988): 19, 22.

———. "Why I Prefer the Tandem Trainer." *Rotorcraft* 31, no. 5 (August 1993): 21–22.

Peck, William C. "Landing Characteristics of an Autogiro," pamphlet. Washington, D.C.: National Advisory Committee for Aeronautics, 1943.

"The People Come First: 9th International PRA Fly-In." *Rotorcraft* 9, no. 5 (September–October 1971): 18–21.

Peters, Max. "Shepherders and Subarus." *Rotorcraft* 29, no. 5 (August 1991): 36.

Piper, William. "The Piper-Marriott Autogyro." *Popular Rotorcraft Flying* 7, no. 6 (November–December 1969): 10–13.

Pitcairn, Harold F. "The Autogiro Answers Its Critics." *Aviation* (April 1932); Reprinted in "The Autogiro." *Autogiro Company of America*, 1932, 73–82.

———. "The Autogiro: Its Characteristics and Accomplishments." *Smithsonian Report for 1930,* Washington, D.C., 1931, 265–277.

———. "The Autogiro as I See It." *Aviation* no. 30 (November 1931): 630–632.

———. "Juan de la Cierva: In Memoriam." Booklet by the Autogiro Company of America, January 9, 1937.

Pitcairn, Stephen. "Flying the Pitcairn PCA-2 Autogiro." *Strut & Axle* 11, no. 2 (1989): 6–15; Also published in *Vintage Airplane* 17, no. 5 (May 1989): 16.

"PRA meets FAA." *Popular Rotorcraft Flying* 2, no. 3 (Summer 1964): 5–6.

"Preservation Report: Shuttleworth's Cierva C.30A G-AHMU." *Aeroplane Monthly* 12, no. 9 (September 1984): 490–491.

Prewitt, Richard H. "Possibilities of the Jump Take-Off Autogiro." *Journal of the Aeronautical Sciences* (November 1938).

"Profile: Wing Commander K H Wallis." *Popular Flying* (April–May 1996): 13–21.

Pynchon, George Jr. "Something About the Autogiro." *Town & Country* 86, no. 4062 (August 15, 1931): 46–47.

"The Queer Birds: Air & Space Model 18-A." *Flying* 76, no. 4 (April 1965): 40–41, 45–46.

Ray, James G. "Is the Autogiro Making a Comeback?" *Flying* 66, no. 1 (January 1960): 34–35, 91–92.

Reed, Lt. Colonel Boardman C. "The Forgotten Rockne Crash." *Vintage Airplane* 17, no. 1 (January 1989): 23–24.

Regnier, Norm. "The French Gyro Connection." *Homebuilt Rotorcraft* 7, no. 11 (November 1994): 4–7.

Renner, Dennis. "The Building of an 'Angel.'" *Rotorcraft* 28, no. 4 (June–July 1990): 36.

"Return of the Flight of the Phoenix." *Popular Flying* (January–February 1971): 10–11.

"The Reverend Igor B. Bensen." *Popular Rotorcraft Flying* 7, no. 4 (July–August 1969): 18.

Reynoso, Fred E. "The Story of a Lonely Gyronaut." *Popular Rotorcraft Flying* 25, no. 6 (December 1987): 16–17.

Riviere, Pierre, and Gerry Beauchamp. "Autogyros At War." *Air Classics Quarterly Review* 3, no. 4 (Winter 1976): 92–97.

"The Road Warrior." *Cinescape* no. 63 (August 2002): 102.

"Roadable Autogiro." *AVIATION* 35, no. 11 (November 1936): 33–34.

Rose, Donald R. "Pitcairn Aircraft." *Aero Digest* 10 (March 1927): 176–177.

"Rotary-wing Aircraft." *Flying* 67, no. 4 (October 1960): 24–26, 100–101.

"Rotorcraft Trainers." *Popular Rotorcraft Flying* (Summer 1963): 12.

"Rotodyne Demonstrates VTOL Features." *Aviation Week* 69, no. 14 (October 6, 1958).

"Salon Stars: A Selection of Types at the November 1936 Paris Salon Aéronautique." *Air Enthusiast* no. 91 (February–January 2001): 2–6.

Sanders, Bill. "The Rebirth of N4353G and N4364G." *Rotorcraft* 28, no. 4 (June–July 1990): 14–19.

Schonheer, Rudolf. "Gyrocopters in Germany." *Rotorcraft* 28, no. 7 (October–November 1990): 50.

Scott, William B. "Hawk 4T Breathes New Life into Gyroplanes." *Aviation Week & Space Technology* 153, no. 19 (November 6, 2000): 54–56.

Sikorsky, Igor. "Commercial and Military Uses of Rotating Wing Aircraft." *Proceedings of the Second Annual Rotating Wing Aircraft Meeting*. Institute of the Aeronautical Sciences, 1939.

Smith, Frank Kingston. "Mr. Pitcairn's Autogiros." *Airpower* 12, no. 2 (March 1983): 28–49.

Sottile, Jim. "RAF Makes Dreams Come True." *Rotorcraft* 32, no. 6 (September 1994): 11–13.

———. "RAF Visits Long Island." *Rotorcraft* 30, no. 4 (June–July 1992): 14–17.

Springer, Marion. "Accident Prone and Macho Personalities." *Homebuilt Rotorcraft* 12, no. 6 (June 1999): 13–14.

———. "Endorsed for Solo." *Rotorcraft* 29, no. 5 (August 1991): 28–30.

———. "Grant Me Patience, Oh Lord, But Hurry!" *Homebuilt Rotorcraft* 7, no. 2 (February 1994): 4–5.

———. "In Pursuit of a Gyroplane Rating." *Rotorcraft* 27, no. 3 (May Extra 1989): 16–19.

———. "No Training Wheels, Please." *Rotorcraft* 28, no. 4 (June–July 1990): 40.

———. "Story of a Blue Angel." *Popular Rotorcraft Flying* 8, no. 1 (January–February 1970): 20–21.

———. "What's Out There?" *Homebuilt Rotorcraft* 12, no. 12 (May 1999): 14–15.

"Staff Interview—Ken Brock—From Long Beach to Kitty Hawk......By Gyrocopter." *Popular Rotorcraft Flying* 9, no. 5 (September–October 1971): 28–31.

Stiles, Tony. "Let's Talk Subaru." *Rotorcraft* 32, no. 3 (May 1994): 24–26.

———. "More Subaru News." *Rotorcraft* 30, no. 1 (February–March 1992): 16–17.

———. "Speaking of Subaru." *Rotorcraft* 32, no. 4 (April 1994): 17–20.

Stone, Tony. "Weekend Training Courses." *Rotorcraft* 29, no. 2 (April 1991): 39–40.

Stump, Michael J. "The Commander Gyroplane—Is It For You?" *Homebuilt Rotortcraft* 10, no. 8 (August 1997): 6–9.

Sutcliffe, Glyn. "Autogyros Big and Small." *Aviation Modeller International* 7, no. 9 (August 2002): 20–23.

Taggart, Ralph E. "How Far Can You Fly." *Rotorcraft* 34, no. 1 (February–March 1996): 33–35.

———. "Making the Most of Part 103: The Gyrobee." *Rotorcraft* 28, no. 7 (October–November 1990): 45–49.

———. "A New Bee on the Block." *Rotorcraft* 37, no. 6 (September 1999): 31–35.

———. "Rotors Over Ohio." *Kitplanes* 11, no. 2 (February 1994): 66–71.

———. "Rotorbyte BBS Update." *Rotorcraft* 30, no. 1 (February–March 1992): 24–25.

———. "The Subaru Story." *Rotorcraft* 29, no. 7 (October–November 1991): 26–28.

———. "What About Those Autogyros." *Rotorcraft* 29, no. 7 (October–November 1991): 32–34.

Tasker, Peter. "Historic Wings X-25A/B." *AUTOGYRO 1/4ly* no. 6 (2000): 29–31.

Temple III, Lt Col L. Parker. "Of Autogyros and Dinosaurs." http://www.airpower.maxwell.af.mil/airchronicles/apj/apj88/temple.html. Accessed January 15, 2003.

Tervamäki, Jukka. "Is An Electric Autogyro a Real Possibility?" *FlyGyro!* no. 2 (November–December 2000): 14–17.

———. "Losing Faith in Autogyros and Gaining It Back Again." *Sport Aviation* 20, no. 5 (May 1971): 40–41.

———. "New Super Sleek Autogyro From Finland." *Popular Rotorcraft Flying* 11, no. 3 (August 1973): 24.

———. "The Sleek New JT-5 From Finland." *Sport Aviation* 23, no. 2 (February 1974): 39–41, 61.

———. "Some Thoughts of Autogyro Design: Part One." *Sport Aviation* 14, no. 11 (November 1965): 6–9.

———. "Some Thoughts of Autogyro Design: Part Two." *Sport Aviation* 15, no. 2 (February 1966): 11–13.

———. "Some Thoughts of Autogyro Design: Part Three." *Sport Aviation* 15, no. 4 (April 1966): 36–37.

———. "Tevarmaki JT-5." *Sport Aviation* 23, no. 2 (February 1974): 22–23.

Tervamäki, Jukka, and A[ulis] Eerola. "The ATE-3 Project." *Popular Rotorcraft Flying* 7, no. 4 (July–August 1969): 20–24.

Thomas, Kas. "From Long Beach to Kitty Hawk......By Gyrocopter." *Popular Rotorcraft Flying* 9, no. 5 (September–October 1971): 10–16.

———. "It's All Yours He Said...Kas Thomas Flys the 18A." *Popular Rotorcraft Flying* 10, no. 2 (March–April 1972): 8–9.

———. "Like a Theodore Bear." *Popular Rotorcraft Flying* 9, no. 3 (May–June 1972): 11–16.

———. "Medical Questions & Answers: Some Commonly-Asked Questions About Medical Certification of Pilots—Part I." *Popular Rotorcraft Flying* 13, no. 5 (October 1975): 11.

———. "Rotary Connection." *Popular Rotorcraft Flying* 13, no. 5 (October 1975): 23.

———. "The Umbaugh Story: Rags to Riches (and Back?)." *Popular Rotorcraft Flying* 10, no. 2 (March–April 1972): 10, 23.

Thomas, Ray. "UK Update." *Rotorcraft* 29, no. 5 (August 1991): 36.

Tichenor, Frank A. "Air—Hot and Otherwise." *Aero Digest* 17, no. 7 (December 1930): 40, 124–134.

Tinsley, Frank. "The Autogyro Joins the Army." *Bill Barnes Air Trails* 6, no. 3 (June 1936): 30–32.

Torres, Alejandro. "Rotors Over Venezuela." *Rotorcraft* 28, no. 2 (April 1990): 16.

Townson, George. "Autogiro Air Mail." *American Helicopter Museum and Education Center Newsletter* 2, no. 3 (Summer 1995): 3.

———. "Autogiro Crop Dusters." *American Helicopter Museum and Education Center Newsletter* 3, no. 1 (Spring 1996): 3–4.

———. "General Information and History of the Autogiro." *American Helicopter Society Newsletter* (March 1961).

———. "History of the Autogiro." *American Helicopter Society Newsletter* 11, no. 4 (March 1961).

———. "The Herrick Convertaplane." *American Helicopter Museum and Education Center Newsletter* 4, no. 3 (Third Quarter 1997): 3–4.

Townson, George, and Howard Levy. "The History of the Autogiro: Part One." *Air Classics Quarterly Review* 4, no. 2 (Summer 1977): 4–18.

Townson, George, and Howard Levy. "The History of the Autogiro: Part Two." *Air Classics Quarterly Review* 4, no. 3 (Fall 1977): 4–19, 110–114.

Trent, Edgar B. "FAA meets PRA." *Popular Rotorcraft Flying* (Fall 1963): 5–7.

———. "Panic in England." *Popular Rotorcraft Flying* 9, no. 1 (January–February 1971): 2, 26.

———. "PRA Fly-In: Let's Do It Again!" *Popular Rotorcraft Flying* (Summer 1963): 3–4.

"TSR2—If Only..." *Aircraft Illustrated* 34, no. 6 (June 2001): 50–54.

"21st Century Rotodyne." *WINGSPAN International* no. 8 (September/October 2001): 65.

Vandewalle, Larry. "Give Me My Training Wheels!" *Rotorcraft* 28, no. 6 (September Extra 1990): 43.

VanVoorhees, John. "The Pitbull Autogyro." *Rotorcraft* 34, no. 3 (May 1966): 16.

Vaz, Mark Cotta. "Rocket Blast." *Cinefex* no. 48 (November 1991): 20–45.

"Vertaplane." *Time* 30, no. 6 (August 9, 1937): 21–22.

"VFW Joins the Parade." *Popular Rotorcraft Flying* 9, no. 4 (July–August 1971): 31.

Viviani, Sonja. "Magni's Day 2001—A 'Wet' Success." *Rotorcraft* 40, no. 1 (February 2002): 4–6.

Wainfan, Barnaby. "CarterCopter: Blazing New Trails in Aviation Technology." *Flight Journal* 7, no. 6 (December 2002): 64–73.

Wallis, Kenneth H. "Autogyro World Records—Past, Present and Future…" *Fly-Gyro!* no. 6 (July–August 2001): 14–18.

———. "From Wing Commander Wallis." *Rotorcraft* 38, no. 8 (November 2000): 12.

———. "I Was 50 Before I Was '007.'" *Rotorcraft* 28, no. 2 (April 1990): 8–14.

———. "The Longest Spin." *Rotorcraft* 33, no. 3 (May 1995): 18–24.

———. "Movie Flying In Brazil." *Rotorcraft* 33, no. 9 (December 1995–January 1996): 10.

———. "World Record! Another One for Ken Wallis." *Rotorcraft* 29, no. 3 (May 1991): 14.

Weisberger, Harry. "Groen Brothers Reinvent a Time-Tested Concept." *Aviation International News* 32, no. 21 (November 2000).

Wheatley, John B. "Lift and Drag Characteristics And Gliding Performance Of An Autogiro As Determined In Flight," Report no. 434. Washington, D.C.: National Advisory Committee for Aeronautics, 1932.

——— "Rotating-Wing Aircraft Compared to Conventional Airplanes," Report. Washington, D.C.: National Advisory Committee for Aeronautics, 1934.

Whiteman, Phillip. "Pilot Profile: Ken Wallis." *Pilot,* January 1998, 22–25.

Wildes, Maxie Wildes. "Maxie Mad Max II Goes Subaru." *Rotorcraft* 30, no. 1 (February–March 1992): 13–15.

Wilson, Jim. "Radical Rotorcraft: Gyrocopters Return in Every Price Range." *Popular Mechanics* 178, no. 8 (August 2001): 56–61.

"Wing Commander Wallis to be Honored." *Popular Rotorcraft Flying* 26, no. 1 (February 1988): 26.

Wood, Roger A. "Cincinnati to Oshkosh: 526 Miles by Gyro." *Popular Rotorcraft Flying* 13, no. 5 (October 1975): 16–18.

———. "PRA Summary Financial Report for 1995–1996." *Rotorcraft* 34, no. 6 (September 1996): 33.

Woods, Harris. "My 7 Gyros." *Popular Rotorcraft Flying* 2, no. 2 (Spring 1964): 6–7, 10–11.

Yaw, Charlie. "The Air Command Story." *Rotorcraft* 26, no. 4 (August–September 1988): 32–33.

Yeatman, H.M. "The Cierva Autogiro." *Aero Digest* (April 1928).

Yoxall, John. "They Who Dared First—No. 2: H.A. Marsh." *Aeroplane Monthly* 18, no. 6: 342–344.

Yulke, Ed. "Gyro Cars for Fun." *Popular Mechanix* 35, no. 1 (November 1945): 74–76, 148.

Zimmerman, Robert. "A Reality at Last: The Family Autogyro." *Popular Mechanics* 131, no. 3 (March 1969): 112–113.

VIDEO/MOVIES/AUDIO

Army-Air Force Newsreels 1941. Traditions Military Videos, www.militaryvideo.com. Accessed November 16, 2002.

A Fairey Rotodyne Storey. Traplet Video Productions, Fairey Aviation Film Unit, Worcestershire, England.

Leslie, Dan. *Bensen Days 2000.* Macon, Georgia: Rotor/Wings Sports TV, 2000. Video.

"Superman on Radio." 1940 radio broadcast. *Smithsonian Historical Performances,* 1997.

INDEX

About the Author

BRUCE H. CHARNOV is Associate Professor and Chairperson of the Management, Entrepreneurship and General Business Department of the Frank G. Zarb School of Business at Hofstra University in Hempstead, Long Island, New York.